## The Conway–Maxwell–Poisson Distribution

While the Poisson distribution is a classical statistical model for count data, the distributional model hinges on the constraining property that its mean equals its variance. This text instead introduces the Conway–Maxwell–Poisson distribution and motivates its use in developing flexible statistical methods based on its distributional form.

This two-parameter model not only contains the Poisson distribution as a special case but, in its ability to account for data over- or under-dispersion, encompasses both the geometric and Bernoulli distributions. The resulting statistical methods serve in a multitude of ways, from an exploratory data analysis tool to a flexible modeling impetus for varied statistical methods involving count data.

The first comprehensive reference on the subject, this text contains numerous illustrative examples demonstrating R code and output. It is essential reading for academics in statistics and data science, as well as quantitative researchers and data analysts in economics, biostatistics and other applied disciplines.

KIMBERLY F. SELLERS is Professor in the Department of Mathematics and Statistics at Georgetown University and a researcher at the U.S. Census Bureau. Her work has contributed to count data research and software for the last 15 years. She is a Fellow of the American Statistical Association and an Elected Member of the International Statistical Institute.

INSTITUTE OF MATHEMATICAL STATISTICS MONOGRAPHS



IMS Monographs are concise research monographs of high quality on any branch of statistics or probability of sufficient interest to warrant publication as books. Some concern relatively traditional topics in need of up-to-date assessment. Others are on emerging themes. In all cases the objective is to provide a balanced view of the field.

Other Books in the Series

1. *Large-Scale Inference*, by Bradley Efron
2. *Nonparametric Inference on Manifolds*, by Abhishek Bhattacharya and Rabi Bhattacharya
3. *The Skew-Normal and Related Families*, by Adelchi Azzalini
4. *Case-Control Studies*, by Ruth H. Keogh and D. R. Cox
5. *Computer Age Statistical Inference*, by Bradley Efron and Trevor Hastie
6. *Computer Age Statistical Inference (Student Edition)*, by Bradley Efron and Trevor Hastie
7. *Stable Lèvy Process via Lamperti-Type Representations*, by Andreas E. Kyprianou and Juan Carlos Pardo

# The Conway–Maxwell–Poisson Distribution

KIMBERLY F. SELLERS
*Georgetown University*

Shaftesbury Road, Cambridge CB2 8EA, United Kingdom

One Liberty Plaza, 20th Floor, New York, NY 10006, USA

477 Williamstown Road, Port Melbourne, VIC 3207, Australia

314–321, 3rd Floor, Plot 3, Splendor Forum, Jasola District Centre, New Delhi – 110025, India

103 Penang Road, #05–06/07, Visioncrest Commercial, Singapore 238467

Cambridge University Press is part of Cambridge University Press & Assessment, a department of the University of Cambridge.

We share the University's mission to contribute to society through the pursuit of education, learning and research at the highest international levels of excellence.

www.cambridge.org
Information on this title: www.cambridge.org/9781108481106

DOI: 10.1017/9781108646437



First published 2023

*A catalogue record for this publication is available from the British Library*

ISBN 978-1-108-48110-6 Hardback

*To those that "count" most in my life:*

*My family, especially my son*

# Contents

# Figures

# Tables

# Preface

Welcome to *The Conway–Maxwell–Poisson Distribution* – the first coherent introduction to the Conway–Maxwell–Poisson distribution and its contributions with regard to statistical theory and methods. This two-parameter model not only serves as a flexible distribution containing the Poisson distribution as a special case but also, in its ability to capture either data over- or under-dispersion, it contains (in particular) two other classical distributions. The Conway–Maxwell–Poisson distribution thereby can effectively model a range of count data distributions that contain data over- or under-dispersion, simply through the addition of one parameter. This distribution's flexibility offers numerous opportunities with regard to statistical methods development. To date, such efforts involve work in univariate and multivariate distributional theory, regression analysis (including spatial and/or temporal models, and cure rate models), control chart theory, and count processes. Accordingly, the statistical methods described in this reference can effectively serve in a multitude of ways, from an exploratory data analysis tool to an appropriate, flexible count data modeling impetus for a variety of statistical methods involving count data.

*The Conway–Maxwell–Poisson Distribution* can benefit a broad statistical audience. This book combines theoretical and applied data developments and discussions regarding the Conway–Maxwell–Poisson distribution and its significant flexibility in modeling count data, where this reference adopts the convention that the counting numbers are the natural numbers including zero, i.e. $\mathbb{N} = \{0, 1, 2, \ldots\}$. Count data modeling research is a topic of interest to the academic audience, ranging from upper-level undergraduates to graduate students and faculty in statistics (and, more broadly, data science). Meanwhile, the compelling nature of this topic and the writing format of the reference intend to draw quantitative researchers and data analysts in applied disciplines, including business and economics, medicine and public health, engineering, psychology, and sociology – broadly anyone interested in its supporting computational discussions and examples using `R`. This reference seeks to assume minimal prerequisite statistics coursework/knowledge (e.g. calculus and a calculus-based introduction to probability and statistics that includes maximum likelihood estimation) throughout the book. More advanced readers, however, will benefit from additional knowledge of other subject areas in some chapters, for example, linear algebra or Bayesian computation.

Along with this reference's discussion of flexible statistical methods for count data comes an accounting of available computation packages in `R` to conduct analyses. Accordingly, preliminary `R` knowledge will also prove handy as this reference brings to light the various packages that exist for modeling count data via the Conway–Maxwell–Poisson distribution through the relevant statistical methods. The Comprehensive R Archive Network (CRAN) regularly updates its system. In the event that any package discussed in this reference is subsequently no longer directly accessible through the CRAN, note that it is archived and thus still accessible for download and use by analysts.

# Acknowledgments

This book is not only the culmination of my years of work and research developing this field but also represents the vastness of contributions in statistical methods and computation by the many authors that are cited in this reference. I am particularly thankful to Dr. Galit Shmueli, the first to introduce me to the Conway–Maxwell–Poisson distribution. That initial collaboration sparked my great interest in this topic where I could easily recognize the broad and diverse areas for potential research, some of which has since come to fruition by myself and/or others. Accordingly, I further thank my colleague and student collaborators with whom we have accomplished a great deal. Thanks as well are extended to other researchers in the larger field of count data modeling with whom I have not only established a camaraderie, but some of whom have also provided discussions and/or analysis associated with their respective models that do not provide an `R` package for direct computation, including Dr. N. Balakrishnan regarding various works on cure-rate models, Dr. Alan Huang regarding the mean-parametrized Conway–Maxwell–Poisson distribution and the `mpcmp` package, Dr. Felix Famoye regarding the bivariate generalized Poisson, Mr. Tong Li for the trivariate-reduced bivariate COM–Poisson, and Dr. S. H. Ong for both Sarmanov-formulated bivariate COM–Poisson models.

Words cannot fully describe how much I thank and appreciate Sir David Cox for his vision and encouragement for me to author this reference; I am only sorry that he was not able to see it come to fruition. I also am forever thankful and appreciative to Dr. Darcy Steeg Morris for our countless, fruitful conversations as she aided with editing the content in this reference that not only helped to shape its vision, but further ensure its accuracy, understanding, and comprehension.

Completing such a project does not happen in a vacuum. I thank family and friends who, through word and deed, kept me encouraged, motivated, and focused to see this project to and through the finish line. I particularly thank Dr. Kris Marsh for our many hours spent together in our self-made writing group. I am also appreciative to Georgetown University for their financial support through avenues including the Senior Faculty Research Fellowship which aided the preparation of this reference.

# 1

# Introduction: Count Data Containing Dispersion

This chapter is an overview summarizing relevant, established, and well-studied distributions for count data that motivate the consideration of the Conway–Maxwell–Poisson (COM–Poisson) distribution. Each of the discussed models provides an improved flexibility and computational ability for analyzing count data; yet associated restrictions help readers to appreciate the need for and usefulness of the COM–Poisson distribution, thus resulting in an explosion of research relating to this model. For completeness of discussion, each of these sections includes discussion of the relevant `R` packages and their contained functionality to serve as a starting point for forthcoming discussions throughout subsequent chapters. Along with the `R` discussion, illustrative examples aid readers in understanding distribution qualities and related statistical computational output. This background provides insights into the real implications of apparent data dispersion in count data models and the need to properly address it.

This introductory chapter proceeds as follows. Section 1.1 introduces the most well-known model for count data: the Poisson distribution. Its probabilistic and statistical properties are discussed, along with `R` tools to perform computations. Section 1.2, however, notes a major limitation of the Poisson distribution – namely its inability to properly model dispersed count data. Focusing first on the phenomenon of data over-dispersion, this section focuses attention on the negative binomial (NB) distribution – the most popular count distribution that allows for data over-dispersion. Section 1.3 meanwhile recognizes the existence of count data that express data under-dispersion and the resulting need for model consideration that can accommodate this property. While several flexible models allowing for data over- or under-dispersion exist in the literature, this section focuses attention on the generalized Poisson (GP) distribution for modeling such data because it is arguably (one of) the most popular option(s) for modeling such data. Section 1.4 offers an overarching perspective about these models as special cases of a larger class of weighted Poisson distributions. Finally, Section 1.5 motivates an interest in the COM–Poisson distribution and summarizes the rest of the book, including the unifying background that will be referenced in subsequent chapters.

## 1.1 Poisson Distribution

The Poisson distribution is the most studied and applied distribution referenced to describe variability in count data. A random variable $X$ with a Poisson($\lambda$) distribution has the probability mass function

$$P(X = x) = \frac{\lambda^x e^{-\lambda}}{x!}, \quad x = 0, 1, 2, \ldots, \tag{1.1}$$

or, on the log-scale,

$$\begin{aligned}\ln\left[P(X = x)\right] &= x \ln \lambda - \ln(x!) - \lambda \\ &= x \ln \lambda - \sum_{j=1}^{x} \ln(j) - \lambda,\end{aligned}$$

where $\lambda$ is the associated intensity parameter; illustrative examples of the distributional form assuming various values of $\lambda$ are provided in Figure 1.1.

Derived as the limiting distribution of a binomial($n, p$) distribution where $n \to \infty$ and $p \to 0$ such that $np = \lambda$, the beauty of this distribution lies in its simplicity. Both its mean and variance equal the intensity parameter $\lambda$; thus, the dispersion index is

$$\mathrm{DI}(X) = \frac{V(X)}{E(X)} = \frac{\lambda}{\lambda} = 1. \tag{1.2}$$

The probability mass function satisfies the recursion

$$\frac{P(X = x - 1)}{P(X = x)} = \frac{x}{\lambda}, \tag{1.3}$$

with its moment generating function $M_X(t) = e^{\lambda(e^t - 1)}$, and the Poisson distribution is a member of the exponential family of the form

$$P(X = x; \theta) = H(x) \exp\left[\eta(\theta) T(x) - \Psi(\theta)\right], \quad x \in \mathbb{N}, \tag{1.4}$$

where, for $\theta = \lambda$, $\eta(\theta) = \ln(\lambda)$, $\Psi(\theta) = \lambda$, $T(x) = x$, and $H(x) = (x!)^{-1}$. The simplicity of the Poisson distribution, however, can also be viewed as theoretically constraining and not necessarily representative of real count

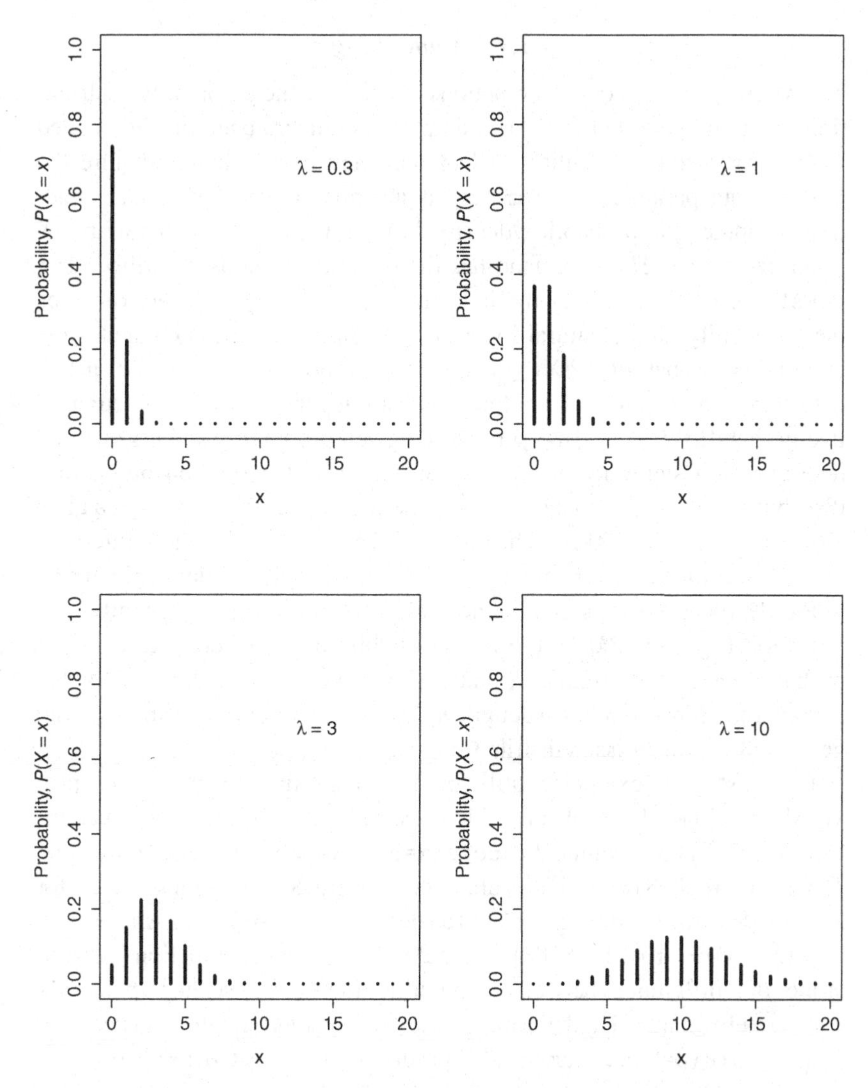


Figure 1.1 Poisson probability mass function illustrations for $\lambda \in \{0.3, 1, 3, 10\}$.

data distributions. Thus, applying statistical methods that are motivated and/or developed by the Poisson model assumption can cause significant repercussions with regard to statistical inference. This matter is discussed in more detail in the subsequent sections in Chapter 1 and throughout this reference.

### *1.1.1* R *Computing*

The `stats` package contains functions to compute the probability, distribution function, quantile function, and random number generation associated with the Poisson distribution. All of the relevant commands require the Poisson rate parameter $\lambda$ (`lambda`) as an input value. The `dpois` function computes the probability/density $P(X = x)$ for a random variable $X$ at observation $x$. The command has the default setting as described (`log = FALSE`), while changing the indicator input to `log = TRUE` computes the probability on the natural-log scale. The `ppois` function computes the cumulative probability $P(X \leq q)$ given a quantile value $q$, while `qpois` determines the quantile $q$ (i.e. the smallest integer) for which the cumulative probability $P(X \leq q) \geq p$ for some given probability $p$. This quantile determination stems from the discrete nature of the Poisson probability distribution. Both commands contain the default settings `lower.tail = TRUE` and `log.p = FALSE`. The condition `lower.tail = TRUE` infers interest regarding the cumulative probability $P(X \leq q)$ while `lower.tail = FALSE` focuses on its complement $P(X > q)$ (i.e. the upper tail). The indicator `log.p = FALSE (TRUE)` meanwhile infers whether to consider probabilities on the original or natural-log scale, respectively. Finally, the `rpois` function produces a length $n$ (`n`) vector of count data randomly generated via the Poisson distribution.

Demonstrative examples utilizing the respective functions are provided in Code 1.1, all of which assume the Poisson rate parameter $\lambda = 3$. The command `dpois(x=5, lambda=3)` determines that $P(X = x) = 0.1008188$; this value is illustrated in Figure 1.1 for $\lambda = 3$. Meanwhile, `dpois(x=5, lambda=3, log = TRUE)` shows that $\ln(P(X = x)) = \ln(0.1008188) = -2.29443$. The `ppois` functions demonstrate the difference between computing the lower versus upper tail, respectively; naturally, the sum of the two results equals 1. The command `qpois(p=0.9, lambda=3)` produces the expected result of 5 because we see that the earlier `ppois(q=5, lambda=3)` result showed that $P(X \leq 5) = 0.9160821 > 0.9$. Meanwhile, one can see that `qpois(p=0.9, lambda=3, lower.tail = FALSE)` produces the value 1 by considering the corresponding `ppois` commands:

`ppois(q=0,lambda=3,lower.tail=FALSE)` produces the result 0.9502129
`ppois(q=1,lambda=3,lower.tail=FALSE)` produces the result 0.8008517.

Recall that the discrete nature of the Poisson distribution requires a modified approach for determining the quantile value; the resulting quantile is

Code 1.1 Examples of R function use for Poisson distributional computing: dpois, ppois, qpois, rpois.

```
> dpois(x=5, lambda=3)
[1] 0.1008188
> dpois(x=5, lambda=3, log = TRUE)
[1] -2.29443
> ppois(q=5, lambda=3)
[1] 0.9160821
> ppois(q=5, lambda=3, lower.tail = FALSE)
[1] 0.08391794
> qpois(p=0.9, lambda=3)
[1] 5
> qpois(p=0.9, lambda=3, lower.tail = FALSE)
[1] 1
> rpois(n=10, lambda=3)
 [1] 3 4 3 5 2 0 5 5 4 3
```

determined such that the cumulative probability of interest is at least as much as the desired probability of interest. This definition suggests that, when considering the upper tail probability, the resulting quantile now implies that the corresponding upper tail probability is no more than the desired probability of interest. As noted above, $P(X > 0) = 0.9502129$ and $P(X > 1) = 0.8008517$; because the desired upper tail probability in the example is 0.9, we see that 0 produces an upper tail probability that is too large for consideration, while the upper tail probability associated with 1 is the first integer that satisfies $P(X > x) \leq 0.9$, thus producing the solution as 1. Finally, for completeness, the `rpois` function produces 10 randomly generated potential observations stemming from a Poisson(3) distribution. Given the probability mass function illustration provided in Figure 1.1 for $\lambda = 3$, these outcomes appear reasonable.

## 1.2 Data Over-dispersion

Over-dispersion (relative to a comparable Poisson model) describes distributions whose variance is larger than the mean, i.e. $\text{DI}(X) > 1$ for a random variable $X$. This is a well-studied phenomenon that occurs in most real-world datasets. Over-dispersion can be caused by any number of situations, including data heterogeneity, the existence of positive correlation between responses, excess variation between response probabilities or counts, and violations in data distributional assumptions. Apparent over-dispersion can also exist in datasets because of outliers or, in the case of regression models, the model may not include important explanatory variables or a sufficient number of interaction terms, or the link relating the response to the explanatory variables may be misspecified. Under such circumstances, over-dispersion causes problems because resulting standard errors associated with parameter estimation may be underestimated, thus producing biased inferences. Interested readers should see Hilbe (2007) for a comprehensive discussion regarding over-dispersion and its causes.

The most popular distribution to describe over-dispersed data is the NB distribution. A random variable $X$ with an NB$(r, p)$ distribution has the probability mass function

$$P(X = x) = \binom{r + x - 1}{x} p^x (1 - p)^r \tag{1.5}$$

$$= \frac{\Gamma(r + x)}{x!\Gamma(r)} p^x (1 - p)^r, \quad x = 0, 1, 2, \ldots, \tag{1.6}$$

and can be viewed as the probability of attaining a total of $x$ successes with $r > 0$ failures in a series of independent Bernoulli$(p)$ trials, where $0 < p < 1$ denotes the success probability associated with each trial. Alternatively, the NB distribution can be derived via a mixture model of a Poisson$(\lambda)$ distribution, where $\lambda$ is gamma distributed[1] with shape and scale parameters, $r$ and $p/(1 - p)$, respectively. The latter approach is a common technique for addressing heterogeneity. Other possible distributions for $\lambda$ include the generalized gamma (which produces a generalized form of the NB distribution (Gupta and Ong, 2004)), the inverse Gaussian, and the generalized inverse Gaussian (which produces the Sichel distribution (Atkinson and Yeh, 1982; Ord and Whitmore, 1986)). Various other mixing distributions have also been considered; see Gupta and Ong (2005) for discussion.

The moment generating function of the NB$(r, p)$ random variable $X$ is

$$M_X(t) = \left( \frac{p}{1 - (1 - p)e^t} \right)^r, \quad t < -\ln(1 - p),$$

[1] For a gamma$(\alpha, \beta)$ distributed random variable $X$ with shape and scale parameters $\alpha$ and $\beta$, respectively, its probability density function (pdf) is $f(x) = \frac{1}{\Gamma(\alpha)\beta^\alpha} x^{\alpha-1} e^{-x/\beta}$ (Casella and Berger, 1990).

which produces a respective mean and variance,

$$\mu \doteq E(X) = \frac{r(1-p)}{p} \quad \text{and} \tag{1.7}$$

$$V(X) = \frac{r(1-p)}{p^2} = \mu + \frac{1}{r}\mu^2, \tag{1.8}$$

where $r > 0$ can be viewed as a dispersion parameter. Given the dispersion parameter $r$, this distribution can be represented as an exponential family (Equation (1.4)), where $\theta = p$, $H(x;r) = \binom{r+x-1}{x}$, $T(x) = x$, $\eta(p) = \ln p$, and $\psi(p;r) = r\ln(1-p)$. Equation (1.8) demonstrates that the NB distribution can accommodate data over-dispersion ($\mathrm{DI}(X) > 1$) because one can clearly see that the distribution's variance is greater than or equal to its mean since $r > 0$. Further, the NB distribution contains the Poisson as a limiting case; as $r \rightarrow \infty$ and $p \rightarrow 1$ such that $r(1-p) \rightarrow \lambda$, $0 < \lambda < \infty$, not only do the NB mean and variance both converge to $\lambda$, but the NB probabilities likewise converge to their respective Poisson counterparts. Figure 1.2 illustrates the distributional convergence of the $\mathrm{NB}(r,p)$ to the Poisson($\lambda = 3$) distribution, where $r \rightarrow \infty$ and $p \rightarrow 1$ such that $r(1-p) = 3$. The NB distribution likewise contains the geometric($p$) as a special case when $r = 1$.

The NB distribution can alternatively be represented as $\mathrm{NB}(r, r/(r+\mu))$ with the probability mass function

$$P(X = x) = \binom{x+r-1}{x}\left(\frac{r}{r+\mu}\right)^x\left(\frac{\mu}{r+\mu}\right)^r, \quad x = 0, 1, 2, \ldots, \tag{1.9}$$

where $r > 0$, $\mu > 0$; this formulation explicitly has a mean $\mu$ and a variance $\mu + \mu^2/r$. The `MASS` package in `R` utilizes this parametrization and defines the dispersion parameter as `theta` such that $V(X) = \mu + \mu^2/\theta$, i.e. $\theta \doteq r$; we will revisit this in Chapter 5. While the NB distribution has been well studied and statistical computational ability is supplied in numerous software packages (e.g. `R` and `SAS`), an underlying constraint regarding the NB distribution leads to its inability to address data under-dispersion (i.e. the dispersion index is less than 1, or the variance is smaller than the mean).

### *1.2.1* `R` *Computing*

The `stats` package provides functionality for determining the probability, distribution function, quantile function and random number generation for the NB distribution. These commands all require the inputs `size` ($r$) and either the success probability $p$ (`prob`) or mean $\mu$ (`mu`), depending on the

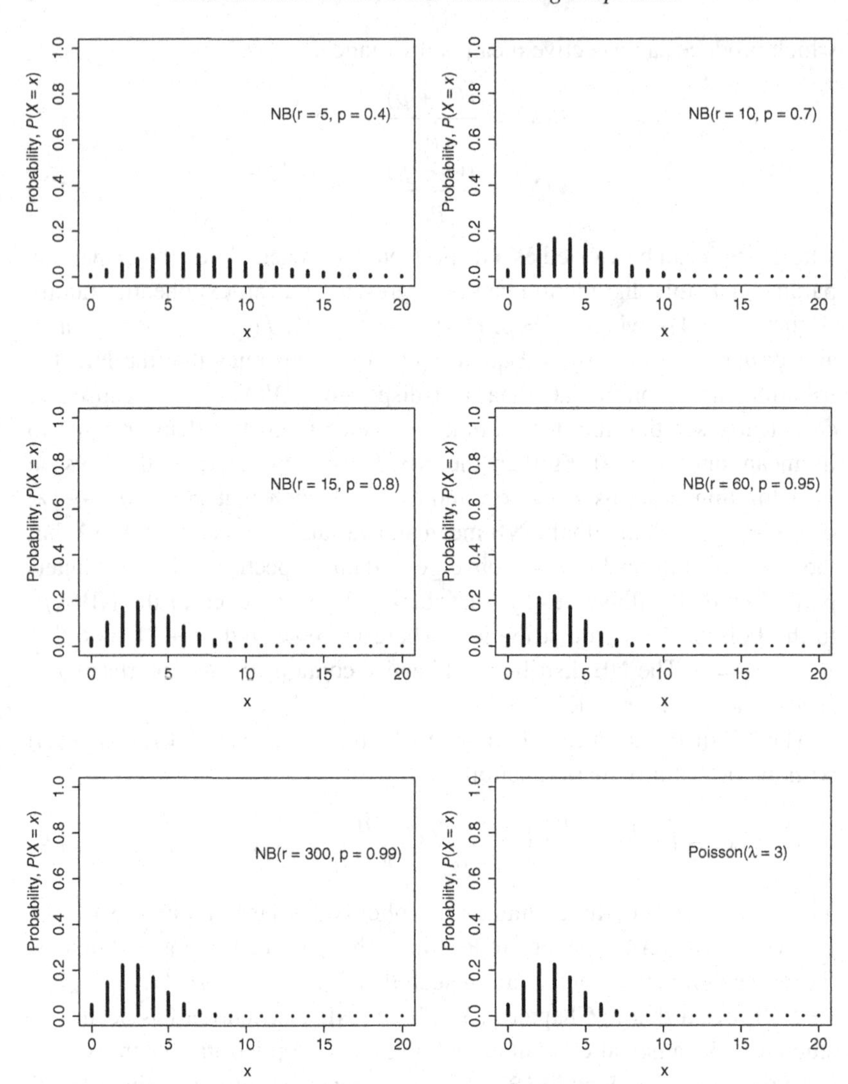


Figure 1.2 Negative binomial distribution illustrations for values of $(r,p) \in \{(5,0.4),(10,0.7),(15,0.8),(60,0.95),(300,0.99)\}$ and the Poisson($\lambda = 3$) probability mass function. This series of density plots nicely demonstrates the distributional convergence of the negative binomial to the Poisson as $r \to \infty$ and $p \to 1$ such that $r(1-p) \to \lambda$.

choice of parametrization. The function `dnbinom` computes the probability $P(X = x)$ for a random variable $X$ at observation $x$, either on the original scale (`log = FALSE`; this is the default setting) or on a natural-log scale

Code 1.2 Examples of R commands for NB distributional computing: dnbinom, pnbinom, qnbinom, rnbinom.

```
> dnbinom(x=5, size=10, prob=0.7)
[1] 0.1374203
> dnbinom(x=5, size=10, prob=0.7, log = TRUE)
[1] -1.984712
> pnbinom(q=5, size=10, prob=0.7)
[1] 0.7216214
> pnbinom(q=5, size=10, prob=0.7, lower.tail = FALSE)
[1] 0.2783786
> qnbinom(p=0.9, size=10, prob=0.7)
[1] 8
> qnbinom(p=0.9, size=10, prob=0.7, lower.tail = FALSE)
[1] 1
> rnbinom(n=10, size=10, prob=0.7)
 [1] 1 8 7 3 5 8 4 2 5 3
```

(`log = TRUE`). For a given quantile value $q$, the pnbinom function determines the cumulative probability $P(X \leq q)$, where the default settings, `lower.tail = TRUE` and `log.p = FALSE`, imply that the resulting cumulative probability is attained by accumulating the probability from the lower tail and on the original probability scale. The command qnbinom meanwhile determines the smallest discrete quantile value $q$ that satisfies the cumulative probability $P(X \leq q) \geq p$ for a given probability $p$. This function likewise assumes the default settings, `lower.tail = TRUE` and `log.p = FALSE`, such that the quantile $q$ is determined from the lower tail and on the original probability scale. For both of these commands, changing the default settings to `lower.tail = FALSE` and `log.p = TRUE`, respectively allows analysts to instead consider quantile determination on the basis of the upper tail probability $P(X > q)$, and via a probability computation on the basis of the natural-log scale. Finally, the rnbinom function randomly generates $n$ (n) observations from an NB distribution with the specified size (size) and success probability (prob).

The NB($r = 10$, $p = 0.7$) distribution is provided in Figure 1.2 and serves as a graphical reference for the illustrative commands featured in Code 1.2. All of the demonstrated functions assume $r = 10$ and $p = 0.7$ as the associated NB size and success probability parameters. The first command (`dnbinom(x=5, size=10, prob=0.7)`) shows that $P(X = x) = 0.1374203$; this probability is shown in the associated plot in Figure 1.2. Meanwhile, `dnbinom(x=5, size=10, prob=0.7, log = TRUE)` shows that $\ln(P(X = x)) = \ln(0.1374203) = -1.984712$.

The `pnbinom` functions show the results when computing the lower versus upper tail, respectively; naturally, the sum of the two computations equals 1. Calling `qnbinom(p=0.9, size=10, prob=0.7)` produces the result 8, while `qnbinom(p=0.9, size=10, prob=0.7, lower.tail = FALSE)` yields the value 1. Finally, the `rnbinom` command produces 10 randomly generated potential observations stemming from an $\text{NB}(r = 10, p = 0.7)$ distribution.

## 1.3 Data Under-dispersion

Where data over-dispersion describes excess variation in count data, under-dispersion describes deficient variation in count data. Data under-dispersion (relative to the Poisson model) refers to count data that are distributed such that the variance is smaller than the mean, i.e. its dispersion index $\text{DI}(X) < 1$ for a random variable $X$.

There remains some measures of debate regarding the legitimacy of data under-dispersion as a real concept. Some researchers attribute under-dispersion to the data generation (e.g. small sample values) or to the modeling process (e.g. model over-fitting), noting that the arrival process, birth–death process, or binomial thinning mechanisms can also lead to under-dispersion (Kokonendji, 2014; Lord and Guikema, 2012; Puig et al., 2016). As an example, for renewal processes where the distribution of the interarrival times has an increasing hazard rate, the distribution of the number of events is under-dispersed (Barlow and Proschan, 1965). Efron (1986), however, argues that "there are often good physical reasons for not believing in under-dispersion, however, especially in binomial and Poisson situations."

Whether real or apparent, examples across disciplines are surfacing with more frequency where data under-dispersion is present; thus there exists the need to represent such data. The most popular model that can accommodate data dispersion (whether over- or under-dispersion) is the GP distribution – a flexible two-parameter distribution that contains the Poisson distribution as a special case (Consul, 1988). A random variable $X$ that is $\text{GP}(\lambda_1, \lambda_2)$ distributed has the probability mass function

$$P(X = x) = \begin{cases} \dfrac{\lambda_1(\lambda_1 + \lambda_2 x)^{x-1}}{x!} \exp(-\lambda_1 - \lambda_2 x), & x = 0, 1, 2, \ldots \\ 0, & x \geq m \text{ where } \lambda_1 + \lambda_2 m \leq 0 \end{cases} \tag{1.10}$$

for $\lambda_1 > 0$ and $-1 < \lambda_2 < 1$ (Consul and Jain, 1973). This distribution has the respective mean and variance,

$$E(X) = \frac{\lambda_1}{1 - \lambda_2} \tag{1.11}$$

$$V(X) = \frac{\lambda_1}{(1 - \lambda_2)^3}, \tag{1.12}$$

and can accommodate any form of data dispersion via $\lambda_2$. The GP($\lambda_1, \lambda_2$) distribution contains the special-case Poisson($\lambda_1$) distribution, where $\lambda_2 = 0$; this is the case of equi-dispersion relative to the Poisson model. Meanwhile, for $\lambda_2 > (<) 0$, the GP distribution accommodates data over-dispersion (under-dispersion). Figure 1.3 illustrates various probability mass functions for different values of $\lambda_1$ and $\lambda_2 \in \{-0.5, 0, 0.5\}$. These choices for $\lambda_1$ and $\lambda_2$ demonstrate the change in shape and skewness for this unimodal distribution and also illustrate the data over- or under-dispersion as a function of $\lambda_2$. The middle column of Figure 1.3 contains the respective Poisson($\lambda_1 = 2, 3, 6$) probability distributions.

The GP distribution allows for over- or under-dispersion; however, extreme under-dispersion can result in probability models that do not satisfy the basic probability axioms (Famoye, 1993). Alternative count distributions exist that allow for data under-dispersion, such as the condensed Poisson, the Gamma count, and the double Poisson distributions; see Sellers and Morris (2017) for discussion regarding these distributions. Nonetheless, the GP distribution maintains its status as a very popular and well-studied count distribution that allows for data dispersion.

### *1.3.1* `R` *Computing*

The GP distribution is a popular model for describing count data that express either over- or under-dispersion, and this is reflected through the multiple `R` packages available for statistical computing. Basic functionality exists in the packages `HMMpa` (Witowski and Foraita, 2018), `LaplacesDemon` (Statisticat and LLC., 2021), and `RNGforGPD` (Li et al., 2020), while commands to conduct GP regression are available in the `VGAM` (Yee, 2008) package.

The `HMMpa` and `LaplacesDemon` packages each contain commands that can compute the probability mass function of a GP distribution. `HMMpa` provides the `dgenpois(x, lambda1, lambda2)` function, where `lambda1` and `lambda2` are $\lambda_1$ and $\lambda_2$ as defined in Equation (1.10). `LaplacesDemon` meanwhile contains the function `dgpois(x, lambda,`

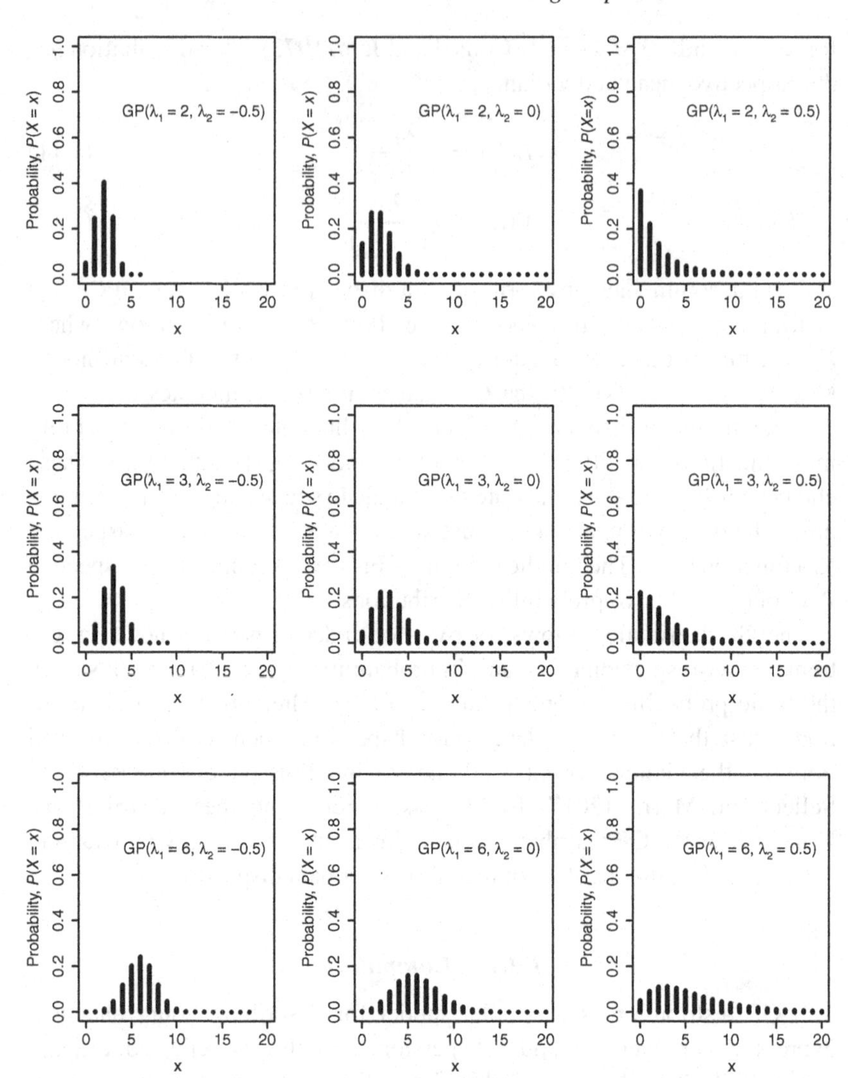


Figure 1.3 Generalized Poisson probability mass function illustrations for values of $\lambda_1 > 0$, and dispersion parameter $\lambda_2 \in \{-0.5, 0, 0.5\}$. For $\lambda_1 > 0$ and $-1 < \lambda_2 < 1$ such that $\lambda_2 > (<)0$ denotes data over-dispersion (under-dispersion), the generalized Poisson distribution has the mean $E(X) = \frac{\lambda_1}{1-\lambda_2}$ and variance $V(X) = \frac{\lambda_1}{(1-\lambda_2)^3}$.

`omega, log=FALSE)` that computes the probability mass function via an alternate parametrization, namely

$$P(X = x) = \frac{\lambda(1-\omega)[\lambda(1-\omega)+\omega x]^{x-1}}{x!} \exp(-\lambda(1-\omega)-\omega x),$$
$$x = 0, 1, 2, \ldots, \tag{1.13}$$

for parameters $\lambda > 0$ and $0 \leq \omega < 1$ (as reported in Statisticat and LLC. (2021)). Under this parametrization, $\omega = 0$ reduces the GP distribution to the Poisson($\lambda$) distribution. Equations (1.10) and (1.13) are equivalent with $\lambda_1 = \lambda(1-\omega)$ and $\lambda_2 = \omega$. The `dgpois` logical input `log` determines whether the probability mass function is provided on the original (`log=FALSE`; this is the default) or natural-log (`log=TRUE`) scale. The two functions `dgenpois` and `dgpois` produce identical outcomes for `lambda1 = lambda(1-omega)` and `lambda2 = omega` for appropriate values of $x$.

While `dgenpois` and `dgpois` both have the capability to compute $P(X=x)$ for a GP Poisson random variable $X$, these functions should be used with caution. The GP parametrization that motivates `dgpois` stems from an applied focus involving claim count data with the argument that such data are not commonly under-dispersed so that distributional focus assumes nonnegative $\omega$ (Ntzoufras et al., 2005). Equation (1.13) thus has a mean and variance

$$E(X) = \lambda \tag{1.14}$$

$$V(X) = \frac{\lambda}{(1-\omega)^2} \tag{1.15}$$

that results in the dispersion index, $\text{DI}(X) = 1/(1-\omega)^2 \geq 1$. The `dgpois` function, however, appears to accurately compute probabilities associated with data under-dispersion (i.e. satisfying $-1 < \omega < 0$); hence analysts can safely maintain $|\omega| < 1$. The `dgenpois` function meanwhile computes the first component of Equation (1.10) (i.e. $\frac{\lambda_1(\lambda_1+\lambda_2 x)^{x-1}}{x!} \exp(-\lambda_1 - \lambda_2 x)$ for $x = 0, 1, 2, \ldots$); however, it does not set $P(X = x) = 0$ for those $x \geq m$, where $\lambda_1 + \lambda_2 m \geq 0$. As a result, the function can compute extraneous output; Figure 1.4 provides an illustrative example. As demonstrated in Figure 1.4(a), because the `dgenpois` function does not properly account for values $x \geq m$, where $\lambda_1 + \lambda_2 m \geq 0$ for some $m$, the resulting outcomes defy the probability axioms. In this illustration, we see that $m = 6$; thus $P(X = x)$ should equal 0 for $x \geq 6$. Reported computations for $x > 6$, however,

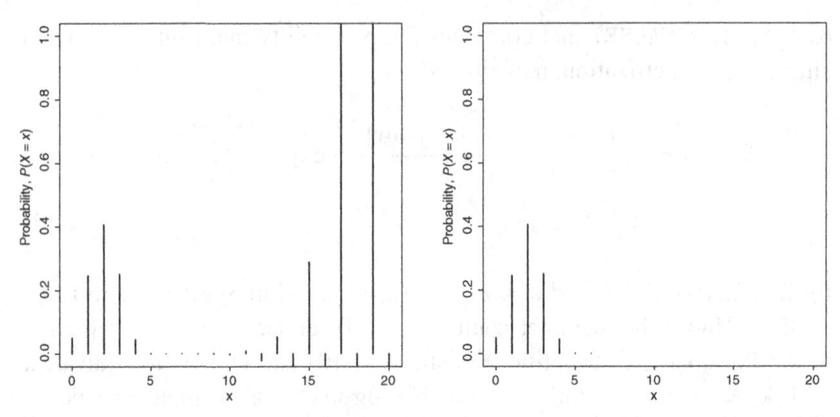


Figure 1.4 The probability mass function $P(X = x)$ created for $x \in \{0, \ldots, 20\}$ for a generalized Poisson distribution (a) via `dgenpois` (`HMMpa`) with $\lambda_1 = 3$, $\lambda_2 = -0.5$; and (b) via `dgpois` (`LaplacesDemon`) with $\lambda = 2$, $\omega = -0.5$. The resulting plots should be identical because $\lambda_1 = \lambda(1 - \omega)$ and $\lambda_2 = \omega$.

instead bifurcate between outcomes that increase in absolute value, whether negative or positive (thus further producing outcomes that are greater than 1); both of these scenarios contradict probability axioms. Thus, in order to get the `dgenpois` function to provide appropriate output, it is important to insert the condition `(lambda1+lambda2*x) >= 0`; see below for illustrative `R` code and output that can produce probabilities as shown in Figure 1.4(b).

```
> x<- 0:20
> lambda1=3
> lambda2=-0.5
> ifelse((lambda1+lambda2*x) >= 0, dgenpois(x, lambda1, lambda2), 0)
 [1] 0.0497870684 0.2462549959 0.4060058497 0.2510214302 0.0459849301
 [6] 0.0009477042 0.0000000000 0.0000000000 0.0000000000 0.0000000000
[11] 0.0000000000 0.0000000000 0.0000000000 0.0000000000 0.0000000000
[16] 0.0000000000 0.0000000000 0.0000000000 0.0000000000 0.0000000000
[21] 0.0000000000
```

The command `dgpois(x=0:20, lambda=2, omega=-0.5)` likewise produces Figure 1.4(b); this is because `dgpois` properly detects the need to set $P(X = x) = 0$ for $x \geq 7$. This function, however, does so by producing a warning and `NaN`s as outcomes for those probabilities $P(X = x), x \geq m_2$ for some $m_2$ such that $\lambda_1 + \lambda_2 m_2 = \lambda(1 - \omega) + \omega m_2 < 0$; see the following illustration for details. The term $\lambda(1 - \omega) + \omega x$ is defined as `lambda.star` in the `dgpois` function and is referenced in the following warning message. In this example, $x \geq 7$ produces `NaN` (i.e. in the eighth position).

```
> dgpois(x=0:20, lambda=2, omega=-0.5)
 [1] 0.0497870684 0.2462549959 0.4060058497 0.2510214302 0.0459849301
 [6] 0.0009477042 0.0000000000          NaN          NaN          NaN
[11]          NaN          NaN          NaN          NaN          NaN
[16]          NaN          NaN          NaN          NaN          NaN
[21]          NaN
Warning message:
In log(lambda.star) : NaNs produced
```

The `dgpois` and `dgenpois` $P(X = x)$ outputs are equivalent for $x = 0, \ldots, 6$ (see Figure 1.4); thus both of these functions are capable of computing the first condition of the GP probability mass function as shown in Equations (1.10) and (1.13). Analysts are thus encouraged to first confirm that the constraint $\lambda_1 + \lambda_2 x = \lambda(1 - \omega) + \omega x \geq 0$ is satisfied in order to ensure proper GP probability computation.

`HMMpa` also contains the functions `pgenpois` and `rgenpois` to conduct cumulative probability computation and random number generation, respectively, based on the GP distribution. Both functions require the parameter inputs `lambda1` and `lambda2`; `pgenpois` needs the added input `q` to determine the cumulative probability $P(X \leq q)$ for a quantile value $q$, while `rgenpois` further requires the value `n` to obtain $n$ randomly generated observations from a GP($\lambda_1, \lambda_2$) distribution. Recognizing the aforementioned issue, however, that the `genpois` functions contained in `HMMpa` do not first constrain the support space for $x$ such that $\lambda_1 + \lambda_2 x \geq 0$, one should ensure that this caveat holds for any subsequent use of `dgenpois` or `pgenpois` in order to have confidence in the resulting output. The `HMMpa` function `rgenpois` appears to operate properly as a random number generator based on the GP distribution; the function selects proper values associated with the true support space. The `RNGforGPD` package offers alternative commands with the ability to randomly generate univariate or multivariate generalized Poisson data. The `GenUniGpois` function generates univariate GP data via one of five methods (inversion, build-up, chop-down, normal-approximation, and branching) selected by the analyst. For the given rate and dispersion parameters, `theta` and `lambda` respectively, and `method`, `GenUniGpois` can generate `n` univariate data from a GP(`theta`=$\lambda_1$, `lambda`=$\lambda_2$) distribution, where we note the aforementioned variable substitutions to adhere to Equation (1.10) for $\lambda_1 > 0$ and $-\lambda_1/4 \leq \lambda_2 < 1$.

As with other GP representations, the `RNGforGPD` package recognizes the Poisson model as a special case of the GP distribution when `lambda` $= \lambda_2 = 0$; under this circumstance, any data-generation method can be specified. Analysts should otherwise be mindful of which method is selected for random number generation as constraints exist in order to ensure performance and/or reliability. The branching method does not work for generating under-dispersed data (thus $\lambda_2 \geq 0$), and the normal approximation approach is not necessarily reliable for $\lambda_1 < 10$ (Demirtas, 2017; Li et al., 2020). The `GenMVGpois` function meanwhile generates data of size `sample.size` from a multivariate GP distribution with the marginal rate and dispersion vectors `theta.vec` and `lambda.vec`, respectively, and the correlation matrix `cmat.star`; see Li et al. (2020) for details.

The `RNGforGPD` package likewise contains the function `Quantile-Gpois` that can determine the quantile $q$ that satisfies the cumulative probability $P(X \leq q) \geq p$ for some percentile $p$ associated with a GP(`theta=`$\lambda_1$, `lambda=`$\lambda_2$) distributed random variable. This function includes the logical input `details`, where `details=FALSE` (the default setting) reports the quantile value, and `details=TRUE` provides the probability $P(X = x)$ and cumulative probability $P(X \leq x)$ for every $x \leq q$. When providing a negative dispersion parameter, it may be helpful to set `details=TRUE` as `RNGforGPD` adjusts the initially provided cumulative probabilities to account for the truncation error, and then lists the adjusted cumulative probabilities.

## 1.4 Weighted Poisson Distributions

The weighted Poisson distribution is a flexible model class for count data that can account for either over- or under-dispersion. Let $X^w$ denote the weighted version of a Poisson random variable $X$ with the probability mass function $P(X = x; \lambda)$ as defined in Equation (1.1); $X^w$ has the probability

$$P(X^w = x; \lambda) = \frac{w(x)P(X = x; \lambda)}{E_\lambda(w(X))}, \quad x = 0, 1, 2, \ldots, \tag{1.16}$$

where $w(\cdot)$ is a nonnegative weight function, and $E_\lambda(w(X)) = \sum_{j=0}^{\infty} w(j)$ $P(X = j; \lambda) > 0$ is the finite expectation. The weighted Poisson is actually a class of distributions that depends on their associated weight functions and does not offer its own general statistical computing packages (e.g. in `R`). Examples of weighted Poisson distributions include the NB and GP distributions; Table 1.1 provides the weight functions that define several examples of weighted Poisson models.

The weighted Poisson distribution has several interesting properties. For a Poisson weight function having an exponential form,

$$w(y) = \exp[rt(y)], \quad y \in \mathbb{N},$$

Table 1.1 *Weight functions associated with various examples of weighted Poisson distributions.*

| Distribution | Weight function, $w(x)$ |
|---|---|
| Poisson | 1 |
| negative binomial | $\Gamma(r+x)$, where $r > 0$ |
| generalized Poisson | $\left(\frac{\lambda_1+\lambda_2 x}{\lambda_1}\right)^{x-1} \exp(-\lambda_2 x)$, where $\lambda_1 > 0$ and $-1 < \lambda_2 < 1$ |

where $r \in \mathbb{R}$ and $y \to t(y)$ is a convex function (that may or may not depend on the original Poisson parameter), $r > 0$ corresponds to a weighted Poisson distribution that is over-dispersed. Similarly, $r = (<)\ 0$ implies that it is equi-dispersed (under-dispersed) (del Castillo and Pérez-Casany, 2005). The random variable $X^w$ is over-dispersed (under-dispersed) if and only if the mean weight function $E_\lambda(w(X;\phi))$ for a weight function $w(x;\phi)$ that does not depend on the Poisson mean $\lambda > 0$ is log-convex (log-concave). Further, $E_\lambda(w(X;\phi))$ has the same direction of log-concavity as $w(x;\phi)$; if $w(x;\phi)$ is log-convex (log-concave), then $E_\lambda(w(X;\phi))$ is likewise log-convex (log-concave). Thus, one can simply assess the shape of $w(x;\phi)$ to determine the direction of dispersion for $X^w$. Accordingly, a positive weight function's log-concavity implies the log-concavity of the weighted Poisson distribution; if the weight function $w(x;\phi)$ is log-concave, then the associated weighted Poisson distribution is likewise log-concave. These concavity results are compelling because they imply other relationships regarding distributional forms. Discrete log-concave distributions have an increasing failure rate and are unimodal, while log-convex distributions have a decreasing failure rate (DFR) and are infinitely divisible, thus implying over-dispersion (Kokonendji et al., 2008). Two weighted Poisson distributions are defined as a pointwise dual pair if their respective positive Poisson weight functions $w_1$ and $w_2$ satisfy $w_1(x)w_2(x) = 1$ for all $x \in \mathbb{N}$. The dual of weighted Poisson distributions is closed if the two distributions have differing dispersion types, i.e. one is over-dispersed (under-dispersed) and the other is under-dispersed (over-dispersed). Further, all natural exponential families of the form

$$P(X = x;\theta,\phi) = \Gamma(x;\phi)\exp\left[\eta(\theta)T(x) - \Psi(\theta;\phi)\right], \quad x \in \mathbb{N}, \tag{1.17}$$

with a fixed $\phi > 0$ are weighted Poisson distributions where the weight function is $w(x;\phi) = x!\Gamma(x;\phi)$, $x \in \mathbb{N}$; however, not all weighted Poisson distributions have the exponential family form. The weighted Poisson distribution is likewise a member of an exponential dispersion family if it satisfies the form

$$P(X = x; \theta, \phi) = H(x; \phi) \exp\left(\frac{(\eta(\theta)T(x) - \Psi(\theta; \phi))w}{\phi}\right). \quad (1.18)$$

Weighted Poisson distributions give rise to a destructive cure rate model framework in survival analysis. Let $M^w$ denote the number of competing causes associated with an event occurrence and have a weighted Poisson distribution as defined in Equation (1.16). Given $M^w$, let

$$D^w = \begin{cases} \sum_{i=1}^{M^w} B_i & M^w > 0 \\ 0 & M = 0, \end{cases} \quad (1.19)$$

where $B_i$ are independent and identically Bernoulli($p$) distributed random variables (independent from $M^w$) noting the presence (1) or absence (0) of Cause $i = 1, \ldots, M^w$. $D^w$ denotes the total number of competing risks or causes that remain viable after eradication or treatment. Accordingly, the destructive weighted Poisson cure rate survival function is

$$S_p(y) = P(Y \geq y) = \sum_{d=0}^{\infty} P(D^w = d)[S(y)]^d,$$

where $Y = \min(W_0, W_1, W_2, \ldots, W_{D^w})$ measures the survival time based on $D^w$ competing risks and their independent and identically distributed survival times $S(y)$ (Rodrigues et al., 2011, 2012). We will revisit these ideas in Chapter 8.

## 1.5 Motivation, and Summary of the Book

The Poisson distribution is a classical statistical model for modeling count data and, because its probability mass function is the simplest distribution for counts, is a "fan favorite" in the statistics community. Its underlying equi-dispersion property, however, is idealistic and constraining such that real data do not typically satisfy this attribute. Over-dispersed data are often modeled via the NB distribution; however, it cannot address data under-dispersion. A distribution that can effectively model data over- or under-dispersion would be convenient for analysts because such a construct could address any exploratory analyses regarding dispersion in a direct sense without a priori knowledge of the dispersion type in the data. More broadly, any statistical methods motivated and/or derived by such a distribution would likewise allow for more flexibility and thus more proper inference. The GP distribution is a popular two-parameter distribution that allows for over- or under-dispersion; however, its distributional complexity and inability to properly model extreme under-dispersion are troubling. Thus, there remains the need to consider an alternate count distribution that can likewise accommodate data over- or under-dispersion.

This book introduces the reader to the COM–Poisson distribution and motivates its use in developing flexible statistical methods based on its form. This two-parameter model not only serves as a flexible distribution containing the Poisson distribution as a special case but, in its ability to capture either data over- or under-dispersion, it contains (in particular) two other classical distributions as special cases (namely, the geometric and Bernoulli distributions). The COM–Poisson distribution thereby can effectively model a range of count data distributions that contain data over- or under-dispersion, from the geometric to the Poisson to the Bernoulli distributions, simply through the addition of one parameter. The statistical methods described in this reference cover a myriad of topics, including distributional theory, generalized linear modeling, control chart theory, and count processes. Chapter 2 describes the COM–Poisson distribution in further detail and discusses its associated statistical properties. It further introduces various proposed parametrizations of the model and offers added discussion regarding the normalizing constant and its approximations. Chapter 3 introduces readers to several distributional extensions of the COM–Poisson distribution and/or other distributions that otherwise associate with the COM–Poisson model. Chapter 4 highlights bivariate and multivariate count distributions that are motivated by the COM–Poisson and discusses their respective statistical properties. Chapter 5 highlights various approaches for COM–Poisson regression under the various parametrizations, including discussions regarding model formulation and estimation approach. It further discusses subsequent advancements, including considerations of observation-level dispersion, additive models, and accounting for excess zeroes and/or data clustering. Chapter 6 introduces the reader to flexible control chart developments for discrete data, including COM–Poisson-motivated generalized control charts, cumulative sum charts, and generalized exponentially weighted moving average control charts. Chapter 7 presents methods for analyzing serially dependent count data via COM–Poisson-motivated stochastic processes, as well as time series and spatio-temporal models. Finally, Chapter 8 presents COM–Poisson-motivated cure rate models that can be used to describe time-to-event data, thus demonstrating the use of this flexible model as a tool in survival analysis. All of the chapters incorporate (where possible) discussions regarding statistical computations via `R`, thus introducing readers to the opportunities for data analysis via the featured R packages and their functionality.

As demonstrated in the subsequent chapters, a great deal of work has emerged where statistical methodologies are motivated by the COM–Poisson distribution. The utility of the COM–Poisson distribution, however, is not limited to these areas. Additional COM–Poisson-related works have emerged in fields, including capture–recapture and other abundance estimation methods (Anan et al., 2017; Wu et al., 2015), and disclosure limitation (Kadane et al., 2006a). Further, the COM–Poisson distribution has been employed in a variety of applications, including biology (Ridout and Besbeas, 2004), linguistics (Shmueli et al., 2005), risk analysis (Guikema and Coffelt, 2008), transportation (Lord and Guikema, 2012; Lord et al., 2008, 2010), and marketing and eCommerce (Boatwright et al., 2003; Borle et al., 2006, 2005, 2007).

Throughout this reference, much of the discussion focuses on parameter-estimation techniques associated with the various statistical method developments. These approaches are relatively thematic, falling in line with one of three approaches: maximum likelihood estimation, generalized quasi-likelihood estimation, and Bayesian estimation (Markov Chain Monte Carlo, Metropolis–Hastings, etc.). This referencc will provide a high-level discussion of the respective approaches as they relate to the featured concepts; however, it assumes that the reader has a prerequisite, rudimentary knowledge of these concepts.

A common theme regarding parameter estimation in this reference centers on its dependence on statistical computation to obtain results because the COM–Poisson distribution does not have a closed form. Various optimization tools exist, however, to aid analysts with such issues. This reference focuses on R tools where existing package functions or analyst-generated codes can utilize optimization tools such as `optim`, `nlm`, or `nlminb` to determine parameter estimates. Details are supplied throughout the manuscript in relation to the respective statistical methodologies under discussion. Meanwhile, hypothesis testing discussions generally center on the likelihood ratio test, while other test statistics (e.g. Rao's score test) can likewise be considered. The likelihood ratio test statistic is

$$\Lambda = \frac{\sup_{\theta \in \Theta_0} L(\theta)}{\sup_{\theta \in \Theta} L(\theta)}, \tag{1.20}$$

where $\theta$ denotes the collection of parameters under consideration, and $\Theta_0$ and $\Theta$ represent the parameter space under the null hypothesis and in general, respectively; as $n \to \infty$, $-2 \ln \Lambda$ converges to a chi-squared

Table 1.2 *Levels of model support based on AIC difference values,* $\Delta_{\mathrm{i}} = AIC_{\mathrm{i}} - AIC_{min}$, *for Model* i *(Burnham and Anderson, 2002).*

| $\Delta_i$ | Empirical support level for Model $i$ |
|---|---|
| $[0, 2]$ | Substantial |
| $[4, 7]$ | Considerably less |
| $(10, \infty)$ | Essentially none |

distribution. Tests about a boundary condition under the null hypothesis meanwhile produce a likelihood ratio test statistic whose asymptotic distribution is based on the equally weighted sum of a point mass and the cumulative probability of a chi-squared distribution (i.e. $0.5 + 0.5\chi^2$) (Self and Liang, 1987). For example, a common interest is to test for statistically significant dispersion where the dispersion parameter may be bounded by 0; this test is introduced in Section 2.4.5 and noted throughout subsequent chapters in this reference as the implications of this test relate to the corresponding chapter content.

Discussions will also include model comparisons to demonstrate and substantiate the COM–Poisson model's importance and flexibility. The Akaike information criterion (AIC) and the Bayesian information criterion (BIC) are two popular measures used for model comparisons, where

$$\text{AIC} = -2\ln(L) + 2k \quad \text{and} \quad \text{BIC} = -2\ln(L) + k\ln(n)$$

for a model's maximized likelihood value $L$, number of parameters $k$, and sample size $n$. For a collection of considered models, the selected model is desired to have the minimum AIC or BIC, respectively. In particular, this reference adopts the Burnham and Anderson (2002) approach for model comparison, where models are compared via the AIC and relative performance is measured via AIC difference values $\Delta_i = \text{AIC}_i - \text{AIC}_{\text{min}}$, where $\text{AIC}_i$ denotes the AIC associated with Model $i$, and $\text{AIC}_{\text{min}}$ is the minimum AIC among the considered models. Table 1.2 supplies the levels of model support based on recommended $\Delta_i$ ranges.

# 2

# The Conway–Maxwell–Poisson (COM–Poisson) Distribution

Conway and Maxwell (1962) developed what is now referred to as the Conway–Maxwell–Poisson (abbreviated as COM–Poisson or CMP in the literature) distribution as a more flexible queuing model to allow for state-dependent service rates. While this is a significant contribution to the field of count distributions, it did not gain notoriety in the statistics community until Shmueli et al. (2005) studied its probabilistic and statistical properties. This chapter defines the COM–Poisson distribution in greater detail, discussing its associated attributes and computing tools available for analysis. Section 2.1 details how the COM–Poisson distribution was derived, highlighting the underlying queuing model under consideration. Section 2.2 describes the probability distribution, and introduces computing functions available in R (R Core Team, 2014) that can be used to determine various probabilistic quantities of interest, including the normalizing constant, probability and cumulative distribution functions, random number generation, mean, and variance. Section 2.3 outlines the distributional and statistical properties associated with this model. Section 2.4 discusses parameter estimation and statistical inference associated with the COM–Poisson model. Section 2.5 describes various processes for generating random data, along with associated available R computing tools. Section 2.6 provides reparametrizations of the density function that serve as alternative forms for statistical analyses and model development. This section will introduce the proposed density representations and their implications on the associated statistical properties. Section 2.7 considers the COM–Poisson as a weighted Poisson distribution. Section 2.8 provides detailed discussion describing the various ways to approximate the COM–Poisson normalizing function. Finally, Section 2.9 concludes the chapter with discussion. Throughout the book, general references regarding Conway–Maxwell–Poisson distributions will be made as "COM–Poisson" when there is no need to distinguish between various model representations. Meanwhile, respective acronyms will be introduced to denote the model parametrizations.

## 2.1 The Derivation/Motivation: A Flexible Queueing Model

Conway and Maxwell (1962) deviate from a queueing model whose service rate is independent of the system state to account for adjustments conducted in actual queues to prevent significant system imbalance; while service centers can be equally loaded, varying service times can cause some service centers to have long lines, while others do not. Accommodations instilled in practice thus serve as motivators to generalize the queueing model, including machine loading, allowances for machine and operator flexibility to prevent service overload, and changes in individual service rates based on the amount of backlog in the queue. Accordingly, they consider a flexible queueing system that allows for dependent arrival and/or service rates that are functions of the current system state, i.e.

$$\pi_y = y^\nu \pi, \tag{2.1}$$

where $\pi_y$ denotes the mean service rate for each of the $y$ units in a system, $1/\pi$ is the mean service time for a unit when that unit is the only one in the system, and $\nu$ is a "pressure coefficient" describing the impact of the system state on the service rate of the system. Two special cases involving the pressure coefficient $\nu$ are immediately evident via Equation (2.1): $\nu = 0$ in Equation (2.1) (i.e. $\pi_y = \pi$ for all $y$) represents the special case of a queueing system where the service rate is independent of the system state and $\nu = 1$ describes the case when the service rate is directly proportional to the number of units in the system. The pressure coefficient can likewise be negative (i.e. $\nu < 0$, denoting when the service center slows in association with an increased workload), between 0 and 1 (i.e. $0 < \nu < 1$, denoting when the constant of proportionality between $\pi_y$ and $\pi$ is less than the number of units in the system), or greater than 1 (i.e. $\nu > 1$, producing a constant of proportionality between $\pi_y$ and $\pi$ that is greater than the number of units in the system).

Conway and Maxwell (1962) consider a single-queue, single-server system with random arrival times and a first-come-first-serve policy for arriving units, where the interarrival times are exponentially distributed with mean $\rho$, and the service times are exponentially distributed with the mean as defined in Equation (2.1). These assumptions associated with this queuing system with a state-dependent service rate construct produce the system of differential equations:

$$P_0'(t) = -\rho P_0(t) + \pi P_1(t) \tag{2.2}$$

$$P_y'(t) = -(\rho + y^\nu \pi)P_y(t) + \rho P_{y-1}(t) + (y+1)^\nu \pi P_{y+1}(t), \text{ for } y > 0, \tag{2.3}$$

where assuming a steady state and letting $\lambda = \rho/\pi$ produce the recursion equations

$$\lambda P_0 = P_1 \tag{2.4}$$

$$(\lambda + y^\nu)P_y = \lambda P_{y-1} + (y+1)^\nu P_{y+1}, \text{ for } y > 0. \tag{2.5}$$

Solving the recursion produces what is now known as the COM–Poisson distribution with the form

$$P_y = P(Y = y) = \frac{\lambda^y}{(y!)^\nu} P_0, \tag{2.6}$$

where $P_0 = P(Y = 0) = \frac{1}{\sum_{y=0}^{\infty} \frac{\lambda^y}{(y!)^\nu}}$. We denote $Z(\lambda, \nu) = 1/P_0$, the normalizing term associated with this distribution. Let this parametrization be referred to as the CMP($\lambda, \nu$) distribution; other parametrizations are introduced in Section 2.6.

Along with the derivation of this distribution, Conway and Maxwell (1962) determine the form of the mean number $\mu$ of the system as

$$\mu = \sum_{y=0}^{\infty} y P_y = \frac{\sum_{y=0}^{\infty} \frac{y\lambda^y}{(y!)^\nu}}{\sum_{y=0}^{\infty} \frac{\lambda^y}{(y!)^\nu}}, \tag{2.7}$$

recognizing that closed-form results emerge for $P_0$ and $\mu$ for special cases involving $\nu$; see Section 2.2 for a more detailed discussion. Finally, the above queuing system model with the state-dependent service rate can likewise be derived in two other ways. One way is to let the arrival rate depend on the system state through the relation, $\rho_y = (y+1)^{-\xi}\rho$, while assuming an exponentially distributed service time. The second approach is to consider the model where the arrival rate is state-dependent ($\rho_y = (y+1)^{-\xi}\rho$) as is the service rate ($\pi_y = y^\nu \pi$). With these assumptions in place along with the added assumption of exponentially distributed service times and interarrival periods, the $(\xi + \nu) = \alpha$ system is consistent with the state-dependent service rate model with $\nu = \alpha$. More precisely, this more flexible state-dependent arrival and service rate model contains the special case of the state-dependent service rate model when $\xi = 0$, and the state-dependent arrival rate model when $\nu = 0$ (Conway and Maxwell, 1962).

## 2.2 The Probability Distribution

The COM–Poisson distribution is a flexible, two-parameter count distribution that allows for data dispersion relative to the Poisson model. Derived by Conway and Maxwell (1962), the CMP parametrization has a probability mass function of the form

$$P(X = x) = \frac{\lambda^x}{(x!)^\nu Z(\lambda, \nu)}, \qquad x = 0, 1, 2, \ldots, \tag{2.8}$$

or, on a log scale,

$$\begin{aligned} \ln[P(X = x)] &= x \ln \lambda - \nu \ln(x!) - \ln[Z(\lambda, \nu)] \\ &= x \ln \lambda - \nu \sum_{j=1}^{x} \ln(j) - \ln[Z(\lambda, \nu)] \end{aligned} \tag{2.9}$$

for a random variable $X$, where the CMP intensity or location parameter $\lambda = E(X^\nu)$ is a generalized form of the Poisson rate parameter; $\nu \geq 0$ is a dispersion parameter such that $\nu = 1$ denotes equi-dispersion, and $\nu > (<)1$ signifies under-dispersion (over-dispersion) relative to the Poisson model; and $Z(\lambda, \nu) = \sum_{j=0}^{\infty} \frac{\lambda^j}{(j!)^\nu}$ normalizes the CMP distribution so that it satisfies the basic probability axioms.[1] The CMP($\lambda$, $\nu$) distribution has the form of an exponential family as defined in Equation (1.17), where $x$ and $\ln(x!)$ denote the joint sufficient statistics for $\lambda$ and $\nu$, respectively. More broadly, given a sample of observations $x_1, \ldots, x_n$ from a CMP($\lambda$, $\nu$) distribution, the likelihood and log-likelihood functions are

$$L(\lambda, \nu; \boldsymbol{x}) = \prod_{i=1}^{n} \frac{\lambda^{x_i}}{(x_i!)^\nu Z(\lambda, \nu)} = \frac{\lambda^{\sum_{i=1}^{n} x_i}}{[Z(\lambda, \nu)]^n \left(\prod_{i=1}^{n} x_i!\right)^\nu} \quad \text{and}$$

$$\ln L(\lambda, \nu; \boldsymbol{x}) = \sum_{i=1}^{n} x_i \ln \lambda - \nu \sum_{i=1}^{n} \ln(x_i!) - n \ln Z(\lambda, \nu). \tag{2.10}$$

The CMP distribution does not, however, have the form of a two-parameter exponential dispersion family (see Equation (1.18)) because the $Z(\lambda, \nu)$ function cannot be factored into separate functions of $\lambda$ and $\nu$. This can be overcome, however, for a CMP distribution with a fixed $\nu$; here, $\eta(\lambda) = \ln \lambda$, $T(x) = x$, $H(x; \nu) = (x!)^{-\nu}$, $\Psi(\lambda; \nu) = \ln(Z(\lambda, \nu))$, $w = 1$, and $\phi = 1$. Similarly, these denotations define the terms for the exponential family form as shown in Equation (1.4) for a given dispersion $\nu$.

[1] $Z(\lambda, \nu) = 1/P_0$, given the notation used in Conway and Maxwell (1962) and described in Section 2.1.

Table 2.1 *Special cases of the CMP parametrization distribution.*

| Constraint | $Z(\lambda, \nu)$ | pmf, $P(X = x)$ | Mean | Distribution |
|---|---|---|---|---|
| $\nu = 1$ | $e^{\lambda}$ | $P(X = x) = \frac{\lambda^x e^{-\lambda}}{x!}, x = 0, 1, 2, \ldots$ | $\lambda$ | Poisson($\lambda$) |
| $\nu = 0$, $\lambda < 1$ | $\frac{1}{1-\lambda}$ | $P(X = x) = (1 - \lambda)\lambda^x, x = 0, 1, 2, \ldots$ | $\frac{\lambda}{1-\lambda}$ | Geometric ($p = 1 - \lambda$) |
| $\nu \to \infty$ | $1 + \lambda$ | $P(X = x) \to \begin{cases} \frac{1}{1+\lambda} & x = 0 \\ \frac{\lambda}{1+\lambda} & x = 1 \\ 0 & x = 2, 3, 4, \ldots \end{cases}$ | $\frac{\lambda}{1+\lambda}$ | Bernoulli $\left(p = \frac{\lambda}{1+\lambda}\right)$ |

The CMP normalizing constant $Z(\lambda, \nu)$ contains three special cases for which the infinite sum can be determined exactly; Table 2.1 summarizes the solutions for $Z(\lambda, \nu)$ and the associated distributions that result. Conway and Maxwell (1962) recognized the two special cases for the normalizing constant and corresponding mean, while the third case is a limiting result. With particular focus on the dispersion parameter $\nu$, we see that

$$Z(\lambda, \nu) = \begin{cases} \sum_{j=0}^{\infty} \frac{\lambda^j}{j!} = e^{\lambda} & \text{for } \nu = 1 \\ \sum_{j=0}^{\infty} \lambda^j = \frac{1}{1-\lambda} & \text{for } \nu = 0, \lambda < 1 \\ 1 + \lambda + \sum_{j=2}^{\infty} \frac{\lambda^j}{(j!)^{\nu}} \to 1 + \lambda & \text{for } \nu \to \infty. \end{cases} \tag{2.11}$$

Note that the special case where $\nu = 0$ requires the additional constraint that $\lambda < 1$ because, for $\nu = 0$ and $\lambda \geq 1$, $Z(\lambda, \nu)$ does not converge. Given these results for the normalizing constant (Equation (2.11)), one can deduce that the CMP distribution contains three classical special-case distributions: the Poisson model with rate parameter $\lambda$ (when $\nu = 1$), the geometric distribution with success probability $1 - \lambda$ (for $\nu = 0, \lambda < 1$), and the Bernoulli distribution with success probability $\frac{\lambda}{1+\lambda}$ (as $\nu \to \infty$); see Table 2.1.

Examples illustrating the flexibility of the CMP model can be found in Figure 2.1. Its subplots consider dispersion values ranging from 0 to 10, where for all $\nu > 0$, $\lambda = 3$ in order to easily assess the impact of the dispersion parameter on the distributional form. The CMP($\lambda = 3$, $\nu = 1$) plot is identical to the Poisson($\lambda = 3$) plot shown in Figure 1.1 (which is likewise shown in Figure 1.2). We can further see larger (smaller) variation in the CMP distribution for $\nu < (>)1$. The first subplot of Figure 2.1 meanwhile considers the special case where $\nu = 0$ and $\lambda = 0.3 < 1$; recall that

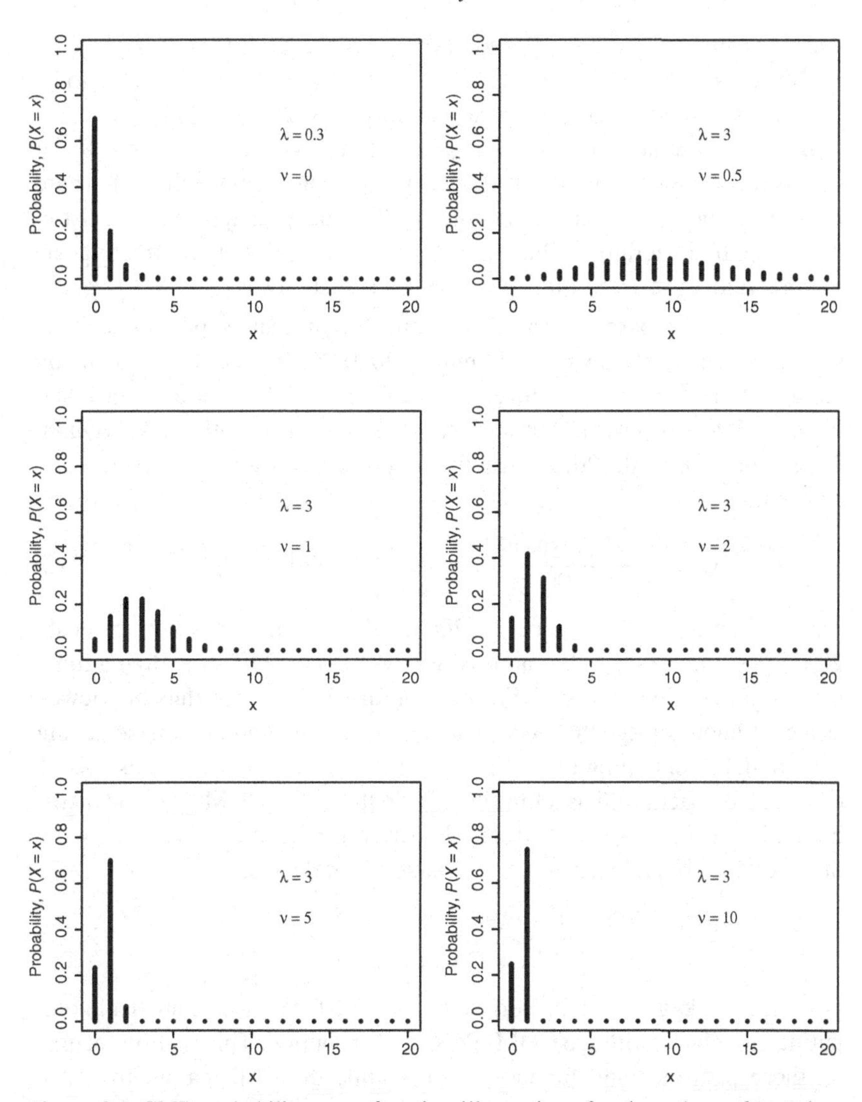


Figure 2.1 CMP probability mass function illustrations for the values of $\lambda$ and $\nu$. Respective illustrative plots define the same value for $\lambda$ when $\nu > 0$ for easy distributional comparisons, while for $\nu = 0$, $\lambda$ must be constrained to be less than 1; $\nu < (>)1$ signifies data over-dispersion (under-dispersion) relative to the Poisson ($\nu = 1$) distribution.

for this special case of $\nu$, $\lambda$ is constrained to be less than 1 so that the normalizing constant can converge. This probability mass function represents a geometric distribution with success probability $(1 - 0.3) = 0.7$. Similarly, the last subplot of Figure 2.1 also appears to reflect the extreme case of data under-dispersion, namely a Bernoulli($p = \lambda/(1+\lambda) = 0.75$) distribution; in fact, we see that, even for $\nu = 10$, the probability at 1 is estimated to be 0.75, while the probability at 0 is approximately 0.25.

Beyond these special cases, $Z(\lambda, \nu)$ does not reduce to a closed form; therefore, one may wish to consider an approximate form for $Z(\lambda, \nu)$, where $\lambda$ and $\nu$ are both nonnegative parameters. Two popular approaches exist in the computing literature. One approach is to consider approximating the infinite sum by a truncation point $M$ such that $Z(\lambda, \nu) = \sum_{j=0}^{\infty} \frac{\lambda^j}{(j!)^\nu} \approx \sum_{j=0}^{M} \frac{\lambda^j}{(j!)^\nu}$, either based on a given value for $M$ that is predicted to be sufficiently large (Sellers and Shmueli, 2010; Sellers et al., 2019) or by taking into account a given level of precision such that, for a given $\lambda$ and $\nu$, the $(M+1)$st term in the summand $\frac{\lambda^M}{(M!)^\nu}$ is sufficiently small (Dunn, 2012; Shmueli et al., 2005). Another approach considers the asymptotic approximation

$$Z(\lambda, \nu) = \frac{\exp(\nu\lambda^{1/\nu})}{\lambda^{(\nu-1)/2\nu}(2\pi)^{(\nu-1)/2}\sqrt{\nu}}\left\{1 + O\left(\frac{1}{\lambda^{1/\nu}}\right)\right\} \tag{2.12}$$

for $\nu \leq 1$ or $\lambda > 10^\nu$ (Minka et al., 2003; Sellers et al., 2019; Shmueli et al., 2005); Section 2.8 describes approximations to the CMP normalizing function in greater detail. The COM–Poisson distribution can thus be viewed as a continuous bridge that accommodates over- and under-dispersed count data models, containing the Poisson and geometric distributions as special cases and the Bernoulli as a limiting distribution. The COM–Poisson distribution is not closed under addition; however, it motivates the development of a COM–Poisson extension; see Chapter 3 for details.

### *2.2.1* `R` *Computing*

Several `R` packages have been developed to perform basic statistical computing associated with the COM–Poisson distribution. This section focuses on those packages and functions that assume the CMP parametrization; see Table 2.2. The `rcomp` (`CompGLM`) and `rcmp` (`COMPoissonReg`) commands both perform random number generation, but they will be discussed in Section 2.5.

The `CompGLM` package (Pollock, 2014a) can determine the CMP probability mass and cumulative probability via its respective functions, `dcomp` and `pcomp`. Both commands require the user to supply the values for $\lambda$ (`lam`) and $\nu$ (`nu`), where both parameters can be supplied as double vectors. An additional input `sumTo` (whose default value is set at 100) defines the truncation point $M$ for the infinite summation so that $Z(\lambda, \nu) \approx \sum_{k=0}^{M} \frac{\lambda^k}{(k!)^\nu}$.

Table 2.2 *Available* R *functions for CMP computing.*

| Package | Function | Computational result |
|---|---|---|
| CompGLM | dcomp | Probability mass function |
| | pcomp | Cumulative distribution function |
| | rcomp | Random number generator |
| compoisson | com.compute.z | Normalizing constant, $Z(\lambda, \nu)$ |
| | com.compute.log.z | $\ln(Z(\lambda, \nu))$ |
| | dcom | Probability mass function |
| | com.log.density | Log-probability |
| | com.loglikelihood | Log-likelihood |
| | com.expectation | Expectation, $E[f(\cdot)]$ |
| | com.mean | Mean |
| | com.var | Variance |
| COMPoissonReg | dcmp | Probability mass function |
| | pcmp | Cumulative distribution function |
| | qcmp | Quantile function |
| | rcmp | Random number generator |

The dcomp and pcomp functions further provide a logical operator to allow for the density function to be computed on a logarithmic scale (logP = TRUE) or not (logP = FALSE); the default setting assumes the original scale. Finally, dcomp requires the integer value y for which the CMP probability is to be computed. pcomp meanwhile contains the additional inputs, q and lowerTail; q is an integer vector[2] for which the CMP cumulative probabilities $P(X \leq q)$ are determined, and lowerTail is a logical term that identifies whether it (lowerTail = TRUE) or its complement (lowerTail = FALSE) should be computed.

The compoisson package (Dunn, 2012) contains the functions, dcom and com.log.density, which respectively compute the CMP probability mass function on the original and natural log scales. Both functions require x, lambda, and nu as inputs in order to compute $P(X = x)$ (on either scale) for some CMP($\lambda, \nu$) random variable $X$. The respective commands further allow for the normalizing constant $Z(\lambda, \nu)$ to be directly supplied by the analyst, but this is not required; dcom identifies the potential input as z, while com.log.density denotes it as log.z.

The compoisson package does not contain a function to directly compute the cumulative probability (i.e. analogous to the pcomp function in CompGLM); however, one can determine it by summing over the relevant

[2] Noninteger-valued terms are coerced to integer form.

`dcom` results from zero to $x$. Meanwhile, the `com.loglikelihood` command computes the log-likelihood of the data from a frequency table that is assumed to stem from a CMP($\lambda, \nu$) distribution. This function requires three inputs: `x`, the frequency table (in a matrix form), and `lambda` and `nu`, the CMP parameters. With the provided inputs, the analyst can obtain the log-likelihood value as provided in Equation (2.10). Meanwhile, because the log-likelihood function is the sum of the log-probabilities shown in Equation (2.9), for a given dataset of values $x_1, \ldots, x_n$, the resulting output provided from `com.loglikelihood` equals the sum of the reported `com.log.density` values associated with $x_1, \ldots, x_n$. The `compoisson` package also contains functions to compute the mean (`com.mean`) and variance (`com.var`), as well as the general expectation (`com.expectation`) of a CMP random variable; these functions will be discussed in more detail in Section 2.3.

The `COMPoissonReg` package (Sellers et al., 2019) is a broader `R` package that allows analysts to conduct a CMP regression associating a count response variable to several explanatory variables. Contained within this package, however, are the commands for the probability (`dcmp`), cumulative probability (`pcmp`), and quantile function (`qcmp`) for the CMP($\lambda, \nu$) distribution. All of the aforementioned functions require the inputs `lambda` and `nu`. Besides these inputs, `dcmp` requires the vector of quantile values `x` for which the function computes $P(X = x)$. This function also offers the option of computing the density function on the log-scale (`log = TRUE`) or not (`log = FALSE`); the function defaults assuming the original scale. `pcmp` likewise requires `x` in order to determine the cumulative probability $P(X \leq x)$. Meanwhile, this is the only package that can perform quantile calculations assuming a CMP parametrization. The `qcmp` function determines the quantile value $x$ such that $P(X \leq x) \geq q$, given some probabilty of interest $q$; `qcmp` includes the inputs `q` (the vector of probabilities) and the logical operator `log.p` that defaults such that the quantile value is determined on the original scale (i.e. `log.p = FALSE`). These functions (as for all in the `COMPoissonReg` package) rely on computing the normalizing constant $Z(\lambda, \nu)$. This package determines $Z(\lambda, \nu)$ in a hybrid fashion where, for $\lambda$ sufficiently large and $\nu$ sufficiently small, the closed-form approximation (Equation (2.12)) is used to estimate $Z(\lambda, \nu)$, while the infinite summation is otherwise Winsorized to meet a desired accuracy level if the conditions for $\lambda$ and $\nu$ are not satisfied.

To illustrate the various packages and their respective functionality, consider computing the (cumulative) probability at $x = 2$ of a CMP($\lambda = 4, \nu$) random variable where $\nu \in \{0.3, 1, 3, 30\}$. Naturally, we expect the corresponding (cumulative) probability computations from the respective packages to equal each other. Table 2.3 provides the respective outcomes obtained using the `compoisson` (Dunn, 2012), `CompGLM` (Pollock, 2014a), and `COMPoissonReg` (Sellers et al., 2019) packages. Rounding computational results to seven significant digits, we obtain equality for all (cumulative) probabilities determined by `CompGLM` and `COMPoissonReg`, while these results differ slightly from `compoisson` by within 0.0001 for all values of $\nu$. In particular, we expect the special cases for the CMP model to hold via statistical computation. For the case where $\nu = 1$, we expect to obtain the same (cumulative) probabilities for the CMP($\lambda, \nu = 1$) and Poisson($\lambda$) distributions. Table 2.4 provides a comparison of the two (cumulative) probabilities, using `compoisson`, `CompGLM`, and `COMPoissonReg` for CMP, and the `stats` package for Poisson. For illustrative purposes, calculations shown assume $\lambda = 4$, and results are rounded to six decimal places. We see that `dcomp` (`CompGLM`) and `dcmp` (`COMPoissonReg`) agree exactly with that from `dpois`, while `dcom` (`compoisson`) produces a result whose difference from `dpois` is more than $4 \times 10^{-5}$. Similarly, the results for `pcomp` (`CompGLM`) and `pcmp` (`COMPoissonReg`) equal that from `ppois`, while summing the `dcom` values from 0 to 2 (thus determining the cumulative probability, $P(X \leq 2)$) differs from `ppois` by more than $6 \times 10^{-5}$. We likewise expect the CMP special case where $\nu = 0$ and $\lambda < 1$ to agree with the geometric distribution with success probability $1 - \lambda$. Table 2.5 compares the CMP (cumulative) probability calculations determined by `compoisson`, `CompGLM`, and `COMPoissonReg` with the geometric distribution calculations attained via the `stats` package; for illustrative purposes, let $\lambda = 0.25$. The resulting calculations are rounded to six decimal places. We again find that the `dcomp` (`CompGLM`) and `dcmp` (`COMPoissonReg`) computations agree exactly with that from `dgeom` (`stats`), while `dcom` (`compoisson`) produces a result whose difference from `dgeom` is more than $1 \times 10^{-5}$. Similarly, the `pcomp` (`CompGLM`) and `pcmp` (`COMPoissonReg`) cumulative probabilities equal that from `pgeom`, while summing over the `dcom` values from 0 to 2 produces a cumulative probability result that differs from `pgeom` by more than $2 \times 10^{-4}$. The differences in which the normalizing function is determined appear to induce the different results. `CompGLM`, for example, uses the default truncation value for the normalizing function; this choice of summation limit proves to be sufficient and satisfactory for this illustration. In contrast, `compoisson` uses a default measure to constrain the amount of allowable error (`log.error = 0.001`). This default value appears to be slightly deficient for this example, thus producing

Table 2.3 *(Cumulative) probability computations via various* R *packages and their respective functions, illustrated assuming a CMP($\lambda = 4, \nu$) random variable* X *evaluated at the value 2 for $\nu \in \{0.3, 1, 3, 30\}$. Functions produce equal calculations when rounded to three decimal places.*

| | | Probability, $P(X = 2)$ | | | |
|---|---|---|---|---|---|
| Package | Function | $\nu = 0.3$ | $\nu = 1$ | $\nu = 3$ | $\nu = 30$ |
| CompGLM | dcomp | $9.314738 \times 10^{-14}$ | 0.1465251 | 0.2733952 | $2.980232 \times 10^{-9}$ |
| compoisson | dcom | $4.339295 \times 10^{-14}$ | 0.1465652 | 0.2733956 | $2.980232 \times 10^{-9}$ |
| COMPoissonReg | dcmp | $9.314738 \times 10^{-14}$ | 0.1465251 | 0.2733952 | $2.980232 \times 10^{-9}$ |
| | | Cumulative probability, $P(X \leq 2)$ | | | |
| Package | Function | $\nu = 0.3$ | $\nu = 1$ | $\nu = 3$ | $\nu = 30$ |
| CompGLM | pcomp | $1.289842 \times 10^{-13}$ | 0.2381033 | 0.9568831 | 1 |
| compoisson | Sum dcom results | $6.008764 \times 10^{-14}$ | 0.2381685 | 0.9568845 | 1 |
| COMPoissonReg | pcmp | $1.289842 \times 10^{-13}$ | 0.2381033 | 0.9568831 | 1 |

Table 2.4 *Probability* P(X = 2) *and cumulative probability* P(X ≤ 2) *computations for the CMP*($\lambda = 4, \nu = 1$) = *Poisson*($\lambda = 4$) *distributed random variable* X. *CMP computations determined using* `compoisson` *(Dunn, 2012),* `CompGLM` *(Pollock, 2014a), and* `COMPoissonReg` *(Sellers et al., 2019); Poisson results obtained using the* `stats` *package. All calculations rounded to six decimal places.*

| Package | Function | $P(X = 2)$ | Function | $P(X \leq 2)$ |
|---|---|---|---|---|
| `compoisson` | `dcom` | 0.146565 | Sum `dcom` results | 0.238169 |
| `CompGLM` | `dcomp` | 0.146525 | `pcomp` | 0.238103 |
| `COMPoissonReg` | `dcmp` | 0.146525 | `pcmp` | 0.238103 |
| `stats` | `dpois` | 0.146525 | `ppois` | 0.238103 |

Table 2.5 *Probability* P(X = 2) *and cumulative probability* P(X ≤ 2) *computations for the CMP*($\lambda = 0.25, \nu = 0$) = *Geom*(p = *0.75*) *distributed random variable* X. *CMP computations determined using* `compoisson` *(Dunn, 2012),* `CompGLM` *(Pollock, 2014a), and* `COMPoissonReg` *(Sellers et al., 2019); geometric results obtained using the* `stats` *package. All calculations rounded to six decimal places.*

| Package | Function | $P(X = 2)$ | Function | $P(X \leq 2)$ |
|---|---|---|---|---|
| `compoisson` | `dcom` | 0.046886 | Sum `dcom` results | 0.984615 |
| `CompGLM` | `dcomp` | 0.046875 | `pcomp` | 0.984375 |
| `COMPoissonReg` | `dcmp` | 0.046875 | `pcmp` | 0.984375 |
| `stats` | `dgeom` | 0.046875 | `pgeom` | 0.984375 |

the computational differences shown in Tables 2.3–2.5. The difference, however, can be argued as sufficiently small such that it may not make any significant difference in subsequent computations. Alternatively, the amount of allowable error can be decreased in order to provide greater accuracy.

To illustrate the relationship between the respective `compoisson` functions, `com.log.density` and `com.loglikelihood`, consider the illustrative frequency table provided in Table 2.6 where we assume a CMP($\lambda = 2$, $\nu = 3$) model. Using the `com.log.density` function, we can determine the log-probabilities at 0, 1, and 2, respectively, as −1.263617, −0.5704694, and −1.956764; see Code 2.1. Meanwhile, using the `com.loglikelihood` function, we provide the frequency table provided in Table 2.6 as a matrix where the count values are supplied in the

Table 2.6 *Hypothetical frequency table for count data. These data are used for illustrative analyses in Code 2.1.*

| Count | Frequency |
|---|---|
| 0 | 6 |
| 1 | 15 |
| 2 | 4 |

first column and the associated frequencies are given in the second column. With this input provided along with the assumed values for $\lambda$ and $\nu$, we find that the associated log-likelihood under this construct equals $-23.96579$. Note that this summand equals the weighted sum of the respective log-probabilities, weighted by their respective frequencies. This is equivalent to summing the reported `com.log.density` values associated with each count value contained in this dataset; see Code 2.1.

Code 2.1 R output comparing the results from the functions `com.log.density` and `com.loglikelihood`. Associated computations assume the CMP($\lambda$, $\nu$) model with parameters `lambda` $= \lambda = 2$ and `nu` $= \nu = 3$.

```
> ex <- matrix(c(0,6,1,15,2,4),byrow=TRUE,nrow=3)
> colnames(ex) <- c("Count","Frequency")
> ex
     Count Frequency
[1,]     0         6
[2,]     1        15
[3,]     2         4
> com.log.density(0,lambda=2,nu=3)
[1] -1.263617
> com.log.density(1,lambda=2,nu=3)
[1] -0.5704694
> com.log.density(2,lambda=2,nu=3)
[1] -1.956764
> 6*com.log.density(0,lambda=2,nu=3)
     + 15*com.log.density(1,lambda=2,nu=3)
     + 4*com.log.density(2,lambda=2,nu=3)
[1] -23.96579
> com.loglikelihood(ex,lambda=2,nu=3)
          [,1]
[1,] -23.96579
```

Table 2.7 *Quantile determinations* x *such that* P(X ≤ x) ≥ *0.9 for the CMP(*$\lambda = 3, \nu$*) distributed random variable* X*, where* $\nu \in \{0.3, 1, 3, 30\}$*. Computations conducted via the* `qcmp` *function (*`COMPoissonReg`*).*

| $\nu$ | $x$ |
|---|---|
| 0.3 | 55 |
| 1.0 | 5 |
| 3.0 | 2 |
| 30.0 | 1 |

For completeness of discussion, Table 2.7 provides the quantile values assuming that $X$ has a CMP($\lambda = 3, \nu$), where $\nu \in \{0.3, 1, 3, 30\}$, and we desire the quantile $x$ such that $P(X \leq x) \geq 0.9$. We see that $x$ decreases as $\nu$ increases; this makes sense because larger $\nu$ implies tighter variation in the data; thus, more probability is contained in a smaller region, implying an increased cumulative probability associated with a value $x$.

## 2.3 Distributional and Statistical Properties

The ratio between probabilities of two consecutive values is

$$\frac{P(X = x - 1)}{P(X = x)} = \frac{x^{\nu}}{\lambda}. \tag{2.13}$$

This nonlinear relationship simplifies to the linear ratio between probabilities of two consecutive values from a Poisson (i.e. $\nu = 1$) model; see Equation (1.3). For $\nu < 1$, successive ratios are flatter than those of a Poisson model; this scenario implies longer tails than the Poisson, i.e. data over-dispersion. Meanwhile, values of $\nu > 1$ demonstrate the opposite effect, hence shorter tails than the Poisson, i.e. data under-dispersion (Shmueli et al., 2005). The CMP is a unimodal distribution whose probability mass function is log-concave, and it has an increasing failure rate and decreasing mean residual life. In fact, as a special case of a weighted Poisson distribution, the CMP is log-concave when $\nu \geq 1$ because, for these values of $\nu$, the weight function $w(x;\nu) = (x!)^{1-\nu}$ is itself log-concave (Gupta et al., 2014; Kokonendji et al., 2008). Finally, the mode of a CMP($\lambda, \nu$) distribution occurs at $\lfloor \lambda^{1/\nu} \rfloor$ if $\lambda^{1/\nu} \notin \mathbb{N}$ or, for $\lambda^{1/\nu} \in \mathbb{N}$, the modes are $\lambda^{1/\nu}$ and $\lambda^{1/\nu} - 1$ (Daly and Gaunt, 2016).

Generating functions associated with the CMP distribution are respectively defined in terms of $Z$ function representations. For a random variable $X$, the probability generating function is $\Pi_X(t) = E(t^X) = \frac{Z(\lambda t, \nu)}{Z(\lambda, \nu)}$, while the moment generating function is $M_X(t) = E(e^{Xt}) = \frac{Z(\lambda e^t, \nu)}{Z(\lambda, \nu)}$. The moment generating function can be utilized to obtain the moments of the CMP distribution (Sellers et al., 2011). Exact formulae for the expected value and variance are

$$E(X) = \frac{\partial \ln Z(\lambda, \nu)}{\partial \ln \lambda} = \lambda \frac{\partial \ln Z(\lambda, \nu)}{\partial \lambda} \text{ and} \tag{2.14}$$

$$V(X) = \frac{\partial^2 \ln Z(\lambda, \nu)}{\partial (\ln \lambda)^2} = \frac{\partial E(X)}{\partial \ln \lambda}. \tag{2.15}$$

More broadly, the following recursive relationship holds for the CMP moments:

$$E(X^{r+1}) = \begin{cases} \lambda[E(X+1)]^{1-\nu} & r = 0 \\ \lambda \frac{\partial}{\partial \lambda} E(X^r) + E(X)E(X^r) & r > 0, \end{cases} \tag{2.16}$$

where $E(X^r) = \sum_{x=0}^{\infty} x^r P(X = x)$ denotes the $r$th moment of a random variable $X$ with probability $P(X = x)$ as defined in Equation (2.8). Meanwhile, letting $(j)_r = j(j-1) \cdot (j-r+1)$ denote a falling factorial, $E[((X)_r)^\nu] = \lambda^r$ for a CMP($\lambda$, $\nu$) distributed random variable $X$ and $r \in \mathbb{N}$, and the $r$th moment has the approximation $E(X^r) \approx \lambda^{r/\nu} \left(1 + O(\lambda^{-1/\nu})\right)$ as $\lambda \to \infty$.

The CMP distribution belongs to the class of two parameter power series distributions, and is a special case of a modified power series distribution (Gupta, 1974, 1975), i.e. a distribution whose probability mass function has the form

$$P(X = x) = \frac{A(x)(g(\theta))^x}{f(\theta)}, \quad x \in B, \tag{2.17}$$

where $B$ is a subset of the set of nonnegative integers, $A(x) > 0$, and $f(\theta)$ and $g(\theta)$ are the positive, finite, differentiable functions; here, $\theta = \lambda$ such that $g(\theta) = g(\lambda) = \lambda, f(\theta) = f(\lambda) = Z(\lambda, \nu)$, and $A(x) = (x!)^{-\nu}$. This form of the modified power series distribution not only confirms that the mean has the form provided in Equation (2.14) along with the recursion for the $r$th moment (Equation (2.16)) but further infers the existence of the following recursion formulae for the $r$th central and factorial moments, respectively, $r = 1, 2, 3, \ldots$:

$$E\left((X-\mu)^{r+1}\right) = \lambda \frac{\partial E\left((X-\mu)^r\right)}{\partial \lambda} + r\lambda E\left((X-\mu)^{r-1}\right) \frac{\partial^2}{\partial \lambda^2}[\ln Z(\lambda, \nu)] \tag{2.18}$$

$$E((X)_{r+1}) = \lambda \frac{\partial E((X)_r)}{\partial \lambda} + \lambda E((X)_r) \frac{\partial}{\partial \lambda}[\ln Z(\lambda, \nu)] - rE((X)_r). \tag{2.19}$$

Results of interest regarding the CMP distribution likewise include the median, cumulants, and other statistical measures. For a CMP($\lambda$, $\nu$) distributed random variable $X$, its median $m$ is approximately $\lambda^{1/\nu}$ as $\lambda \to \infty$; more precisely, $m \approx \lambda^{1/\nu} + O(\lambda^{1/2\nu})$ as $\lambda \to \infty$. The $c$th cumulant $\kappa_c$ is approximated as $\kappa_c \approx \frac{1}{\nu^{c-1}}\lambda^{1/\nu} + O(1)$ as $\lambda \to \infty$. Further, as $\lambda \to \infty$, the skewness $\gamma_1$ and kurtosis $\gamma_2$ are approximately

$$\gamma_1 \approx \frac{1}{\sqrt{\nu}}\lambda^{-1/2\nu} + O(\lambda^{-3/2\nu}) \tag{2.20}$$

$$\gamma_2 \approx \frac{1}{\nu}\lambda^{-1/\nu} + O(\lambda^{-2/\nu}). \tag{2.21}$$

Finally, the mean deviation of $X^\nu$ is

$$E|X^\nu - \lambda| = \frac{2}{Z(\lambda, \nu)} \frac{\lambda^{\lfloor \lambda^{1/\nu} \rfloor + 1}}{\lfloor \lambda^{1/\nu} \rfloor !}. \tag{2.22}$$

The CMP mean and variance (Equations (2.14) and (2.15)) have the respective approximations

$$E(X) \approx \mu \doteq \lambda^{1/\nu} - \frac{\nu - 1}{2\nu} \quad \text{and} \tag{2.23}$$

$$V(X) \approx \frac{1}{\nu}\lambda^{1/\nu} = \frac{\mu}{\nu} + \frac{\nu - 1}{2\nu^2}, \tag{2.24}$$

where these approximations are determined via the asymptotic approximation for $Z(\lambda, \nu)$ as described in Section 2.8, and they hold for $\nu \leq 1$ or $\lambda > 10^\nu$ (Minka et al., 2003; Shmueli et al., 2005; Sunecher et al., 2020). Other methods for approximating the mean of the distribution include estimating the mean via the distribution mode, summing the first terms of $Z$ when $\nu$ is large, or bounding the mean when $\nu$ is small (Francis et al., 2012; Lord et al., 2008). Given the approximations for the mean and variance (Equations (2.23) and (2.24)), one can see that, given $\nu$, the CMP mean increases (decreases) as $\lambda$ increases (decreases). The variance likewise increases (decreases) with $\lambda$ for any data that are equi- or over-dispersed. Meanwhile, for given $\nu > 1$, the variance increases (decreases) as $\lambda$ decreases (increases) (Alevizakos and Koukouvinos, 2022). The approximations provided in Equations (2.23) and (2.24) further motivate two reparametrizations of the COM–Poisson distribution: the approximate COM–Poisson (ACMP) parametrization (Guikema and Coffelt, 2008) which assumes $\mu_* = \lambda^{1/\nu}$ as a measure of center (that refraining from referring to this as a mean) and a mean-parametrized COM–Poisson (MCMP2) (Ribeiro Jr. et al., 2019) where the mean $\mu$ is estimated as done in Equation (2.23). See Section 2.6 for more details regarding these and other reparametrizations.

The CMP dispersion index is defined and approximated as

$$\mathrm{DI}(X) = \frac{\partial^2 \ln Z(\lambda, \nu)/\partial(\ln \lambda)^2}{\partial \ln Z(\lambda, \nu)/\partial \ln \lambda} \approx \frac{\lambda^{1/\nu}/\nu}{\lambda^{1/\nu} - (\nu - 1)/2\nu} \approx \frac{1}{\nu}. \tag{2.25}$$

Thus, for $\nu = 1$, we find that the data are equi-dispersed (the Poisson model) since the mean and variance equal, while for $\nu < ( > )1$, we see that the data are over- (under-)dispersed as the variance is greater (less) than the mean (Anan et al., 2017). The dispersion index, along with other indexes regarding zero-inflation (ZI) and heavy-tail (HT), namely

$$\mathrm{ZI}(X) = 1 + \frac{\ln P(X = 0)}{E(X)} \quad \text{and} \quad \mathrm{HT}(X) = \frac{P(X = x + 1)}{P(X = x)}, \quad \text{for } x \to \infty, \tag{2.26}$$

depend on the value of $\mu$ and become stable for large $\mu$; they confirm that the relationship between the mean and variance is proportional to the dispersion. Finally, as the mean increases, the expected number of zeros decreases (Brooks et al., 2017), and the moment

$$E(\ln(X!)) = -\frac{d \ln Z(\lambda, \nu)}{d\nu} \approx \frac{1}{2\nu^2} \ln(\lambda) + \lambda^{1/\nu}\left(\frac{\ln(\lambda)}{\nu} - 1\right) + \frac{1}{2\nu} + \frac{\ln(\pi)}{2}, \tag{2.27}$$

where the sum of the first two approximation terms in Equation (2.27) reasonably estimates the result for $\nu \leq 1$ or $\lambda > 10^{\nu}$ (Gupta et al., 2014; Minka et al., 2003).

Additional properties regarding the CMP distribution hold. Daly and Gaunt (2016) note that, for a CMP($\lambda, \nu$) distributed random variable $X$ and function $f{:}\mathbb{Z}^+ \to \mathbb{R}$ such that $E|f(X+1)| < \infty$ and $E|X^{\nu} f(X)| < \infty$,

$$E[\lambda f(X+1) - X^{\nu} f(X)] = 0, \tag{2.28}$$

and its converse is likewise true. They further have results regarding stochastic ordering and other orderings relating to the CMP distribution. Meanwhile, Gupta et al. (2014) show that the CMP random variable is smaller than the corresponding Poisson random variable with respect to the likelihood ratio order, hazard ratio order, and mean residual life order.

For two random variables, $X_P$ and $X$, that are Poisson($\lambda$) and CMP($\lambda, \nu$) distributed respectively,

$$\frac{P(X_P = x)}{P(X = x)} = \frac{(x!)^{\nu-1} Z(\lambda, \nu)}{e^{\lambda}} \tag{2.29}$$

increases with $x$; hence, $X$ is smaller than $X_P$ with regard to the likelihood ratio order. The latter two results are inferred given the likelihood ratio ordering result.

### *2.3.1* R *Computing*

As noted in Table 2.2, the `compoisson` package can calculate several of the distributional and statistical measures described in Section 2.3. The `com.expectation` function computes the expected value of a function of the CMP random variable of interest, $E[f(X)]$. The inputs associated with this function are `f`, the function on which the random variable (a single argument) is applied; `lambda` and `nu`, the CMP parameters; and `log.error`, i.e. the allowable amount of error (on the log scale) for approximating the logarithm of the expectation. The default setting for this error is `log.error = 0.001`, where the logarithm of the expected value is determined exactly (i.e. not based on the moment approximation formula). Given the general ability of `com.expectation`, this function can be used to determine the CMP moments provided in Equation (2.16). In particular, it is naturally applied in `com.mean` and `com.var` to compute the mean and variance of the CMP distribution. The `com.mean` and `com.var` functions, however, only allow `lambda` and `nu` (i.e. the CMP parameters) as inputs; the `log.error` input is presumed to remain as provided in the default settings.

To illustrate these commands, consider a CMP distributed random variable with $\lambda = 4$ and $\nu = 0.5$ and suppose we want to compute the associated mean and variance. Using `com.mean` and `com.var`, we find that the mean and variance are 16.50979 (`com.mean(lambda = 4, nu = 0.5)`) and 32.00716 (`com.var(lambda = 4, nu = 0.5)`), respectively. These results make sense in that $0 \leq \nu < 1$ implies that the distribution is over-dispersed, i.e. the distribution's variance is greater than its mean (as reflected here). In contrast, if we consider a CMP($\lambda = 4, \nu = 5.0$) random variable, we instead find the associated mean and variance to equal 0.9122332 and 0.2706445, respectively. Here, the variance is considerably smaller than the mean of the distribution. Finally, let us consider a CMP($\lambda = 4, \nu = 1$) model. By definition, this represents the special case of a Poisson model with rate parameter, $\lambda = 4$; accordingly, `com.mean` and `com.var` should both produce values equaling 4. When performing these functions, however, we obtain

```
> com.mean(lambda = 4,nu = 1)
[1] 4
> com.var(lambda = 4,nu = 1)
[1] 4.00079
```

The result for the mean is as expected; however, the variance computation shows error due to the numerical approximation. This difference in result is presumed to be due to the default choice of an error threshold. A tighter threshold value for the error should resolve this issue, thus producing computations under the Poisson assumption such that the mean and variance computations equal.

## 2.4 Parameter Estimation and Statistical Inference

Three approaches exist for parameter estimation associated with a COM–Poisson model: (1) combining a graphical technique with a least-squares method, (2) maximum likelihood estimation, and (3) Bayesian estimation.

### *2.4.1 Combining COM–Poissonness Plot with Weighted Least Squares*

Shmueli et al. (2005) note this as a simple and computationally efficient method where the graphical procedure allows analysts to determine if the COM–Poisson distribution is a reasonable model for the dataset in question and, if so, to estimate $\lambda$ and $\nu$ via the method of least squares. Utilizing the relationship between successive probabilities from the CMP distribution (Equation (2.13)), taking the logarithm of both sides yields a linear model of the form

$$\ln\left(P(X = x-1)/P(X = x)\right) = -\ln\lambda + \nu\ln x, \tag{2.30}$$

where the respective probabilities on the left-hand side of Equation (2.30) can be estimated by their relative frequencies at $x-1$ and $x$. Plotting the ratio of successive probabilities (where positive counts exist) against $\ln(x)$ will demonstrate that the CMP is a reasonable model if the points follow the shape of a line. This "COM–Poissonness plot" is similar to the Ord plot (Ord, 1967) that associated the ratios of successive probabilities with quantile values from a power series distribution. In particular, the COM–Poissonness plot will illustrate that a Poisson model is reasonable if the shape of the line follows a 45° line.

The parameters, $-\ln\lambda$ and $\nu$, can be estimated more precisely by considering a linear model that regresses $\ln(P(X = x-1)/P(X = x))$ on $\ln x$. Two assumptions associated with an ordinary, Gaussian regression model, however, are violated; this model construct contains heteroskedasticity in the data, and the data are not independent. These issues can be overcome by conducting a weighted least-squares approach toward parameter estimation, where the inverse weight matrix has a tri-diagonal form with the variances on the diagonal and the one-step covariances on the off-diagonals. This method is found to be effective, particularly if there are not too many low values with zero counts (Shmueli et al., 2005).

### 2.4.2 Maximum Likelihood Estimation

Using the log-likelihood determined in Equation (2.10), the maximum likelihood estimates (MLEs) of $\lambda$ and $\nu$ satisfy the score equations

$$\begin{cases} \frac{\partial \ln L}{\partial \lambda} = \frac{\sum_{i=1}^{n} x_i}{\lambda} - \frac{n\frac{\partial Z(\lambda,\nu)}{\partial \lambda}}{Z(\lambda,\nu)} = 0 \\ \frac{\partial \ln L}{\partial \nu} = -\sum_{i=1}^{n} \ln(x_i!) - \frac{n\frac{\partial Z(\lambda,\nu)}{\partial \nu}}{Z(\lambda,\nu)} = 0, \end{cases} \tag{2.31}$$

where

$$\frac{\partial Z(\lambda,\nu)}{\partial \lambda} = \sum_{j=0}^{\infty} \frac{j\lambda^{j-1}}{(j!)^{\nu}} = \frac{Z(\lambda,\nu)}{\lambda} \sum_{j=0}^{\infty} j \cdot \frac{\lambda^j}{(j!)^{\nu} Z(\lambda,\nu)} = \frac{Z(\lambda,\nu)E(X)}{\lambda} \tag{2.32}$$

$$\begin{aligned} \frac{\partial Z(\lambda,\nu)}{\partial \nu} &= -\sum_{j=0}^{\infty} \frac{\lambda^j \ln(j!)}{(j!)^{\nu}} = -Z(\lambda,\nu) \sum_{j=0}^{\infty} \ln(j!) \cdot \frac{\lambda^j}{(j!)^{\nu} Z(\lambda,\nu)} \\ &= -Z(\lambda,\nu) E(\ln(X!)). \end{aligned} \tag{2.33}$$

Substituting Equations (2.32) and (2.33) into Equation (2.31) produces

$$\begin{aligned} E(X) &= \lambda \overline{X} \\ E(\ln(X!)) &= \overline{\ln(X!)}; \end{aligned}$$

these equations cannot be solved directly and do not have a closed form. A numerical iterative procedure (e.g. a Newton–Raphson-type optimization method) can instead be used to optimize the log-likelihood where, at each iteration, the respective expectations and variances, and covariance of $X$ and $\ln(X!)$, are determined by plugging in the previous step's estimates into the general equation for an expectation, $E(f(X))$, where $f(X)$ is as appropriate for the required computation (e.g. expectation, variance, or covariance), and the infinite sum is approximated by a finite sum that produces the desired level of precision (Minka et al., 2003; Shmueli et al., 2005). Gupta et al. (2014) instead suggest determining MLEs via the simulated annealing algorithm as a means to conduct numerical optimization, where the normalizing constant is truncated such that $Z(\lambda, \nu) \leq 1 \times 10^{50}$ and calculated with double precision.

### 2.4.3 Bayesian Properties and Estimation

The CMP distribution belongs to the exponential family (see Section 2.2), so a distribution of the form

$$h(\lambda, \nu) = \lambda^{a-1} \exp(-\nu b)[Z(\lambda, \nu)]^{-c} \kappa(a, b, c) \tag{2.34}$$

is a conjugate prior, where $\lambda > 0$, $\nu \geq 0$, and

$$\kappa^{-1}(a, b, c) = \int_0^\infty \int_0^\infty \lambda^{a-1} e^{-b\nu} [Z(\lambda, \nu)]^{-c} d\lambda d\nu \tag{2.35}$$

is the normalizing constant. Accordingly, given a dataset and priors for $\lambda$ and $\nu$, the posterior distribution maintains the same form as the prior distribution, with $a$, $b$, and $c$ respectively updated to $a' = a + \sum_{i=1}^n X_i$, $b' = b + \sum_{i=1}^n \ln(X_i!)$, and $c' = c + n$. Equation (2.34) can be viewed as an extended bivariate gamma distribution where it is necessary and sufficient for the hyperparameters $a, b$, and $c$ to be closed under sampling with

$$b/c > \ln(\lfloor a/c \rfloor !) + (a/c - \lfloor a/c \rfloor) \ln(\lfloor a/c \rfloor + 1) \tag{2.36}$$

in order for Equation (2.34) to be a proper density function; see Kadane et al. (2006b) for details.

Given the values of the hyperparameters, the predictive probability function is

$$P(X = x \mid a, b, c) = \frac{\kappa(a, b, c)}{\kappa(a + y, b + \ln(x!), c + 1)}, \tag{2.37}$$

where $\kappa(\cdot)$ is as defined in Equation (2.35) and can be computed through the aid of numerical procedures and computational software. The double integrals can be computed by using a nonequally spaced grid over the support space for $\lambda$ and $\nu$, where the computational procedures produce more robust representations of the parameter space when transformed to the log-scale in both parameters (i.e. for $\lambda^* = \ln(\lambda)$ and $\nu^* = \ln(\nu)$); see Kadane et al. (2006b) for details.

Given Equation (2.34), the respective marginal densities of $\lambda$ and $\nu$ are

$$h_1(\lambda) = \lambda^{a-1}\kappa(a,b,c)\int_0^\infty e^{-b\nu}[Z(\lambda,\nu)]^{-c}d\nu \text{ and} \tag{2.38}$$

$$h_2(\nu) = e^{-b\nu}\kappa(a,b,c)\int_0^\infty \lambda^{a-1}[Z(\lambda,\nu)]^{-c}d\lambda. \tag{2.39}$$

Accordingly, the conditional density of $\nu$ given $\lambda$ is

$$h(\nu \mid \lambda) \propto e^{-b\nu}[Z(\lambda,\nu)]^{-c}, \tag{2.40}$$

while the conditional density of $\lambda$ given $\nu$ is

$$h(\lambda \mid \nu) \propto \lambda^{a-1}[Z(\lambda,\nu)]^{-c}. \tag{2.41}$$

The latter conditional distribution (Equation (2.41)) simplifies to three special cases: (1) for $\nu = 0$, the conditional distribution of $\lambda$ on $\nu$ is Beta($a$,$c + 1$); (2) for $\nu = 1$, the conditional distribution simplifies to a gamma distribution with parameters $a$ and $c$; and (3) as $\nu$ goes to infinity, the conditional distribution of $\lambda$ given $\nu$ is $F(2a, 2(c-a))$ distributed for $c > a$, where the last case is recognized because $\frac{\lambda}{1+\lambda}$ has a Beta($a$, $c - a$) distribution when $\nu = \infty$ (Kadane et al., 2006b).

It is important to note that assuming a conjugate distribution is just one avenue for pursuing Bayesian estimation. While the presented conjugate forms provide welcomed properties, other priors can be used, for example, with Markov chain Monte Carlo (MCMC)-type estimation in order to conduct Bayesian estimation. This matter is pursued moreso with discussion in the context of COM–Poisson regression and hence addressed in greater detail in Chapter 5.

### *2.4.4* `R` *Computing*

Statistical computing and estimation for constants $\lambda$ and $\nu$ in `R` focus on maximum likelihood estimation and estimation via the COM–Poissonness plot. The `R` packages `compoisson`, `COMPoissonReg`, and `CompGLM` each contain functions that conduct maximum likelihood estimation. No package supplies functionality to conduct estimation via the COM–Poissonness plot; however, associated estimates for $\lambda$ and $\nu$ are directly attainable in `R`. Both approaches are discussed in greater detail below in this section. Meanwhile, basic Bayesian estimation was historically addressed via WinBugs (e.g. Lord and Guikema, 2012) and hence not addressed within this reference. There do exist two `R` packages that allow for Bayesian estimation; however, they are not pursued here. The `mpcmp` package (Fung et al., 2020) is equipped to conduct Bayesian COM–Poisson regression as described in Huang and Kim (2019), but it requires a plug-in that is only available through the first author. The `combayes` package (Chanialidis, 2020) also conducts Bayesian estimation in `R`; however, this package does not currently supply sufficient documentation in order to determine how to directly estimate $\lambda$ and $\nu$; see Chapter 5 for package discussion in the regression format.

Several functions exist to allow an analyst to perform maximum likelihood estimation and related statistical computation assuming a CMP model. The `com.fit` function (contained in the `compoisson` package (Dunn, 2012)) computes the maximum likelihood estimators for the CMP parameters and estimated frequencies associated with respective count values. The only input required for this function is the frequency table `x` (in a matrix form) that contains the count levels in the first column and the associated frequencies in the second column. As a result, `com.fit` returns the MLEs $\hat{\lambda}$ and $\hat{\nu}$ for the CMP parameters (`lambda` and `nu`, respectively), the value of the normalizing constant (`z`), the estimated frequencies associated with each count level in `x` (`fitted.values`), and the maximum log-likelihood value (`log.likelihood`). Meanwhile, `com.confint` (also in the `compoisson` package) computes bootstrap confidence intervals for $\lambda$ and $\nu$. The analyst provides the inputs `data`, a frequency table of counts; `level`, the confidence level of interest (as a decimal); `B`, the number of bootstrap repetitions; and `n`, the bootstrap sample size.

`COMPoissonReg` (Sellers et al., 2019) and `CompGLM` (Pollock, 2014b) are two `R` packages that conduct CMP regression via generalized linear models of the forms $\ln(\boldsymbol{\lambda}) = \boldsymbol{X\beta} = \beta_0 + \beta_1 X_1 + \cdots + \beta_{p_1-1} X_{p_1-1}$ and $\ln(\boldsymbol{\nu}) = \boldsymbol{G\gamma} = \gamma_0 + \gamma_1 G_1 + \cdots + \gamma_{p_2-1} G_{p_2-1}$. Details regarding the respective packages and their capabilities are provided in Chapter 5; however, they both allow for parameter estimation via the method of maximum likelihood on intercept-only models in order to obtain $\hat{\lambda} = \exp(\hat{\beta}_0)$ and $\hat{\nu} = \exp(\hat{\gamma}_0)$. The `COMPoissonReg` function `glm.cmp` calls an intercept-only model via the input `formula.lambda =  1` in order to obtain $\hat{\lambda}$; `formula.nu =  1` already serves as the default construct, thus estimating $\hat{\nu}$ as a constant. Similarly, the `CompGLM` function `glm.comp` considers intercept-only models for both parameters (i.e. `lamFormula= 1` and `nuFormula= 1`) to estimate $\lambda$ and $\nu$. In both cases, the estimators are determined via the `optim` function in `R` – a general-purpose optimization procedure that seeks to minimize a function of interest (in this case, the negated log-likelihood). By minimizing the negated log-likelihood, the analyst maximizes the log-likelihood, thus identifying the maximum likelihood estimators. The resulting output for `glm.cmp` (`COMPoissonReg`) includes the usual coefficient table for $\beta_0$ and $\gamma_0$, along with another table that provides the transformed estimates and standard errors for $\hat{\lambda}$ and $\hat{\nu}$. Additional output includes a chi-squared test for equi-dispersion (see Section 2.4.5 for details), and other reports including the log-likelihood, and AIC and BIC values. The `glm.comp` function (`CompGLM`) similarly reports a coefficient table,[3] along with the resulting log-likelihood and AIC values. While both `glm.cmp` (`COMPoissonReg`) and `glm.comp` (`CompGLM`) produce similar displays (`glm.cmp` being more inclusive), analysts are cautioned against using `glm.comp` for estimating the constants $\lambda$ and $\nu$. The coefficient estimates are reported correctly (thus the estimates themselves are useable); however, its reporting of the standard errors associated with coefficient estimates is located in swapped positions in the coefficient table, which can introduce incorrectly reported $t$-values and $p$-values.

To illustrate these function capabilities, we consider a classic count data example where Bailey (1990) studies the number of occurrences of the articles "the," "a," and "an" in five-word samples from "Essay on Milton" by Macaulay (1923) (Oxford edition) as a means to measure Macaulay's writing style. Table 2.8 provides the observed number of occurrences in 100

Table 2.8 *Observed versus CMP estimated frequency of occurrences of the articles "the," "a," and "an" in five-word samples from "Essay on Milton" by Macaulay (1923) (Oxford edition). CMP estimated frequencies obtained based on the maximum likelihood estimators $\hat{\lambda} \approx 1.0995$ and $\hat{\nu} \approx 3.2864$.*

| No. of articles | Observed | CMP estimated |
|---|---|---|
| 0 | 45 | 44.90 |
| 1 | 49 | 49.37 |
| 2 | 6 | 5.56 |
| >2 | 0 | 0.17 |
| Total | 100 | 100.00 |

[3] `glm.comp` (`CompGLM`) denotes the coefficients associated with $\nu$ as `zeta` instead of `gamma` in the equation $\ln(\boldsymbol{\nu}) = \boldsymbol{G\gamma}$.

Code 2.2 R codes and output illustrating the use of the compoisson (Dunn, 2012) package to model observed data provided in Bailey (1990) via the COM–Poisson distribution.

```
> word5 <- matrix(c(0,45,1,49,2,6),nrow=3,byrow=TRUE)
> com.fit(word5)
$lambda
[1] 1.099509

$nu
[1] 3.286388

$z
[1] 2.227142

$fitted.values
[1] 44.90060 49.36859  5.56350

$log.likelihood
           [,1]
[1,] -87.95295
```

such samples; this dataset is under-dispersed with a mean number of occurrences equaling 0.61 while the variance equaling approximately 0.362. Code 2.2 provides the example code where the observed frequency is first defined in R as word5 and then inputted into the com.fit (compoisson) function for model fitting. This example nicely illustrates the ability of the COM–Poisson distribution (here, via the CMP parametrization) to describe the variation in count data that express some measure of data dispersion. Using com.fit, we find that the CMP distribution with $\hat{\lambda} \approx 1.0995$ and $\hat{\nu} \approx 3.2864$ optimally models the observed distribution; Table 2.8 further provides the resulting estimated frequencies associated with the CMP($\hat{\lambda}$, $\hat{\nu}$) model. These estimates maximize the log-likelihood at approximately $-87.9530$ and produce a CMP normalizing constant, $Z(\hat{\lambda}, \hat{\nu}) \approx 2.2271$. The estimated dispersion parameter $\hat{\nu} \approx 3.2864 > 1$ verifies the apparent under-dispersion present in this data example. This output, however, does not provide any context surrounding the resulting estimates in that no quantified measure of variation associated with $\hat{\lambda}$ and $\hat{\nu}$ is provided.

The glm.cmp (COMPoissonReg) function instead provides a greater context and understanding by supplying a more elaborate coefficient table for $\beta_0$ and $\gamma_0$; see Code 2.3. The function output (primarily) references the resulting coefficient estimates $\hat{\beta}_0 = \ln(\hat{\lambda})$ and $\hat{\gamma}_0 = \ln(\hat{\nu})$. Solving for $\hat{\lambda}$ and $\hat{\nu}$ shows that the estimates equal those determined in Code 2.2;

glm.cmp provides the estimates for $\hat{\lambda}$ and $\hat{\nu}$ explicitly (along with the associated standard errors) in the "Transformed intercept-only parameters" section. The direct call to glm.cmp produces the coefficient table supplying the estimates, standard errors, test statistic values, and $p$-values. The additional outputs can be used to assess the fit of the CMP distribution and will be discussed in Chapter 5.

The glm.comp (CompGLM) function likewise reports analysis regarding $\lambda$ and $\nu$ via the coefficient table for $\beta_0$ and $\gamma_0$. The command for glm.comp is identical to that provided in Code 2.3; a subsequent model object summary() call reports the resulting coefficient table. The table lists identical estimates to those provided in the glm.cmp (COMPoissonReg) output; however, their corresponding standard errors are swapped such that the standard error associated with $\beta_0$ is actually the standard error for $\gamma_0$ and vice versa. While this error does not ultimately influence perceived statistical inference in this example, the error is nonetheless noted, thus cautioning analysts for using this command for generally estimating constants $\lambda$ and $\nu$ or any associated statistical inference.

Alternatively, one can utilize the COM–Poissonness plot to assess the appropriateness of the CMP model to these data, creating a scatterplot of data points, where $\ln x$ and $\ln(P(X = x - 1)/P(X = x))$ denote the

Code 2.3 R codes and output illustrating the use of the COMPoissonReg (Sellers et al., 2019) package to model observed data provided in Bailey (1990) via the CMP–distribution.

```
> word5 <- c(rep(0,45),rep(1,49),rep(2,6))
> glm.cmp(word5 ~ 1)
CMP coefficients
                 Estimate     SE z.value   p.value
X:(Intercept)      0.0949 0.2051  0.4626    0.6436
S:(Intercept)      1.1898 0.2142  5.5555 2.768e-08
--
Transformed intercept-only parameters
       Estimate     SE
lambda   1.0995 0.2255
nu       3.2864 0.7038
--
Chi-squared test for equidispersion
X^2 = 14.7160, df = 1, p-value = 1.2498e-04
--
Elapsed Sec: 0.01   Sample size: 100   SEs via Hessian
LogLik: -87.9530   AIC: 179.9059   BIC: 185.1163
Optimization Method: L-BFGS-B   Converged status: 0
```

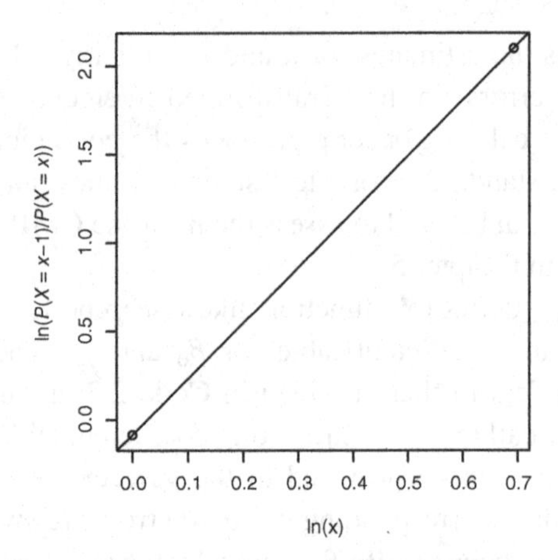


Figure 2.2 COM–Poissonness plot associated with Macaulay (1923) data.

explanatory and response variables, respectively. Given the small support space for the number of articles in this example (namely, {0,1,2}), the resulting scatterplot has only two points, i.e. $(\ln(1) = 0, \ln(0.45/0.49) \approx -0.0852)$ and $(\ln(2) \approx 0.6931, \ln(0.49/0.06) \approx 2.1001)$, and naturally a linear relationship is reasonable; see Figure 2.2. Given that there are only two data points, one can fit a line exactly with the intercept $-\widetilde{\ln\lambda} \approx -0.0852$, implying that $\tilde{\lambda} \approx 1.0889$, and slope equaling $\tilde{\nu} = 3.1527 > 1$, again indicating data under-dispersion. The COM–Poissonness plot estimates for $\lambda$ and $\nu$ are close to their respective maximum likelihood estimators. The programs conducting COM–Poisson estimation typically use Poisson estimates as starting values for running the underlying optimizations; however, this illustration demonstrates that the COM–Poissonness plot can likewise potentially be utilized for starting value determination with optimization tools (e.g. maximum likelihood estimation).

As a second example, consider data from Winkelmann (1995) that provide sample observations from the second wave of the 1985 German Socio-Economic Panel. The random sample comprises 1,243 women over 44 years old and in their first marriages, where the information collected includes the number of children born to this sample of women. For now, attention focuses on the variation in the number of children that ranges from 0 to 11 with a mean and variance equaling approximately 2.384 and 2.330, respectively. Given that the variance is smaller than the mean, the distribution representing the number of children is recognized as being at least mildly under-dispersed. Utilizing `com.fit` (`compoisson`) to analyze these data first requires data restructuring via the `table` function in `R`; this

Table 2.9 *Observed versus CMP estimated frequency of the number of children born to a random sample of females (Winkelmann, 1995).*

| No. of children | Observed | CMP estimated |
|---|---|---|
| 0 | 76 | 94.6238 |
| 1 | 239 | 270.8529 |
| 2 | 483 | 345.3846 |
| 3 | 228 | 274.4449 |
| 4 | 118 | 155.9054 |
| 5 | 44 | 68.2681 |
| 6 | 30 | 24.1661 |
| 7 | 10 | 7.1466 |
| 8 | 8 | 1.8086 |
| 9 | 3 | 0.3989 |
| 10 | 3 | 0.0778 |
| 11 | 1 | 0.0136 |
| Total | 1243 | 1243.0910 |

tool tabulates the frequency associated with each of the possible number of children representing the support space in the dataset. Again, the CMP model is able to recognize the under-dispersion present in these data. We find that the CMP distribution with $\hat{\lambda} \approx 2.862$ and $\hat{\nu} \approx 1.1665 > 1$ optimally models the observed distribution; Table 2.9 provides the fitted values associated with the CMP model. These MLEs maximize the log-likelihood at approximately $-2182.141$ and produce a CMP normalizing constant, $Z(\hat{\lambda}, \hat{\nu}) \approx 13.13623$.

The `glm.cmp` (`COMPoissonReg`) function can likewise be used to model the data via an intercept-only model such that $\hat{\lambda} = \exp(\hat{\beta}_0)$ and $\hat{\nu} = \exp(\hat{\gamma}_0)$ for the coefficient estimates $\hat{\beta}_0$ and $\hat{\gamma}_0$. Code 2.4 provides the `glm.cmp` output, including the coefficient table to not only assess the statistical significance associated with the two estimates (both deemed to be statistically significant, given the small $p$-values) but also explicitly providing $\hat{\lambda} = \exp(1.0533) = 2.8671$ and $\hat{\nu} = \exp(0.1555) = 1.1683$. `glm.comp` (`CompGLM`) likewise supplies the summary output associated with the coefficients, producing nearly identical estimates for $\beta_0$ and $\gamma_0$; however, we again note that the reported standard errors appear to swap those provided in the `glm.cmp` output. Thus, analysts are cautioned against using `glm.comp` for the constants $\lambda$ and $\nu$ estimation or inference.

Code 2.4 R codes and output illustrating the use of the COMPoissonReg (Sellers et al., 2019) package to model observed child data provided in Winkelmann (1995) via the CMP distribution.

```
> child.cmp <- glm.cmp(child ~ 1, data=Child)
> child.cmp
CMP coefficients
                 Estimate     SE z.value   p.value
X:(Intercept)      1.0533 0.0658 16.0019 1.238e-57
S:(Intercept)      0.1555 0.0493  3.1535  0.001613
—
Transformed intercept-only parameters
          Estimate     SE
lambda      2.8671 0.1887
nu          1.1683 0.0576
—
Chi-squared test for equidispersion
X^2 = 9.0823, df = 1, p-value = 2.5810e-03
—
Elapsed Sec: 0.21   Sample size: 1243   SEs via Hessian
LogLik: -2182.2347   AIC: 4368.4693   BIC: 4378.7199
Optimization Method: L-BFGS-B   Converged status: 0
Message: CONVERGENCE: REL_REDUCTION_OF_F <= FACTR*EPSMCH
```

### 2.4.5 Hypothesis Tests for Dispersion

Does a dataset exhibit a statistically significant amount of (over- or under-) dispersion? This is an obvious question that deserves attention in that analysts often question the need to consider alternative models to the ever-popular Poisson distribution when modeling count data, and the presence of statistically significant dispersion provides an insight into improved analysis via the COM–Poisson model. The CMP($\lambda$, $\nu$) distribution reduces to the Poisson($\lambda$) distribution when the dispersion parameter $\nu = 1$. Since the dispersion parameter describes the level and type of data dispersion, the likelihood ratio test can be used to test the null hypothesis $H_0 : \nu = 1$ against the alternative hypothesis $H_1 : \nu \neq 1$ to simply note potential data dispersion (in either direction). The resulting likelihood ratio test statistic is as defined in Equation (1.20), where, for this test, $-2 \ln \Lambda$ converges to a chi-squared distribution with one degree of freedom from which the appropriate critical values or $p$-value can be determined to draw inference regarding $\nu$ (Sellers and Raim, 2016; Sellers and Shmueli, 2010); see Section 1.5. The proposed (two-sided) test does not address the direction of dispersion; however, one can resolve the matter either by comparing the

MLE for $\nu$ to 1 to determine the type of data dispersion (recalling that over-dispersion (under-dispersion) is detected when $\hat{\nu} < ( > )1$). Alternatively, the hypothesis test can be modified to consider $\nu$ being strictly greater than or less than 1, as appropriate. Analogous likelihood ratio tests can likewise be considered for the respective boundary cases ($H_0$:$\nu = 0$ or $H_0$:$\nu \to \infty$ versus $H_1$: otherwise). Under these respective premises, the interested analyst can determine the appropriate likelihood ratio test statistic, bearing in mind that the corresponding $p$-value is now based on a mixture of a point mass and the $\chi_1^2$ distribution, i.e. $0.5 + 0.5\chi_1^2$ (Self and Liang, 1987).

Rao's score test is an alternative procedure for hypothesis testing. Here, the test statistic for $H_0$ versus $H_1$ is $R = SI^{-1}S'$, where

$$S = \left( \frac{\partial \ln L}{\partial \lambda} \frac{\partial \ln L}{\partial \nu} \right) \text{ and} \tag{2.42}$$

$$I = -\begin{pmatrix} E\left(\frac{\partial^2 \ln L}{\partial \lambda^2}\right) & E\left(\frac{\partial^2 \ln L}{\partial \nu \partial \lambda}\right) \\ E\left(\frac{\partial^2 \ln L}{\partial \nu \partial \lambda}\right) & E\left(\frac{\partial^2 \ln L}{\partial \nu^2}\right) \end{pmatrix} \tag{2.43}$$

denote the score vector and Fisher information matrix, respectively, associated with the log-likelihood (Equation (2.10)); this statistic likewise converges to a chi-squared distribution with one degree of freedom. Through various data simulation studies, the statistical power associated with each of the two tests are very close when the sample size is large (i.e. over 500), whether the simulated data reflect over- or under-dispersion. Meanwhile, when applying either test to simulated equi-dispersed data, both tests achieve reasonable estimated empirical levels that are close to the prespecified significance level (Gupta et al., 2014).

More broadly, analysts can also consider a Wald test to assess the statistical significance of the coefficient $\gamma_0$ associated with $\nu$. Considering the generalized linear model $\ln(\nu) = \gamma_0$, analysts can determine the Wald statistic $W_\gamma = \frac{\hat{\gamma}_0}{\text{se}(\hat{\gamma}_0)}$, where $\hat{\gamma}_0$ and $\text{se}(\hat{\gamma}_0)$, respectively, denote the maximum likelihood estimator and its corresponding standard error for $\gamma_0$; $W_\gamma$ has an asymptotic standard normal distribution from which its $p$-value can be determined. As an aside, analysts can likewise utilize the Wald test to draw statistical inference regarding the coefficient $\beta_0$ associated with $\lambda$ via the model $\ln(\lambda) = \beta_0$; the Wald statistic is now $W_\beta = \frac{\hat{\beta}_0}{\text{se}(\hat{\beta}_0)}$, which likewise has an asymptotic standard normal distribution.

### R *Computing for Dispersion Tests*

For the constant dispersion case, the `glm.cmp` (`COMPoissonReg`) function conducts both a likelihood ratio test and a Wald test for equi-dispersion, including both corresponding outputs in its report. The likelihood ratio test statistic is provided as $-2 \ln \Lambda$ and thus reported as a chi-squared statistic with one degree of freedom; accordingly, the test statistic value, its degree of freedom, and the associated $p$-value are provided in the output. The Wald test output is meanwhile contained within the resulting coefficient table produced in the "`CMP coefficients`" output provided via `glm.cmp`; the information regarding $\gamma_0$ is provided in the "`S: (Intercept)`" line. As illustrative examples, one can revisit the data analyses provided in Section 2.4.4; the respective `glm.cmp` outputs are provided in Codes 2.3 and 2.4. Both test examples demonstrate reported statistically significant data under-dispersion, directly reporting test statistics with corresponding $p$-values that are sufficiently small when compared to the traditionally assumed significance levels in the "`Chi-squared test for equidispersion`" output. Meanwhile, both examples likewise report Wald test statistics (listed in the output as `z.value`) and corrsponding $p$-values that demonstrate that $\gamma_0$ is statistically significantly different from zero. While the test for equi-dispersion only infers that some form of statistically significant data dispersion exists deeming the Poisson model unreasonable, that alone does not provide insight regarding the direction of the dispersion. That additional knowledge is attained via the reported dispersion estimate, both of which are greater than 1 in these examples; see Codes 2.3 and 2.4. Accordingly, analysts considering an initial model structure to describe variation in these respective datasets should consider a CMP rather than a Poisson distribution.

## 2.5 Generating Data

Random number generation is an important aspect of distributional study as one can benefit from data simulations to gain a better understanding of statistical properties. Two popular approaches have been considered for generating COM–Poisson data: the inversion, and rejection sampling methods.

### *2.5.1 Inversion Method*

Minka et al. (2003) provide a simple algorithm for generating data from a CMP distribution via the inversion method: to generate a dataset with $n$ observations,

1. simulate $n$ Uniform(0,1) observations;
2. starting from $P(X = 0)$, sum the CMP probabilities until the total exceeds the value of the associated simulated observation from Step 1;
3. $X = x$ is the observation such that Step 2 is satisfied.

The probability $P(X = 0) = [Z(\lambda, \nu)]^{-1}$ can be determined via methods described in Section 2.8. Subsequent probabilities can be easily determined via the recursion,

$$P(X = x) = P(X = x - 1)\frac{\lambda}{(x)^{\nu}}. \tag{2.44}$$

Analogous algorithms can be constructed for random number generation via the other COM–Poisson parametrizations; see Section 2.6 for the alternative COM–Poisson constructs.

### 2.5.2 Rejection Sampling

Rejection sampling is another approach used to generate independent samples from a distribution of interest. The appeal behind this approach is that COM–Poisson sampling can be determined without computing the normalizing constant. To date, two rejection samplers have been proposed, and they are each performed in relation to the ACMP parametrization such that $q_f(x) \doteq q_f(x \mid \mu_*, \nu) = \left(\frac{\mu_*^x}{x!}\right)^{\nu}$ denotes the nonnormalized ACMP function; see Section 2.6 for details regarding this parametrization.

Chanialidis et al. (2017) establish a rejection sampler whose rejection envelope is based on a piecewise geometric distribution

$$q_g(x) = \begin{cases} q_f(m-s)\cdot\left(\frac{m-s}{\mu_*}\right)^{\nu(m-s-x)} & x = 0, \ldots, m-s \\ q_f(m-1)\cdot\left(\frac{m-1}{\mu_*}\right)^{\nu(m-1-x)} & x = m-s+1, \ldots, m-1 \\ q_f(m)\cdot\left(\frac{\mu_*}{m+1}\right)^{\nu(x-m)} & x = m, \ldots, m+s-1 \\ q_f(m+s)\cdot\left(\frac{\mu_*}{m+s+1}\right)^{\nu(x-m-s)} & x = m+s, m+s+1, \ldots \end{cases}$$

with normalizing constant

$$Z_{1g}(\mu_*, \nu) = q_f(m-s)\frac{1-\left(\frac{m-s}{\mu_*}\right)^{\nu(m-s+1)}}{1-\left(\frac{m-s}{\mu_*}\right)^{\nu}} + q_f(m-1)\frac{1-\left(\frac{m-1}{\mu_*}\right)^{\nu(s-1)}}{1-\left(\frac{m-1}{\mu_*}\right)^{\nu}}$$
$$+ q_f(m)\frac{1-\left(\frac{\mu_*}{m+1}\right)^{\nu s}}{1-\left(\frac{\mu_*}{m+1}\right)^{\nu}} + q_f(m+s)\frac{1}{1-\left(\frac{\mu_*}{m+s+1}\right)^{\nu}}$$

such that an observation $x$ is drawn from the proper probability mass function $g(x) = q_g(x)/Z_{1g}(\mu, \nu)$ whose acceptance probability for inclusion as part of an ACMP sample is $\frac{q_f(x)}{q_g(x)}$. Benson and Friel (2017) instead introduce a rejection sampler whose envelope distribution depends on the dispersion parameter $\nu$; for $\nu \geq 1$, the Poisson($\mu_*$) distribution envelopes the ACMP,

Given the ACMP parameters $\mu_*$ and $\nu$, determine if $\nu \geq (<)1$.
If $\nu \geq 1$,

1. Sample $\tilde{x}$ from a Poisson($\mu_*$) distribution using the Ahrens and Dieter (1982) algorithm.
2. Calculate $B_{f/g}$ (Equation (2.45)) and determine the acceptance probability, $\alpha = \frac{(\mu_*^{\tilde{x}}/\tilde{x}!)^\nu}{B_{f/g}(\mu_*^{\tilde{x}}/\tilde{x}!)}$.
3. Generate $u \sim$ Unif(0, 1). If $u \leq \alpha$, then return $\tilde{x}$; otherwise, reject it and repeat.

If $\nu < 1$,

1. Sample $\tilde{x}$ from a geometric distribution with success probability $p = \frac{2\nu}{2\mu_*\nu+1+\nu}$. To do this, sample $u_{(0)} \sim$ Unif(0, 1) and return $\left\lfloor \frac{\ln(u_{(0)})}{\ln(1-p)} \right\rfloor$.
2. Calculate $B_{f/g}$ (Equation (2.45)) and determine the acceptance probability, $\alpha = \frac{(\mu_*^{\tilde{x}}/\tilde{x}!)^\nu}{B_{f/g}(1-p)^{\tilde{x}}p}$.
3. Generate $u \sim$ Unif(0, 1). If $u \leq \alpha$, then return $\tilde{x}$; otherwise, reject it and repeat.

**Algorithm 1** Benson and Friel (2017) rejection sampler for the ACMP($\mu_*$, $\nu$) distribution. For multiple ACMP($\mu_*$, $\nu$) draws, calculate the appropriate $B_{f/g}$ bound once (whether for $\nu \geq (<)1$) to use in the algorithm.

while, for $\nu < 1$, they use the geometric distribution with success probability $p$; see Algorithm 1 for details. Considering the Poisson or geometric distribution as an envelope distribution implies that the corresponding enveloping probability mass function is

$$g(x \mid \gamma) = \begin{cases} p(1-p)^x & \text{for } \nu < 1 \\ \frac{\mu_*^x e^{-\mu_*}}{x!} & \text{for } \nu \geq 1 \end{cases}$$

and their corresponding enveloping bound is

$$B_{f/g} = \begin{cases} \frac{1}{p} \frac{\mu_*^{\left(\nu \left\lfloor \frac{\mu_*}{(1-p)^{1/\nu}} \right\rfloor\right)}}{(1-p)^{\left(\left\lfloor \frac{\mu_*}{(1-p)^{1/\nu}} \right\rfloor\right)} \left(\left\lfloor \frac{\mu_*}{(1-p)^{1/\nu}} \right\rfloor !\right)^\nu} & \text{for } \nu < 1 \\ \left(\frac{\mu_*^{\lfloor \mu_* \rfloor}}{\lfloor \mu_* \rfloor !}\right)^{\nu-1} & \text{for } \nu \geq 1 \end{cases} \quad (2.45)$$

such that the acceptance probability is $\frac{q_f(x)}{q_g(x)B_{f/g}}$.

The Benson and Friel (2017) envelope selections make use of the special-case distributions in a nice way, recognizing their ability to envelope the ACMP distribution under certain cases of dispersion. When the ACMP has $\nu \geq 1$, the distribution reflects equi- or under-dispersion. Accordingly, the Poisson distribution will either completely envelope the ACMP when $\nu > 1$ because the Poisson variance is larger (thus producing an acceptance probability that is less than 1), or the Poisson envelope will match the ACMP distribution when $\nu = 1$, producing an acceptance probability that equals 1. Meanwhile, when the ACMP has $\nu < 1$ (over-dispersion), the geometric distribution envelopes the ACMP model for all $0 < \nu < 1$, thus producing an acceptance probability less than 1 under those circumstances, or produces an acceptance probability equaling 1 when $\nu = 0$ (at least, in theory). The ACMP reparametrization, however, appears to limit the perceived choice of success probability $p$ associated with the geometric distribution. Benson and Friel (2017) select $p$ by matching the geometric distribution mean to the ACMP approximated mean, i.e.

$$\frac{1-p}{p} = \mu_* + \frac{1}{2\nu} - \frac{1}{2} \quad \Leftrightarrow \quad p = \frac{2\nu}{2\mu_*\nu + 1 + \nu}. \tag{2.46}$$

Under this approach, Benson and Friel (2017) report that their sampler obtains an acceptance probability between 50% and 100% when $\nu \geq 1$, while that acceptance probability is 30%–80% when $\nu < 1$. Notice that, in Equation (2.46) however, $\nu = 0$ implies that $p = 0$. This contradicts the $p > 0$ constraint associated with the geometric distribution; yet we know that the geometric($p$) distribution is a special case of the COM–Poisson distribution (but not under the ACMP parametrization; see Section 2.6 for discussion). This matter presumably contributes toward explaining the reduced acceptance probability range for $\nu < 1$.

### *Generating Data from* $h(\lambda \mid \nu)$ *When* $c > a$

Along with random number generation regarding the COM–Poisson distribution itself, one can consider simulating data from the corresponding conjugate distribution described in Equation (2.34) or the conditional distribution $h(\lambda \mid \nu)$. Motivated by the special-case results described in Section 2.4.3 for $h(\lambda \mid \nu)$, Kadane et al. (2006b) motivate an algorithm to generate data from $h(\lambda \mid \nu)$ when $c > a$ by recognizing that the conditional distribution for $\lambda$ on $\nu$ has shorter tails than the F distribution; see Algorithm 2. An "F-type dominating curve" can then aid in developing Algorithm 2 for $h(\lambda \mid \nu)$ when $c > a$. Meanwhile, for $\nu < 1$, the special cases imply that $h(\lambda \mid \nu)$ has a shorter tail than the gamma distribution; thus,

1. Choose $\lambda_0$ and compute $q(\lambda_0)$, where $q(\lambda_0) = \frac{dZ(\lambda,\nu)}{d\lambda}\mid_{\lambda=\lambda_0}$.
2. Draw $\lambda$ from the $F$ distribution proportional to $\lambda^{a-1}[q(\lambda_0)(\lambda - \lambda_0) + Z(\lambda_0, \nu)]^{-c}$.
3. Draw a uniform variate $u \sim U(0, 1)$ and accept $\lambda$ if $u \leq \frac{[Z(\lambda,\nu)]^{-c}}{[q(\lambda_0)(\lambda-\lambda_0)+Z(\lambda_0,\nu)]^{-c}}$.
4. If $\lambda$ is rejected, repeat from Step 2.

**Algorithm 2** Kadane et al. (2006b) rejection sampling algorithm to generate data from $h(\lambda \mid \nu)$ when $c > a$, assuming a motivating CMP distribution. See Section 2.4.3 for prerequisite discussion and variable notation.

1. Choose $\lambda_0$ and compute $q(\lambda_0)$, where $q(\lambda_0) = \frac{dZ(\lambda,\nu)}{d\lambda}\mid_{\lambda=\lambda_0}$.
2. Draw $\lambda$ from a gamma distribution of the form $p(\lambda \mid \nu) \propto \lambda^{a-1} \exp(-cq(\lambda_0)\lambda)$.
3. Draw a uniform variate $u \sim U(0, 1)$ and accept $\lambda$ if $u \leq \frac{[Z(\lambda,\nu)]^{-c}}{\exp(-cq(\lambda_0)(\lambda-\lambda_0))[Z(\lambda_0,\nu)]^{-c}}$
4. If $\lambda$ is rejected, repeat from Step 2.

**Algorithm 3** Kadane et al. (2006b) rejection sampling algorithm to generate data from $h(\lambda \mid \nu)$ when $\nu < 1$, assuming a motivating CMP distribution. See Section 2.4.3 for prerequisite discussion and variable notation. This algorithm holds without the restriction, $c > a$.

a gamma-type dominating curve can be constructed and used to develop a rejection sampling scheme that holds even without the restriction, $c > a$; see Algorithm 3.

### *2.5.3* R *Computing*

The `compoisson` (Dunn, 2012), `CompGLM` (Pollock, 2014b), `COMPoissonReg` (Sellers et al., 2019), and `combayes` (Chanialidis, 2020) packages all provide analysts with the ability to randomly generate CMP data. All functions naturally require the analyst to provide the sample size `n`. Meanwhile, the `compoisson`, `CompGLM`, and `COMPoissonReg` packages need the associated parameter values for $\lambda$ and $\nu$ from the analyst; `rcomp` (`CompGLM`) defines these inputs as `lam` and `nu`, while `rcom` (`compoisson`) and `rcmp` (`COMPoissonReg`) refer to these inputs as `lambda` and `nu`. The `rcmpois` function (`combayes`) instead requests the parameter inputs as `mu` and `nu`, where `mu` denotes $\mu_*$ in the ACMP parametrization. The `rcomp` command further allows the analyst to supply the input `sumTo` that denotes the number of summation terms used to approximate $Z(\lambda, \nu)$; the default setting for this input is 100. The `rcom` function meanwhile allows the analyst to input `log.z`, the natural logarithm of the normalizing constant.

The `rcmp` function (`COMPoissonReg`) uses the inversion method with the CMP parametrization, while `rcmpois` (`combayes`) uses rejection sampling with the ACMP structure. It is unclear what methods are used for the other packages; however, the inversion method is presumed.

## 2.6 Reparametrized Forms

Some researchers argue that the CMP parametrization assumed thus far is limited in its usefulness (e.g. as a generalized linear model) because neither $\lambda$ nor $\nu$ provide a clear centering parameter, i.e. the CMP distribution is not directly parametrized by the mean. Barriga and Louzada (2014) (for example) note that, while $\lambda$ is close to the distributional mean for $\nu$ close to 1, the mean differs considerably from $\lambda$ when $\nu$ is significantly different from 1. Huang (2017) further notes that the approximation to the mean (provided in Equation (2.14)) is accurate for $\nu \leq 1$ or $\lambda > 10^{\nu}$; yet it can be inaccurate for scenarios outside of that range, particularly when analyzing under-dispersed count data (unless $\lambda > 10^{\nu}$ is very large). This arguably makes the regression model based on the original CMP formulation difficult to interpret and use for dispersed data (this issue is revisited in Chapter 5).

To circumvent this problem for over-dispersed data (i.e. $\nu < 1$), Guikema and Coffelt (2008) propose the reparametrization

$$\mu_* = \lambda^{1/\nu} \tag{2.47}$$

as an approximation for the "center" of the COM–Poisson distribution; note that Equation (2.47) truncates the CMP mean approximation (Equation (2.23)) and offers an alternative representation of the constraint $\lambda > 10^{\nu}$ (i.e. $\mu_* = \lambda^{1/\nu} > 10$) such that Equation (2.23) holds. The substitution provided in Equation (2.47) produces a reparametrization ACMP reparametrization) that restructures the CMP probability mass function in Equation (2.8) to

$$P(X = x) = \frac{1}{Z_1(\mu_*, \nu)} \left( \frac{\mu_*^x}{x!} \right)^{\nu}, \quad x = 0, 1, 2, \ldots, \tag{2.48}$$

where $Z_1(\mu_*, \nu) = \sum_{j=0}^{\infty} \left( \frac{\mu_*^j}{j!} \right)^{\nu}$ is the modified form of the normalizing constant. Given this representation, the ratio of successive probabilities now has the form

$$\frac{P(X = x - 1)}{P(X = x)} = \left(\frac{x}{\mu_*}\right)^{\nu}.$$

Further, the mean and variance of $X$ are now

$$E(X) = \frac{1}{\nu}\frac{\partial \ln Z_1(\mu_*, \nu)}{\partial \ln \mu_*} \approx \mu_* + \frac{1}{2\nu} - \frac{1}{2} \tag{2.49}$$

$$V(X) = \frac{1}{\nu^2}\frac{\partial^2 \ln Z_1(\mu_*, \nu)}{\partial(\ln \mu_*)^2} = \frac{1}{\nu}\frac{\partial E(Y)}{\partial \ln \mu_*} \approx \frac{\mu_*}{\nu}, \tag{2.50}$$

where the asymptotic approximation is accurate for $\mu_* = \lambda^{1/\nu} > 10$ (Barriga and Louzada, 2014; Lord et al., 2008; Zhu, 2012). The ACMP distribution mode is $\lfloor \mu_* \rfloor$ (Chanialidis et al., 2017) for general $\mu_*$ and for the special case where $\mu_* \in \mathbb{N}$; the ACMP has consecutive modes at $\mu_*$ and $\mu_* - 1$ (Benson and Friel, 2017). Accordingly, "the integral part of $\mu_*$ is now the mode leaving $\mu_*$ as a reasonable approximation of the mean" while maintaining $\nu$'s role as the dispersion parameter and hence a shape parameter (Lord et al., 2008; Zhu, 2012). Chanialidis et al. (2014) confirm that $\mu_*$ closely approximates the mean; however, they further note that this is only true when both $\mu_*$ and $\nu$ are not small. Guikema and Coffelt (2008) likewise stress that these approximations may not be accurate for $\nu > 1$ or $\mu_* < 10$. This distribution is undefined for $\nu = 0$ and is hence restricted to $\nu > 0$; thus, this parametrization does not include the geometric special case (Zhu, 2012).

Huang (2017), however, argues that the ACMP parametrization still does not offer a closed form for the mean in that the $\mu_* = \lambda^{1/\nu}$ designation instead redresses the (say) location parameter; yet this value does not equal the mean nor provide a good approximation for it under any $\nu$. The mean-parametrized COM–Poisson (MCMP1) distribution instead reparametrizes the CMP probability mass function (Equation (2.8)) as

$$P(X = x \mid \mu, \nu) = \frac{\lambda(\mu, \nu)^x}{(x!)^{\nu} Z(\lambda(\mu, \nu), \nu)}, \quad x = 0, 1, 2, \ldots, \tag{2.51}$$

where $\mu \doteq E(X)$ is the mean of the distribution; accordingly, $\lambda(\mu, \nu)$ solves the equation

$$\sum_{x=0}^{\infty} (x - \mu)\frac{\lambda^x}{(x!)^{\nu}} = 0,$$

and $\nu = 1$ maintains the definition of equi-dispersion, while $\nu < (>) 1$ indicates over-dispersion (under-dispersion). The special-case distributions associated with the MCMP1 distribution still hold, namely the MCMP1 distribution reduces to the geometric distribution with success probability $p = 1/(\mu + 1) < 1$ when $\nu = 0$, the Poisson($\mu$) distribution when $\nu = 1$, and converges to the Bernoulli($p = \mu$) distribution as $\nu \to \infty$.

MCMP1 likewise has the form of a two-parameter exponential family, and a one-parameter exponential dispersion family forms with respect to $\mu$ when $\nu$ is fixed. Under this construct, the mean and dispersion parameters $\mu$ and $\nu$ are also orthogonal. This proves beneficial in a regression setting because the resulting estimated regression coefficients linking the covariates to the mean are asymptotically efficient and independent of the dispersion estimate $\hat{\nu}$; see Chapter 5 for details. To estimate the MCMP1 parameters $\mu, \nu$ via maximum likelihood estimation, the log-likelihood associated with one random variable $X$ is

$$\ln L(\mu, \nu \mid x) = x \ln(\lambda(\mu, \nu)) - \nu \ln(x!) - \ln Z(\lambda(\mu, \nu), \nu), \qquad (2.52)$$

which is the reparametrized version of the log-likelihood provided in Equation (2.10). Thus, for a random sample of observations, the respective MLEs of the MCMP1 mean and dispersion parameters are $\hat{\mu} = \bar{X}$ and $\hat{\nu}$, which solve the equation $E_{\bar{X},\nu}(\ln(x!)) = \frac{1}{n}\sum_{i=1}^{n} \ln(x_i!) \doteq \overline{\ln(X!)}$. The closed form for $\hat{\mu}$ provides a direct estimate for $\mu$, thus reducing the overall computation time for the MLEs.

Equation (2.34) not only serves as a conjugate prior for the CMP parametrization, but its essence likewise serves as a conjugate form for the MCMP1 distribution; the Huang and Kim (2019) supplemental discussion maintains the Kadane et al. (2006b) form of the prior, updating $\lambda$ with $\lambda(\mu, \nu)$. Accordingly, the form of the MCMP1 prior distribution becomes

$$h_{+}(\mu, \nu) = \lambda(\mu, \nu)^{a-1} \exp(-\nu b)[Z(\lambda(\mu, \nu), \nu)]^{c} \kappa_{+}(a, b, c), \quad \mu > 0, \nu \geq 0,$$

where

$$\kappa_{+}^{-1}(a, b, c) = \int_0^{\infty} \int_0^{\infty} \lambda(\mu, \nu)^{a-1} e^{-b\nu} [Z(\lambda(\mu, \nu), \nu)]^{-c} d\mu d\nu$$

is now the normalizing constant, and $a$, $b$, and $c$ maintain the restrictions described in Section 2.4.3. This posterior form likewise maintains conjugacy with the updates $a'$, $b'$, and $c'$ as defined in Section 2.4.3.

Ribeiro Jr. et al. (2019) instead work directly with the mean approximation provided in Equation (2.23) and backsolve for $\lambda$ so that $\lambda = \left(\mu + \frac{\nu-1}{2\nu}\right)^{\nu}$ and $\phi = \ln(\nu)$ represent the intensity and dispersion parameter, respectively; $\phi = 0$ captures the Poisson model, and $\phi < (>) 0$ recognizes the data as being over- (under-)dispersed relative to the Poisson. Accordingly, they introduce a second mean-parametrized COM–Poisson (hereafter MCMP2) distribution with parameters $\mu$ and $\phi$ which has the probability mass function

$$P(X = x \mid \mu, \phi) = \left(\mu + \frac{e^{\phi} - 1}{2e^{\phi}}\right)^{xe^{\phi}} \frac{(x!)^{-e^{\phi}}}{Z_2(\mu, \phi)}, \quad x = 0, 1, 2, \ldots, \quad (2.53)$$

for $\mu > 0$, where $Z_2(\mu, \phi) = \sum_{j=1}^{\infty} \left(\mu + \frac{e^{\phi}-1}{2e^{\phi}}\right)^{je^{\phi}} / (j!)^{e^{\phi}}$. Ribeiro Jr. et al. (2019) likewise note that this parametrization provides orthogonality between $\mu$ and $\phi$; yet their parametrization is simpler than that from Huang (2017) because $\mu$ is obtained via "simple algebra." Further, this parametrization offers a more direct parametrization based on the approximate mean than what is achieved via the ACMP parametrization. The asymptotic approximations for the mean and variance of the MCMP2 distribution are shown to be accurate "for a large part of the parameter space"; yet the mean approximation can be inaccurate when the count data are small and strongly over-dispersed (Ribeiro Jr. et al., 2019). Given the ZI and HT indexes described in Equation (2.26), the MCMP2 can handle limited ZI when over-dispersion exists, while in cases of under-dispersion, the MCMP2 can accommodate zero-deflation. Further, studies regarding the HT indexes show that MCMP2 is generally a light-tailed distribution (Ribeiro Jr. et al., 2019).

While all of these parametrizations are technically valid in that they are intended to theoretically produce the same probabilities, differences can result when different parametrizations are used in statistical methodology and computation and thus in their resulting interpretation and inference. Francis et al. (2012) conducted various simulation studies assessing maximum likelihood estimation accuracy, prediction bias, and accuracy of the approximated asymptotic mean for the ACMP parametrization. With regard to maximum likelihood estimation accuracy, under most scenarios, the confidence interval contained the true parameter with the expected percentage of occasions. This, however, was not the case for the dispersion parameter. Accordingly, "the true value of the dispersion parameter may be difficult to ascertain with the level of confidence assumed by the $\alpha$-level selected." In fact, studies further showed that, for data that are over-dispersed with a large mean, the dispersion parameter is overestimated. Regarding the bias of the predicted values, the intercept showed a considerable bias when the data were under- or equi-dispersed, but the bias in the intercept decreased with the mean; further, the largest bias occurred when the data were over-dispersed.

While the Shmueli et al. (2005) constraints imply that the ACMP mean approximation to be suitable for $\mu_* > 10$, Francis et al. (2012) instead argue that simulation studies show that the asymptotic mean approximates the true mean even when $5 < \mu_* < 10$; however, the approximation's accuracy diminishes for $\mu_* < 5$. Further, the asymptotic approximation for the mean is shown to hold well for all dispersion types when $\mu_*$ is moderate to large, while it is further accurate for small $\mu_*$ when the data are under-dispersed. When $\mu_*$ is small, however, the asymptotic accuracy is lost when working with over- or equi-dispersed data. Francis et al. (2012) agree that the true mean and its asymptotic approximation are close when $\mu_* > 10$. In contrast, however, Ribeiro Jr. et al. (2019) counter the Shmueli et al. (2005) claims regarding the constraints required for the approximations to hold, thus making their argument for MCMP2 parametrization. Through numerical studies, the authors determine that the mean approximation is accurate, while the variance approximation experiences its largest errors when $\nu \leq 1$, and that the reported constraint region $\lambda > 10^{\nu}$ appears unnecessary as there does not appear to exist a relationship between their reparametrized parameters in their impact on variance; see Ribeiro Jr. et al. (2019) for details. These assessments, however, rely on several assumptions. The authors estimate the true mean and variance by computing the sum of the first 500 terms from the infinite sum; yet they too concede that the infinite sum can take considerable time to converge. Thus, it is unclear whether $M = 500$ suffices for such a study. Meanwhile, the constraint space for these approximations to hold stems from one's ability to approximate the normalizing function; the approximations for the mean and variance are the by-products of this result. Nonetheless, it is surprising that Ribeiro Jr. et al. (2019) obtain results that contradict the over-dispersion constraint $\nu \leq 1$. Other works (e.g. Guikema and Coffelt, 2008; Lord et al., 2008) provide over-dispersed examples where their ACMP reparametrization works well where their approximation of the mean is a further truncated representation of that provided by Ribeiro Jr. et al. (2019). Meanwhile, assuming a positive approximate mean implies that $\lambda^{1/\nu} > \frac{\nu-1}{2\nu}$ so that the space corresponding to small and under-dispersed count data is prohibited. However, the constraints for which the normalizing constant can be approximated imply that $\lambda^{1/\nu} > 10$.

### `R` *Computing*

While no available `R` package appears to exist for statistical computing via the MCMP2 model, such packages exist assuming either the ACMP or MCMP1 parametrizations; see Table 2.10. The `combayes` package (Chanialidis, 2020) performs ACMP computations and Bayesian regression. Unfortunately, this package does not currently provide any help pages

Table 2.10 *Available* R *functions based on reparametrized COM–Poisson models.*

| Package (Parametrization) | Function | Computational result |
|---|---|---|
| | `dcmpois` | Probability mass function |
| `combayes` (ACMP) | `logzcmpois` | Normalizing constant, $Z_1(\mu_*, \nu)$ (log-scale) |
| | `rcmpois` | Random number generator |
| | `dcomp` | Probability mass function |
| `mpcmp` (MCMP1) | `pcomp` | Cumulative distribution function |
| | `qcomp` | Quantile function |
| | `rcomp` | Random number generator |

for its functions; however, some information regarding package capabilities can be ascertained from the `Github` site and its illustrative examples. The `dcmpois`, `logzcmpois`, `rcmpois` functions respectively compute the density, normalizing constant (on the log-scale), and random number generation associated with the ACMP($\mu_*, \nu$) distribution for given centering and dispersion parameters, $\mu_*$ and $\nu$ (noted as `mu` and `nu` in the `combayes` package). All of the aforementioned functions require `mu` and `nu` as inputs because they aid in detailing the parametric form of the ACMP distribution. Beyond these inputs, `dcmpois` requires the input `x` for the quantile of interest as defined in Equation (2.48), while `rcmpois` instead needs the input `n` to specify the desired number of generated data values (i.e. the sample size).

The `mpcmp` package (Fung et al., 2020) performs MCMP1 computations and regression analysis. The `dcomp`, `pcomp`, `qcomp`, and `rcomp` respectively compute the density, distribution function, quantiles, and random number generation associated with the MCMP1($\mu, \nu$) distribution,[4] although analysts can supply either `mu` or `lambda` (referring to the CMP($\lambda, \nu$) parametrization). All of these functions have the default settings that `nu=1` (i.e. MCMP1 with $\nu = 1$, which is the Poisson model); `lambdalb = 1e-10` and `lambdaub = 1900`, implying that $(1 \times 10^{-10}) \leq \lambda \leq 1900$ are the parameter bounds, `maxlambdaiter = 1000` is the number of iterations allowed to estimate $\lambda$, and `tol = 1e-06` is the tolerance threshold to assess convergence. The `dcomp` and `pcomp` functions, respectively, compute the probability $P(X = x)$ and cumulative probability $P(X \leq x)$, and `qcomp` determines the quantile $x$ such that $P(X \leq x) \geq p$, where, for all of these functions, the additional input `log.p` is a logical operator

[4] The `CompGLM` package also contains functions named `dcomp`, `pcomp`, and `rcomp`; these functions assume the CMP parametrization.

(defaulted as FALSE), noting whether to compute the probability on the log-scale. pcomp and qcomp further offer a logical input lower.tail that determines whether the cumulative density function (i.e. =TRUE) or its complement (i.e. FALSE) is used to determine the output of interest. The function rcomp generates a sample size of $n$ observations (n) from an MCMP1($\mu, \nu$) = CMP($\lambda, \nu$) distribution when providing nu and either mu or lambda. Finally, the mpcmp package contains the function glm.cmp to perform MCMP1 regression via a generalized linear model.[5] Details regarding this function will be discussed in more detail in Chapter 5; however, we can use it here to compute $\hat{\mu}$ and $\hat{\nu}$ via an intercept-only model. Further discussion of the combayes package, particularly its ability to conduct Bayesian ACMP regression, is likewise provided in Chapter 5.

To illustrate the usefulness of these functions, Table 2.11 contains the density calculations from combayes for the ACMP($\mu_* = \lambda^{1/\nu} = 4^{1/\nu}, \nu$), and the mpcmp density and cumulative probabilities for an MCMP1($\mu, \nu$) = CMP($\lambda = 4, \nu$) distributed random variable for $\nu \in \{0.3, 1, 3, 30\}$. The resulting calculations are consistent with the analogous CMP computations provided in Table 2.3 from CompGLM, compoisson, and COMPoissonReg, particularly close to dcomp (CompGLM) and dcmp (COMPoissonReg) when $\nu \ll 1$. Similarly, the density computations shown in Table 2.11 via dcmpois and dcomp equal each other up to at least five decimal places for all considered $\nu$. When trying to compare mpcmp with the other functions for the special case associated with the geometric distribution (Table 2.5), however, the dcomp and pcomp functions produce NaN. This is because the mpcmp package requires mu, lambda, and nu to all be positive; thus, an equivalent calculation for the geometric distribution cannot occur because that requires setting $\nu = 0$. The combayes package meanwhile tries to produce output for when $\nu = 0$ but appears stuck churning, trying to determine this computation. Theoretically, however, we can already recognize this as being unattainable since $\nu = 0$ implies that $\mu_* = \lambda^{1/\nu}$ is undefined; thus, we are unable to determine any density associated with the geometric special case via the ACMP parametrization and dcmpois. Finally (comparing qcomp with results from Table 2.7), the qcomp function likewise determines the quantile $q$ such that $P(X \leq q) \geq p$ for some probability of interest $p$. Assuming that $X$ has an MCMP1($\mu, \nu$) = CMP($\lambda = 3, \nu$), where $\nu \in \{0.3, 1, 3, 30\}$, and $P(X \leq q) \geq 0.9$, qcomp obtains identical quantiles as those presented in Table 2.7. The combayes package does not currently

[5] The COMPoissonReg package also contains a function glm.cmp, but it assumes a CMP parametrization.

Table 2.11 *Probability computations (to six decimal places) via* `dcmpois` (`combayes`) *and* `dcomp` (`mpcmp`) *for ACMP($\mu_* = \lambda^{1/\nu} = 4^{1/\nu}, \nu$) and MCMP1($\mu, \nu$) = CMP($\lambda = 4, \nu$), respectively, and cumulative probability* `pcomp` *for the MCMP1 distribution.*

| | $P(X = 2)$ | | $P(X \leq 2)$ |
|---|---|---|---|
| $\nu$ | `dcmpois` (`combayes`) | `dcomp` (`mpcmp`) | `pcomp` (`mpcmp`) |
| 0.3 | $4.312012 \times 10^{-14}$ | $9.772194 \times 10^{-14}$ | $1.353188 \times 10^{-13}$ |
| 1.0 | 0.146525 | 0.146525 | 0.238104 |
| 3.0 | 0.273395 | 0.273395 | 0.956883 |
| 30.0 | $2.980834 \times 10^{-9}$ | $2.980232 \times 10^{-9}$ | 1 |

provide help pages to know whether an analogous function is contained in its package.

Revisiting the 1985 German Socio-Economic Panel data regarding the number of children born to a random sample of 1,243 women (Winkelmann, 1995; Section 2.4.4), we can use `glm.cmp` (`mpcmp`) to independently obtain the coefficients $\hat{\beta}_0$ and $\hat{\gamma}_0$ such that $\hat{\lambda} = \exp(\hat{\beta}_0)$ and $\hat{\nu} = \exp(\hat{\gamma}_0)$. Code 2.5 supplies the resulting `R` code and output obtained via the `mpcmp::glm.cmp` function. Here, we obtain $\hat{\nu} = 1.168$ and $\hat{\mu} = \exp(0.8687) \approx 2.3821$. While we can directly compare this estimated dispersion parameter to those obtained via `compoisson`, `CompGLM`, and `COMPoissonReg` (where we see that the values are nearly identical; see Section 2.4.4), we cannot directly compare $\hat{\mu}$ and $\hat{\lambda}$ because these estimates stem from different parametrizations that are not easily transformable.

## 2.7 COM–Poisson Is a Weighted Poisson Distribution

As discussed in Chapter 1, weighted Poisson distributions are flexible generalizations of the Poisson distribution that allow for either over- or under-dispersion; see Section 1.4 for a detailed discussion. The COM–Poisson distribution is a special case of the class of weighted Poisson distributions, where the weight function is $w(x;\nu) = (x!)^{1-\nu} = [\Gamma(x+1)]^{1-\nu}, x \in \mathbb{N}$ (Kokonendji et al., 2008). This weight function can be viewed as $w(x;\nu) = \exp[(1-\nu)\ln(x!)]$, i.e. $r = 1 - \nu$ and $t(x) = \ln(x!)$; accordingly, the COM–Poisson distribution with $\nu = 1$ (i.e. the Poisson distribution) is equi-dispersed, $\nu < 1$ is over-dispersed, and $\nu > 1$ is under-dispersed. The COM–Poisson weight function is also log-convex (log-concave) for $0 \leq \nu < 1$ ($\nu > 1$); this result is consistent with the previously stated

Code 2.5 R codes and output illustrating the use of the mpcmp (Fung et al., 2020) package to model observed child data provided in Winkelmann and Zimmermann (1995) via the MCMP1 distribution.

```
> child.mcmp <- glm.cmp(child ~ 1, data=Child)
> summary(child.mcmp)

Call: glm.cmp(formula = child ~ 1, data = Child)

Deviance Residuals:
    Min        1Q    Median        3Q       Max
-2.2702   -1.0707   -0.2711    0.4077    4.3332

Linear Model Coefficients:
            Estimate Std.Err Z value Pr(>|z|)
(Intercept)   0.8687  0.0173    50.2   <2e-16 ***
---
(Dispersion parameter for Mean-CMP estimated to be
   1.168)

    Null deviance: 1343.2  on 1242 degrees of freedom
Residual deviance: 1343.2  on 1242 degrees of freedom

AIC: 4368.469
```

results relating the constrained space for $\nu$ and the type of data dispersion associated with the COM–Poisson distribution. Gupta et al. (2014) further apply relations of the COM–Poisson log-concavity (log-convexity) to discrete life distributions having an increasing (decreasing) failure rate. Finally, the family of COM–Poisson distributions is closed by pointwise duality for $0 \leq \nu \leq 2$; accordingly, for any COM–Poisson distribution whose dispersion parameter is $0 \leq \nu \leq 2$, there exists a corresponding dual COM–Poisson distribution whose dispersion is $0 \leq 2 - \nu \leq 2$, thus having the same magnitude in the other direction (Kokonendji et al., 2008).

## 2.8 Approximating the Normalizing Term, $\mathbf{Z}(\lambda, \nu)$

As noted in Section 2.2, the CMP normalizing function $Z(\lambda, \nu) = \sum_{j=0}^{\infty} \frac{\lambda^j}{(j!)^\nu}$ is an infinite sum that does not generally have a closed-form solution; Table 2.1 lists the special cases such that $Z(\lambda, \nu)$ has a closed form. The infinite sum diverges when $\nu = 0$ and $\lambda \geq 1$ but can likewise cause computational overflow in those scenarios where $\nu$ is small while $\lambda$ is large (Ribeiro Jr. et al., 2019). Ability to compute the infinite sum, however, is important because it is required for computing the probabilities, moments, and other quantities. Throughout this reference, irrespective of the COM–Poisson parametrization and associated normalization term, there exists a consistent need to determine the normalizing constant in a computationally efficient manner, thus reducing computational time. One way to address the matter is to code an `R`-executable `C`-program that further incorporates parallel computing (Wu et al., 2013). Another way is to properly and efficiently approximate this term. Much of the ensuing discussion addresses this matter with focus on the CMP parametrization and its associated normalization term, $Z(\lambda, \nu)$; however, analogous results hold true for the other COM–Poisson parametrizations and their respective representations of the normalizing constant.

While there does not necessarily exist a closed form of $Z(\lambda, \nu)$ for arbitrary $\lambda$ and $\nu$, there exist various methods to approximate it to any level of precision. Minka et al. (2003) noted computational issues and provided useful approximations and upper bounds for both $Z(\lambda, \nu)$ and other related quantities by noting that, for any $\lambda > 0$ and $\nu > 0$, the series $\frac{\lambda^j}{(j!)^\nu}$ converges because (implementing the ratio test for converging series)

$$\lim_{j\to\infty} \frac{\lambda^{j+1}/[(j+1)!]^\nu}{\lambda^j/(j!)^\nu} = \lim_{j\to\infty} \frac{\lambda}{[(j+1)!]^\nu} = 0.$$

In practice, the infinite sum can be approximated by truncation. As noted in Section 2.2, one can approximate the infinite sum by defining some truncation point $M$ such that $Z(\lambda, \nu) \approx \sum_{j=0}^{M} \frac{\lambda^j}{(j!)^\nu}$. The upper bound on the truncation error from using only the first $M + 1$ counts ($j = 0, 1, 2, \ldots, M$) is $\frac{\lambda^{M+1}}{[(M+1)!]^\nu(1-\epsilon_M)}$, where $\epsilon_M > \lambda(k+1)^\nu$ for all $k > M$ (Minka et al., 2003). Several `R` packages (e.g. `CompGLM` (Pollock, 2014a), `multicmp` (Sellers et al., 2017a), `cmpprocess` (Zhu et al., 2017a), `CMPControl` (Sellers and Costa, 2014)) utilize this approach with $M = 100$ as the default truncation; however, this value can be supplied by the analyst. Earlier versions of `COMPoissonReg` (Sellers et al., 2019) (i.e. before 0.5.0) likewise used this truncation method to compute $Z(\lambda, \nu)$. Alternatively, $M$ can be determined such that a desired precision level is attained, i.e. such that $\frac{\lambda^M}{(M!)^\nu}$ is sufficiently small; the `compoisson` package (Dunn, 2012) computes $Z(\lambda, \nu)$ (and $\ln(Z(\lambda, \nu))$) in this manner. The `com.compute.z` and `com.compute.log.z` functions include the input `log.error` whose default value equals 0.001 but can be supplied by the analyst.

When the data are under-dispersed (i.e. $\nu > 1$), $M$ is relatively small because the terms $\frac{\lambda^j}{(j!)^\nu}$ decrease quickly. For over-dispersed data (i.e. $\nu < 1$), however, these terms decline more slowly, thus requiring the value of $M$ to be larger. Accordingly, as demonstrated in Section 2.2.1, while the choice of default truncation (i.e. $M = 100$) is oftentimes sufficient for use in the associated R functions, scenarios exist that require $M > 100$ to sufficiently approximate the infinite sum when the data are over-dispersed (Ribeiro Jr. et al., 2019). This approach can thus require increased computational time to approximate the normalization term. Further, scenarios that do not allow for approximating the normalizing constant can lead to "computationally infeasible Bayesian models" (Wu et al., 2013). As a compromise, one can directly code an approximation function that combines Equation (2.12) with the truncation approach via a program (e.g. C) that can be called from R, e.g. in Wu et al. (2013). Recent versions of `COMPoissonReg` (i.e. after 0.5.0), for example, take this approach where, for $\lambda$ and $\nu$ satisfying $\nu \leq 1$ or $\lambda > 10^{\nu}$, the asymptotic approximation (Equation (2.12)) is applied, while the truncation approach is otherwise utilized based on a pre-determined level of desired precision. Nonetheless, these issues have motivated researchers to determine alternative approaches to approximate $Z(\lambda, \nu)$ for any $\lambda, \nu$.

Shmueli et al. (2005) propose an asymptotic approximation, say $\tilde{Z}(\lambda, \nu)$ (Equation (2.12)), when $\nu \leq 1$ or $\lambda > 10^{\nu}$. This result is derived by noting that the identity $\frac{1}{2\pi}\int_{-\pi}^{\pi} e^{e^{ia}} e^{-iaj} da = \frac{1}{j!}$ implies that

$$\frac{1}{2\pi}\int_{-\pi}^{\pi} e^{e^{ia}} Z(\lambda e^{-ia}, \nu) da = \sum_{j=0}^{\infty} \frac{\lambda^j}{(j!)^{\nu}} \frac{1}{2\pi} \int_{-\pi}^{\pi} e^{e^{ia}} e^{-iaj} da = Z(\lambda, \nu + 1), \tag{2.54}$$

i.e. $Z(\lambda, \nu + 1)$ can be represented as an integral over $Z(\lambda, \nu)$. Applying this result repeatedly produces the representation

$$Z(\lambda, \nu) = \frac{1}{(2\pi)^{\nu-1}} \int \cdots \int \exp\left( \sum_{k=1}^{\nu-1} \exp(ia_k) + \lambda \exp\left( - \sum_{k=1}^{\nu-1} ia_k \right) \right) da_1 \cdot da_{\nu-1} \tag{2.55}$$

for $Z(\lambda, \nu)$, as a multiple integral where the change of variable ($ia_j = ib_j + \frac{1}{\nu} \ln(\lambda)$) and application of Laplace's method result in Equation (2.12). Minka et al. (2003) prove this approximation accurate when $\lambda > 10^{\nu}$ only for integer-valued $\nu$; yet they demonstrate empirically that the approximation holds for any real $\nu > 0$. Olver (1974) likewise obtained and proved this approximation years earlier but only for $0 < \nu \leq 4$. Then, the result was determined by approximating the infinite summation with a contour integral around the nonnegative integers where the proportional error to the approximation converges to 0. This proof, however, fails for $\nu > 4$.

Gillispie and Green (2015) instead consider a different contour integral such that the error integral tends to infinity for large $\lambda$ given any fixed $\nu$ but at such a rate that, compared to $Z(\lambda, \nu)$, the ratio of the error term to the approximation to $Z(\lambda, \nu)$ tends to zero so that the approximation holds for all $\nu > 0$. Gillispie and Green (2015) further discuss the behavior of $Z(\lambda, \nu)$, noting the following properties: $Z(0, \nu) = 1$, $\frac{\partial Z}{\partial \lambda} > 0$ and $\frac{\partial^2 Z}{\partial \lambda^2} > 0$ for all $\nu$; hence, $Z(\lambda, \nu)$ increases at an increasingly faster rate with respect to $\lambda$. Meanwhile, $\frac{\partial Z}{\partial \nu} < 0$ implies that (for fixed $\lambda$) $Z(\lambda, \nu)$ decreases as $\nu$ increases. Thus, $Z(\lambda, \nu)$ increases sharply starting at 1 for $\nu < 1$ and then flatter for $\nu > 1$, where, as $\nu \to \infty$, $Z(\lambda, \nu)$ becomes completely linear with respect to $\lambda$ with a slope of 1 (i.e. $\lim_{\nu\to\infty} Z(\lambda, \nu) = 1 + \lambda$).

Gillispie and Green (2015) show that the behavior of $\tilde{Z}(\lambda, \nu)$ does not behave in the same manner for all $\nu$. At $\nu = 1$, the approximation is exact (i.e. $\tilde{Z}(\lambda, 1) = Z(\lambda, 1) = e^{\lambda}$). While $\lim_{\lambda\to 0} \tilde{Z}(\lambda, \nu) = 0$, for $0 < \nu < 1$,

$$\frac{\partial \tilde{Z}}{\partial \lambda} = \frac{\nu^{-3/2}}{2(2\pi)^{(\nu-1)/2}}(2\nu\lambda^{1/\nu} + 1 - \nu)\lambda^{(1-3\nu)/2\nu} e^{\nu\lambda^{1/\nu}} \tag{2.56}$$

has an infinite slope at zero for $\nu \in \left(\frac{1}{3}, 1\right)$, a positive constant slope at $\nu = \frac{1}{3}$, and a zero slope for $\nu \in \left(0, \frac{1}{3}\right)$. Higher derivatives with respect to $\lambda$ show that $\frac{\partial \tilde{Z}}{\partial \lambda}$ becomes positive for any $\nu$; therefore, $\tilde{Z}$ increases. Meanwhile, for $\nu > 1$, $\lim_{\lambda\to 0} \tilde{Z}(\lambda, \nu) \to \infty$ and $\lim_{\lambda\to 0} \frac{\partial \tilde{Z}}{\partial \lambda} \to -\infty$, falling rapidly before $\lambda = \left(\frac{1}{2}\right)^{\nu}$, and then approaching $Z(\lambda, \nu)$ from below as $\lambda$ increases. Recall that the performance of $\tilde{Z}(\lambda, \nu)$ is considered only for $\nu \in (0, \infty)$ because $Z(\lambda, \nu)$ has a closed form when $\nu \in \{0, 1, \infty\}$; see Table 2.1.

The asymptotic approximation has alternatively been proven via probabilistic arguments. Şimşek and Iyengar (2016) consider the ACMP reparametrization and show that

$$Z_1(\mu_*, \nu) = \frac{e^{\nu\mu_*}}{(2\pi\mu_*)^{(\nu-1)/2}\sqrt{\nu}}[1 + O(\mu_*^{-1})] \text{ as } \mu_* \to \infty. \tag{2.57}$$

While this can be obtained directly by applying Equation (2.47) to Equation (2.12), the authors prove this result by expressing the normalizing constant as an expectation of a function of a Poisson random variable and approximate the function via Stirling's formula. From there, they use distribution theory results relating a Poisson random variable to a normal random variable as $\mu_*$ gets large, expand various expressions, and take the expectation of the penultimate result to obtain Equation (2.57). Şimşek and Iyengar (2016) further advance the connection between the ACMP distribution and generalized hypergeometric functions (initiated by Nadarajah (2009)) to refine the approximation of the normalizing constant for integer values of $\nu$.

Noting that the normalizing constant equals a modified Bessel function of the first kind when $\nu = 2$, they apply an asymptotic expansion of the modified Bessel function to determine a more detailed approximation of the normalizing constant, namely

$$Z_1(\mu_*, 2) \approx \frac{e^{2\mu_*}}{\sqrt{4\pi\mu_*}}\left(1 + \frac{1}{16\mu_*} + \frac{9}{512\mu_*^2} + O(\mu_*^{-3})\right). \tag{2.58}$$

Because the additional terms are both positive, Equation (2.58) improves the under-estimation of the normalizing constant approximation provided in Equation (2.57). Extending the approach described above but with more terms in the expansions, the asymptotic approximation is refined to be

$$Z_1(\mu_*, \nu) = \frac{e^{\nu\mu_*}}{(2\pi\mu_*)^{(\nu-1)/2}\sqrt{\nu}}\left[1 + (\nu - 1)\left(\frac{8\nu^2 + 12\nu + 3}{96\mu_*\nu^2} + \frac{1 + 6\nu}{144\mu_*^2\nu^3}\right)\right.$$
$$\left. + O(\mu_*^{-3})\right] \text{ as } \mu_* \to \infty \tag{2.59}$$

(Şimşek and Iyengar, 2016). Gaunt et al. (2016) further extend this result by providing the form of the lower order terms in the asymptotic expansion. Given $\nu > 0$,

$$Z(\lambda, \nu) = \frac{\exp(\nu\lambda^{1/\nu})}{\lambda^{(\nu-1)/2\nu}(2\pi)^{(\nu-1)/2}\sqrt{\nu}} \sum_{k=0}^{\infty} c_k(\nu\lambda^{1/\nu})^{-k} \text{ as } \lambda \to \infty, \tag{2.60}$$

where the expansion coefficients $c_k$ are uniquely determined by

$$(\Gamma(t+1))^{-\nu} = \frac{\nu^{\nu(t+1/2)}}{(2\pi)^{(\nu-1)/2}} \sum_{j=0}^{\infty} \frac{c_j}{\Gamma(\nu t + (1+\nu)/2 + j)}. \tag{2.61}$$

## 2.9 Summary

The COM–Poisson distribution is a flexible count distribution that accommodates for data over- or under-dispersion and contains the Poisson, geometric, and Bernoulli distributions as special cases. Derived nicely from a more flexible queuing system model, it satisfies several eloquent properties, including having the form of an exponential family and a (modified) power series distribution, and, for a given dispersion $\nu$, it satisfies the form of an exponential dispersion family. Computational tools provide analysts with the ability to model frequency tables with the COM–Poisson distribution, thus providing estimators and confidence intervals for $\lambda$ and $\nu$, as well as fitted values to determine estimated frequencies based on a COM–Poisson model.

The flexibility of the COM–Poisson distribution motivates its application in various statistical models, including multivariate distributions (Chapter 4), regression analysis (Chapter 5), control chart theory (Chapter 6), stochastic processes and time series models (Chapter 7), and survival analysis (Chapter 8) Meanwhile, Chapter 3 introduces the various distributional extensions that are attained and introduced (to date) from a COM–Poisson model.

# 3

# Distributional Extensions and Generalities

The COM–Poisson (with specific parametrizations defined as necessary) distribution has garnered interest in and development of other flexible alternatives to classical distributions. This chapter will introduce various distributional extensions and generalities motivated by functions of COM–Poisson random variables. Section 3.1 describes a generalization of the Skellam distribution, namely the Conway–Maxwell–Skellam (COM–Skellam or CMS) distribution. Section 3.2 introduces the sum-of-Conway–Maxwell–Poissons (sCMP) distribution. Section 3.3 introduces the reader to Conway–Maxwell inspired generalizations of the binomial distribution, particularly the COM–binomial (CMB) and COM–multinomial (CMM) distributions. Section 3.4 highlights several works that seek to generalize the negative binomial (NB) distribution, including the generalized COM–Poisson (GCMP), the COM-type NB (COMtNB), the extended CMP (ECMP), and the COM–NB (COMNB) distributions. Section 3.5 extends the Katz class of distributions to introduce the Conway–Maxwell–Katz (COM–Katz) class. Section 3.6 describes two flexible series system life-length distributions: the exponential-CMP (ExpCMP) and Weibull–CMP (WCMP) distributions, respectively. Section 3.7 discusses three generalizations of the negative hypergeometric (NH) distribution – two named as a Conway–Maxwell–negative hypergeometric (abbreviated as COMNH and CMNH, respectively), and the COM–Poisson-type negative hypergeometric (CMPtNH) distributions. All of the distributions described in this chapter are derived in relation to CMP parametrized random variables. That said, one can likewise consider analogous representations of these distributions based on the alternative COM–Poisson parametrizations described in Chapter 2 (Section 2.6). Finally, Section 3.8 concludes the chapter with discussion.

## 3.1 The Conway–Maxwell–Skellam (COM–Skellam or CMS) Distribution

The Skellam distribution (Skellam, 1946) is a discrete model based on the difference of two Poisson random variables. Its form has been applied in various contexts, e.g. in image background correction (Hwang et al., 2007) and measuring the difference in scores for competing soccer teams (Karlis and Ntzoufras, 2008). The underlying assumed data equi-dispersion that motivates the development of the Skellam distribution, however, is constraining. The COM–Skellam or CMS distribution is a flexible generalization of the Skellam model to allow for significant data over- or under-dispersion; its structure is based on the difference of two COM–Poisson random variables. Consider $S = X_1 - X_2$, where $X_i$, $i = 1, 2$, are independent CMP($\lambda_i, \nu$) distributed random variables; accordingly, $S$ has a CMS($\lambda_1, \lambda_2, \nu$) distribution with probability

$$P(S = s) = \frac{1}{Z(\lambda_1, \nu)Z(\lambda_2, \nu)} \left(\frac{\lambda_1}{\lambda_2}\right)^{s/2} I_{|s|}^{(\nu)}(2\sqrt{\lambda_1\lambda_2}), \qquad s \in \mathbb{Z},$$

where $Z(\lambda_i, \nu)$, $i = 1, 2$, denote the respective normalizing constants of the CMP($\lambda_i, \nu$) distributions as described in Section 2.2, and $I_\alpha^{(\nu)}(z) \doteq \sum_{m=0}^{\infty} \frac{1}{[\Gamma(m+\alpha+1)m!]^\nu} \left(\frac{z}{2}\right)^{2m+\alpha}$ is a modified Bessel function of the first kind (Sellers, 2012a). This probability mass function simplifies to that for the Skellam distribution when $\nu = 1$, while $\nu < ( > )1$ corresponds to over-dispersion (under-dispersion) relative to the Skellam distribution. The CMS distribution likewise contains the difference of geometric and Bernoulli distributions, respectively, as special cases.

The moment generating function of $S$ is $M_S(t) = \frac{Z(\lambda_1 e^t, \nu)}{Z(\lambda_1, \nu)} \cdot \frac{Z(\lambda_2 e^{-t}, \nu)}{Z(\lambda_2, \nu)}$, from which the moments of interest can be derived. In particular, the mean and variance are, respectively,

$$E(S) = E(X_1) - E(X_2) \approx \lambda_1^{1/\nu} - \lambda_2^{1/\nu} \text{ and} \tag{3.1}$$

$$V(S) = V(X_1) + V(X_2) \approx \frac{1}{\nu}\left(\lambda_1^{1/\nu} + \lambda_2^{1/\nu}\right), \tag{3.2}$$

where $E(X_i)$ and $V(X_i)$ and their approximations are defined in Chapter 2 with the approximations deemed reasonable for $\nu \leq 1$ or $\lambda > 10^\nu$ (Minka et al., 2003; Shmueli et al., 2005). Figure 3.1 illustrates the CMS variability via various CMS($\lambda_1 = \lambda_2 = 5, \nu$) probability mass functions for $\nu \in \{0.25, 0.5, 1, 2, 4, 10\}$; in particular, the CMS($\lambda_1 = \lambda_2 = 5, \nu = 1$) represents the special case of the Skellam($\lambda_1 = \lambda_2 = 5$) distribution. These plots show that, as $\nu$ increases, the distribution contracts about the

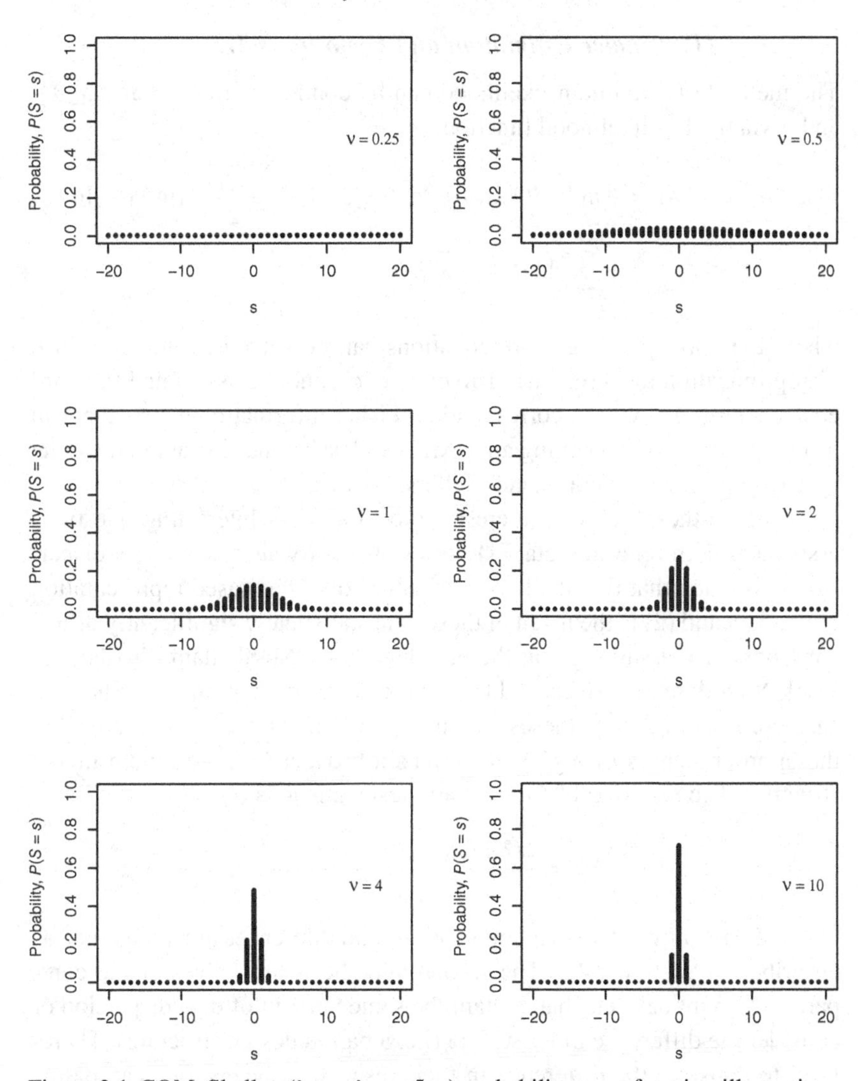


Figure 3.1 COM–Skellam($\lambda_1 = \lambda_2 = 5, \nu$) probability mass function illustrations for the values of $\nu \in \{0.25, 0.5, 1, 2, 4, 10\}$.

mean, consistent with the reduced dispersion in the underlying CMP random variables; thus, we see that the CMS distribution is over-dispersed (under-dispersed) relative to the Skellam distribution when $\nu < (>)1$. Differences between $\lambda_1$ and $\lambda_2$ would meanwhile shift the distribution center as described in Equation (3.1).

### *Parameter Estimation and Hypothesis Testing*

The method of maximum likelihood can be conducted to estimate $\lambda_1$, $\lambda_2$, and $\nu$ via the log-likelihood function,

$$\ln L(\lambda_1, \lambda_2, \nu \mid \mathbf{s}) = -n[\ln Z(\lambda_1, \nu) + \ln Z(\lambda_2, \nu)] + \frac{\sum_{i=1}^{n} s_i}{2}(\ln \lambda_1 - \ln \lambda_2) + \sum_{i=1}^{n} \ln I_{|s_i|}^{\nu}(2\sqrt{\lambda_1 \lambda_2}),$$

where the corresponding score equations can be solved numerically in `R` via optimization tools (e.g. `nlminb` or `optim`), and the associated standard errors determined via the corresponding Fisher information matrix that can be computed via the resulting approximate Hessian matrix supplied via the selected optimization command (Sellers, 2012a).

Two hypothesis tests of interest serve to address interesting questions associated with relevant data: (1) does statistically significant data dispersion exist such that the Skellam distribution offers a biased representation of the data and (2) is the mean of these data statistically significantly different from zero, assuming that the data have a COM–Skellam distribution. A likelihood ratio test can address each of these hypotheses. The first case considers the hypotheses $H_0\colon \nu = 1$ versus $H_1\colon \nu \neq 1$ to compare the appropriateness of a Skellam versus a broader COM–Skellam model structure. The resulting likelihood ratio test statistic is

$$\Lambda_\nu = \frac{L(\hat{\lambda}_{(1,H_0)}, \hat{\lambda}_{(2,H_0)}, \nu_{H_0} = 1)}{L(\hat{\lambda}_1, \hat{\lambda}_2, \hat{\nu})},$$

where the resulting statistical implications and inferences are maintained as described in Section 2.4.5. The second hypothesis test allows one to compare two count datasets that contain the same amount of data dispersion or consider the difference of those data (i.e. a paired design structure). Therefore, to consider the difference in their respective means, one can define $H_0\colon E(S) = 0$ and $H_1\colon E(S) \neq 0$ (which is equivalent to $H_0\colon \lambda_1 = \lambda_2$ versus $H_1\colon \lambda_1 \neq \lambda_2$). The associated likelihood ratio test statistic is now

$$\Lambda_{\mu_S} = \frac{L(\hat{\lambda}_{1,H_0} = \hat{\lambda}_{2,H_0} = \hat{\lambda}, \hat{\nu}_{H_0})}{L(\hat{\lambda}_1, \hat{\lambda}_2, \hat{\nu})},$$

where $-2\ln \Lambda_{\mu_S}$ converges to a chi-squared distribution with one degree of freedom. For both tests, analysts can modify the proposed two-sided tests to be one-sided in order to properly study direction. Analogous to Section 2.4.5, alternative hypothesis test statistics can likewise be derived.

## 3.2 The Sum-of-COM–Poissons (sCMP) Distribution

The sCMP $(\lambda, \nu, m)$ class of distributions can be viewed as a three-parameter structure with CMP intensity parameter $\lambda$, dispersion parameter $\nu$, and number of underlying CMP random variables $m \in \mathbb{N}$. Given a random sample of CMP$(\lambda, \nu)$ distributed random variables $X_i$, $i = 1, \ldots, m$, a random variable $Y = \sum_{i=1}^{m} X_i$ has a sCMP$(\lambda, \nu, m)$ distribution with probability mass function

$$P(Y = y \mid \lambda, \nu, m) = \frac{\lambda^y}{(y!)^\nu Z^m(\lambda, \nu)} \sum_{\substack{x_1, \ldots, x_m = 0 \\ x_1 + \cdots + x_m = y}}^{y} \binom{y}{x_1 \cdots x_m}^{\nu}, \quad y = 0, 1, 2, \ldots, \tag{3.3}$$

where $Z^m(\lambda, \nu)$ is the $m$th power of the CMP$(\lambda, \nu)$ normalizing constant, and $\binom{y}{x_1 \cdots x_m} = \frac{y!}{x_1! \cdots x_m!}$ is the multinomial coefficient. Special cases of the sCMP class include the Poisson$(m\lambda)$ distribution when $\nu = 1$, the NB$(m, 1 - \lambda)$ distribution when $\nu = 0$ and $\lambda < 1$, and the binomial$(m, p)$ distribution as $\nu \to \infty$ with success probability $p = \frac{\lambda}{\lambda+1}$. Accordingly, the sCMP$(\lambda, \nu, m)$ distribution is over-dispersed (under-dispersed) relative to the Poisson$(m\lambda)$ distribution when $\nu < (>)1$. The CMP$(\lambda, \nu)$ distribution naturally is contained as a special case when $m = 1$; this implies that the special cases of the CMP distribution are likewise contained here, namely the Poisson$(\lambda)$ $(\nu = 1)$, geometric (with success probability $1 - \lambda$ for $\nu = 0$ and $\lambda < 1$), and Bernoulli (with success probability $\frac{\lambda}{1+\lambda}$, when $\nu \to \infty$) distributions (Sellers et al., 2017b). The sCMP generating functions include its moment generating function $M_Y(t) = \left(\frac{Z(\lambda e^t, \nu)}{Z(\lambda, \nu)}\right)^m$, probability generating function $\Pi_Y(t) = \left(\frac{Z(\lambda t, \nu)}{Z(\lambda, \nu)}\right)^m$, and characteristic function $\phi_Y(t) = \left(\frac{Z(\lambda e^{it}, \nu)}{Z(\lambda, \nu)}\right)^m$, respectively. Given the same $\lambda$ and $\nu$, the sum of independent sCMP distributions is invariant under addition; however, this result does not generally hold. While this definition assumes $m \in \mathbb{N}$, $m$ need not be integer-valued since the generating functions are valid for all $m \geq 0$.

Figure 3.2 contains various sCMP$(m, \lambda = 1, \nu)$ probability density illustrations for the values of $m \in \{2, 3, 4, 5\}$ and $\nu \in \{0.5, 1, 2\}$. As demonstrated in this illustration, the sCMP distribution can effectively represent discrete data that contain data dispersion. We see that, for a fixed $m$, the distribution contracts closer to zero as $\nu$ increases. This makes sense because an increasing $\nu$ associates with decreased dispersion. Meanwhile, for a fixed $\nu$, we see that the distribution expands as $m$ increases. This also makes sense because an increased $m$ can be thought of as summing more CMP

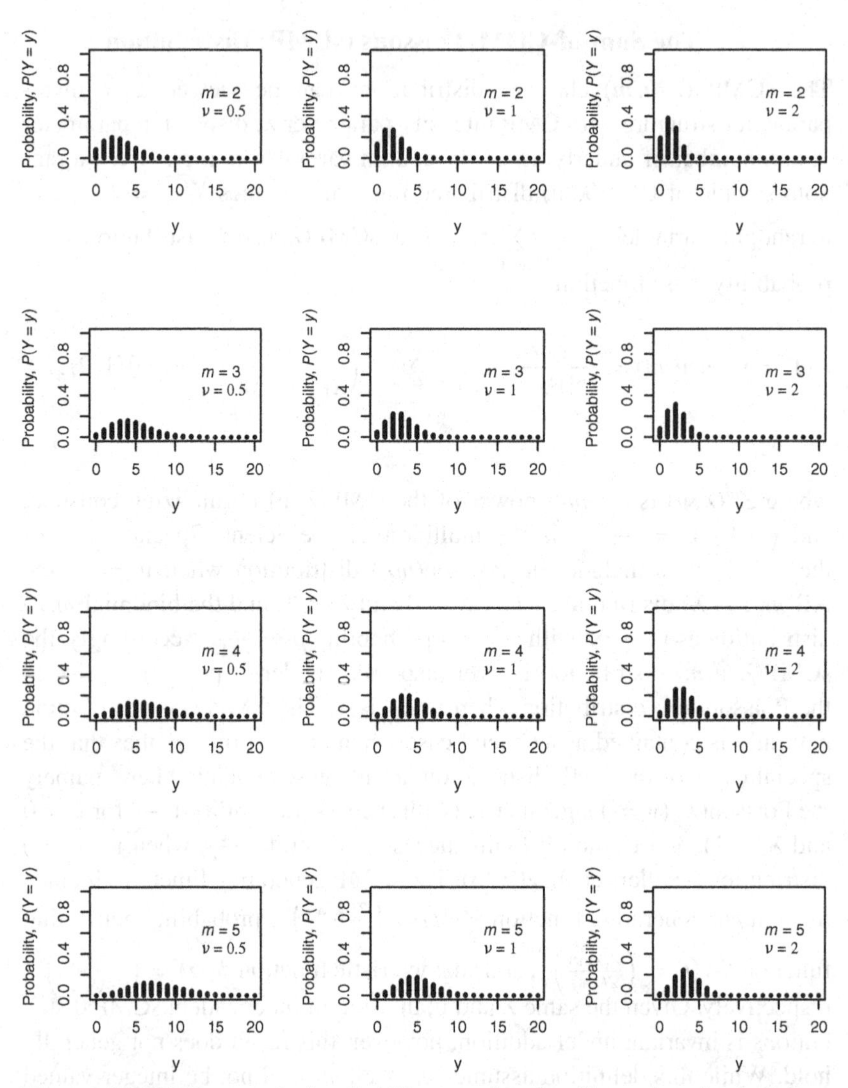


Figure 3.2 sCMP($m, \lambda = 1, \nu$) probability mass function illustrations for the values of $m \in \{2, 3, 4, 5\}$ and $\nu \in \{0.5, 1, 2\}$.

variables. The middle column contains the distribution plots when $\nu = 1$, and note that the sCMP($\lambda = 1, \nu = 1, m$) equals the Poisson($m$) distribution. For example, we see that the sCMP($\lambda = 1, \nu = 1, m = 3$) probability distribution equals the Poisson($\lambda = 3$) model shown in Figure 1.1.

***Parameter Estimation***

Sellers et al. (2017b) consider a profile likelihood approach for parameter estimation to estimate $\lambda$ and $\nu$ for a given $m$ via maximum likelihood estimation. Consider the log-likelihood

$$\ln L(\lambda, \nu \mid m) = \sum_{i=1}^{n} \ln P(Y_i = y_i \mid \lambda, \nu, m) \tag{3.4}$$

for a random sample $Y_1, \ldots, Y_n$, where $P(Y_i = y_i \mid \lambda, \nu, m)$, $i = 1, \ldots, n$ is defined in Equation (3.3). For each considered $m$, the respective conditionally maximized log-likelihood results $\ln L(\hat{\lambda}, \hat{\nu} \mid m)$ are obtained and compared to determine the combination of estimators $(\hat{\lambda}, \hat{\nu}, \hat{m})$ that produce the maximum log-likelihood. The standard errors associated with $\hat{\lambda}$ and $\hat{\nu}$ are obtained by taking the square root of the diagonal elements from the inverse of the information matrix,

$$I(\lambda, \nu) = -n \cdot E \begin{pmatrix} \frac{\partial^2 \ln P(Y=y)}{\partial \lambda^2} & \frac{\partial^2 \ln P(Y=y)}{\partial \lambda \partial \nu} \\ \frac{\partial^2 \ln P(Y=y)}{\partial \lambda \partial \nu} & \frac{\partial^2 \ln P(Y=y)}{\partial \nu^2} \end{pmatrix}. \tag{3.5}$$

The log-likelihood (Equation (3.4)) produces score equations that do not have a closed-form solution; thus, the MLEs $\hat{\lambda}$ and $\hat{\nu}$ are determined numerically. Sellers et al. (2017b) use the `nlminb` function (contained in the `stats` package) to determine the parameter estimates, while they use the `hessian` function (in the `numDeriv` package (Gilbert and Varadhan, 2016)) to determine the information matrix provided in Equation (3.5).

## 3.3 Conway–Maxwell Inspired Generalizations of the Binomial Distribution

The binomial distribution is a popular discrete distribution that models the number of outcomes in either of two categories, while the multinomial distribution is its most popular and well-studied generalization of the binomial distribution, modeling the number of outcomes contained in each of the $k \geq 2$ categories. The common thread regarding these distributions is that the data stem from independent and identically distributed trials. This section instead considers various Conway–Maxwell inspired generalizations of these distributions. Section 3.3.1 introduces readers to the CMB distribution, while Section 3.3.2 discusses a broader generalization of the CMB distribution. Section 3.3.3 meanwhile acquaints readers to the broader CMM distribution. Both the CMB and CMM distributions can be described as a sum of associated Bernoulli random variables; Section 3.3.4

details that discussion. Finally, Section 3.3.5 describes statistical computing tools in R for modeling data and computing probabilities assuming a CMB or CMM distribution.

### 3.3.1 The Conway–Maxwell–binomial (CMB) Distribution

The CMB distribution (also referred to as the Conway–Maxwell–Poisson–binomial distribution) is a three-parameter generalization of the binomial distribution. A random variable $Y$ is CMB$(m, p, \nu)$ distributed with the probability mass function

$$P(Y = y) = \frac{\binom{m}{y}^{\nu} p^{y}(1-p)^{m-y}}{S(m, p, \nu)}, \quad y = 0, \ldots, m, \tag{3.6}$$

where $0 \leq p \leq 1$, $\nu \in \mathbb{R}$, and $S(m, p, \nu) = \sum_{y=0}^{m} \binom{m}{y}^{\nu} p^{y}(1-p)^{m-y}$ is a normalizing constant (Borges et al., 2014; Kadane, 2016; Shmueli et al., 2005). This model satisfies the recursion property,

$$P(Y = y) = \frac{p}{1-p}\left(\frac{m+1-y}{y}\right)^{\nu} P(Y = y-1). \tag{3.7}$$

The binomial$(m, p)$ distribution is the special case of the CMB$(m, p, \nu)$, where $\nu = 1$, while $\nu > (<) 1$ corresponds to under-dispersion (over-dispersion) relative to the binomial model. For $\nu \to \infty$, the probability distribution is concentrated on the point $mp$, while for $\nu \to -\infty$, it is concentrated at 0 or $m$. For $\lambda = m^{\nu} p$, the CMB$(m, p, \nu)$ distribution converges to the CMP$(\lambda, \nu)$ as $m \to \infty$, i.e. $\lim_{m\to\infty} P(Y = y \mid m, p, \nu) = \frac{\lambda^{y}}{(y!)^{\nu} Z(\lambda, \nu)}$, $y = 0, 1, 2, \ldots$ (Borges et al., 2014; Daly and Gaunt, 2016). Meanwhile, the CMB can be derived as the conditional distribution of one CMP random variable given that the sum of it and another (independent) CMP random variable (with the same dispersion parameter) equals an outcome $m$. For two independent CMP$(\lambda_i, \nu)$ distributed random variables $X_i, i = 1, 2$,

$$(X_1 \mid (X_1 + X_2 = m)) \sim \text{CMB}\left(m, \frac{\lambda_1}{\lambda_1 + \lambda_2}, \nu\right). \tag{3.8}$$

Daly and Gaunt (2016) present some general properties regarding the CMB distribution. Letting $(j)_k = j(j-1)\cdots(j-k+1)$ denote the falling factorial for a number $j$,

$$E\left(((Y)_k)^{\nu}\right) = \frac{S(m-k, p, \nu)}{S(m, p, \nu)}((m)_k)^{\nu} p^{k}$$

for $k = 1, \ldots, m-1$. Meanwhile, the mode of the distribution is $\lfloor a \rfloor$ if

$$a = \frac{m+1}{1 + \left(\frac{1-p}{p}\right)^{1/\nu}}$$

is not integer-valued; otherwise, the modes are $a$ and $a-1$. Finally, analogous to Equation (2.28), for a CMB$(m, p, \nu)$ distributed random variable $Y$ and function $f\colon \mathbb{Z} \to \mathbb{R}$ such that $E|f(Y+1)| < \infty$ and $E|Y^\nu f(Y)| < \infty$,

$$E[p(m-Y)^\nu f(Y+1) - (1-p)Y^\nu f(Y)] = 0. \tag{3.9}$$

Results also exist regarding stochastic ordering and other orderings relating to the CMB distribution and also regarding other convergence and approximation results for this distribution; see Daly and Gaunt (2016) for details.

The CMB$(m, p, \nu)$ probability function (Equation (3.6)) can be reparametrized in terms of the odds $\theta = \frac{p}{1-p}$ as

$$\begin{aligned} P(Y = y \mid m, \theta, \nu) &= \frac{1}{\tau(m,\theta,\nu)} \binom{m}{y}^\nu \theta^y \quad \text{or} \quad \frac{1}{\xi(m,\theta,\nu)} \frac{\theta^y}{[y!(m-y)!]^\nu} \\ &= \frac{\exp\left[\nu \ln \binom{m}{y} + y \ln(\theta)\right]}{\tau(m,\theta,\nu)} \quad \text{or} \quad \frac{\exp\left(y \ln(\theta) - \nu \ln(y!(m-y)!)\right)}{\xi(m,\theta,\nu)}, \end{aligned}$$

for $y = 0, 1, 2, \ldots, m$, where $\tau(m,\theta,\nu) = \sum_{j=0}^{m} \binom{m}{j}^\nu \theta^j$ and $\xi(m,\theta,\nu) = \sum_{j=0}^{m} \frac{\theta^j}{[j!(m-j)!]^\nu}$ (Borges et al., 2014; Kadane, 2016). These parametrizations produce analogous representations for the respective generating functions (i.e. probability, moment, and characteristic function), namely

$$\Pi_Y(t) = \frac{\tau(m,\theta t,\nu)}{\tau(m,\theta,\nu)} = \frac{\xi(m,\theta t,\nu)}{\xi(m,\theta,\nu)}, \quad \text{for } 0 \le t \le 1, \tag{3.10}$$

$$M_Y(t) = \frac{\tau(m,\theta e^t,\nu)}{\tau(m,\theta,\nu)} = \frac{\xi(m,\theta e^t,\nu)}{\xi(m,\theta,\nu)}, \quad \text{and} \tag{3.11}$$

$$\phi_Y(t) = \frac{\tau(m,\theta e^{it},\nu)}{\tau(m,\theta,\nu)} = \frac{\xi(m,\theta e^{it},\nu)}{\xi(m,\theta,\nu)}. \tag{3.12}$$

Whether from the moment generating function or a direct representation of the $r$th moment of $Y$ (via the latter parametrization, e.g.),

$$E(Y^r) = \frac{1}{\xi(m,\theta,\nu)} \sum_{x=0}^{m} y^r \frac{\theta^y}{[y!(m-y)!]^\nu},$$

for $r \in \mathbb{R}$, the mean and variance of $Y$ are, respectively,

$$E(Y) = \theta \frac{d \ln (\xi(m, \theta, \nu))}{d\theta} \tag{3.13}$$

$$V(Y) = \theta \frac{dE(Y)}{d\theta}. \tag{3.14}$$

The CMB distribution is a member of the exponential family with the joint probability mass function

$$P(\mathbf{y} \mid p, \nu) \propto (1-p)^{mn} \prod_{i=1}^{n} \left(\frac{p}{1-p}\right)^{y_i} \frac{m!^{n\nu}}{[y_i!(m-y_i)!]^{\nu}}$$
$$\propto \exp\left(\mathcal{A}_{*1} \ln\left(\frac{p}{1-p}\right) - \nu \mathcal{A}_{*2}\right),$$

for a random sample $\mathbf{y} = \{y_1, \ldots, y_n\}$, where $\mathcal{A}_{*1} = \sum_{i=1}^{n} y_i$ and $\mathcal{A}_{*2} = \sum_{i=1}^{n} \ln [y_i!(m-y_i)!]$ are sufficient statistics for $p$ and $\nu$. As an exponential family, however, it cannot be represented as a mixture of binomials for all sample sizes; hence, the CMB family is not marginally compatible (Kadane, 2016). The CMB distribution also belongs to the family of weighted Poisson distributions as described in Section 1.4 where

$$w(y; \nu) = \begin{cases} \frac{[y!(m-y)!]^{1-\nu}}{(m-y)!} & \text{if } y \leq m \\ 0 & \text{if } y > m, \end{cases}$$

and is a generalized power series distribution,[1] where $\eta = \nu$ and

$$a(y; \nu) = \begin{cases} \frac{1}{[y!(m-y)!]^{\nu}} & \text{if } y \leq m \\ 0, & \text{if } y > m. \end{cases}$$

The CMB distribution has a conjugate prior family of the form

$$h(\theta, \nu) = \theta^{a_1 - 1} e^{-\nu a_2} \xi^{-a_3}(\theta, \nu) \psi(a_1, a_2, a_3), \qquad 0 < \theta < \infty, \;\; 0 < \nu < \infty,$$

where $\xi(\theta, \nu) = \sum_{y=0}^{m} \theta^y / [y!(m-y)!]^{\nu}$ and $\psi^{-1}(a_1, a_2, a_3) = \int_0^\infty \int_0^\infty \theta^{a_1 - 1} e^{-\nu a_2} \xi^{-a_3}(\theta, \nu) d\theta d\nu < \infty$; its updates occur as $a_1' = a_1 + y$, $a_2' = a_2 + \ln(y!(m-y)!)$ and $a_3' = a_3 + 1$, and this conjugate prior is symmetric about 0 if and only if $a_1/a_3 = m/2$ (Kadane, 2016).

[1] A generalized power series distribution has the form $P(Y = y; \eta, \theta) = \frac{a(y;\eta)\theta^y}{A(\eta;\theta)}$, $y = 0, 1, 2, \ldots$, where $A(\eta; \theta) = \sum_{j=0}^{\infty} a(j; \eta)\theta^j$.

The maximum likelihood estimators of the CMB parameters $\theta = \frac{p}{1-p}$ and $\nu$ can be determined for a random sample of $n$ CMB$(\theta, \nu)$ random variables, $Y_1, \ldots, Y_n$, via the log-likelihood

$$\ln L(\theta, \nu, ; \mathbf{y}) = \left( \ln(\theta) \sum_{i=1}^{n} y_i - \nu \sum_{i=1}^{n} \ln(y_i!(m-y_i)!) - n \ln(\xi(m, \theta, \nu)) \right)$$

$$= n \left[ \ln(\theta) t_1 - \nu t_2 - \ln(\xi(m, \theta, \nu)) \right],$$

where $t_1 = \frac{1}{n} \sum_{i=1}^{n} y_i$ and $t_2 = \frac{1}{n} \sum_{i=1}^{n} \ln(y_i!(m - y_i)!)$ denote the sample mean and sample log-geometric mean, respectively (Borges et al., 2014; Kadane, 2016). The resulting score equations produce

$$\begin{cases} t_1 = \theta \frac{d}{d\theta} \ln(\xi(m, \theta, \nu)) \\ t_2 = -\frac{d}{d\nu} \ln(\xi(m, \theta, \nu)), \end{cases} \tag{3.15}$$

where, because the CMB is a member of the exponential family, $T_1 = \frac{1}{n} \sum_{i=1}^{n} Y_i$ and $T_2 = \frac{1}{n} \sum_{i=1}^{n} \ln(Y_i!(m - Y_i)!)$ are minimal sufficient statistics, and the maximum likelihood estimators $\hat{\theta}$ and $\hat{\nu}$ are unique solutions to Equation (3.15). As has been the case with other distributions and constructs, an iterative method (e.g. the Newton–Raphson method) is required to determine these solutions. This can be accomplished via statistical computing in `R` (R Core Team, 2014) using, for example, the `optim` function or in `SAS` via the `PROC NLMIXED` procedure.

### 3.3.2 The Generalized Conway–Maxwell–Binomial Distribution

The generalized Conway–Maxwell–binomial (gCMB) distribution broadens the CMB distribution, where the support space remains between 0 and $m$ but allows for even greater flexibility relative to the binomial distribution. The probability mass function associated with a gCMB$(p, \nu, m, n_1, n_2)$ distributed random variable $X_*$ is

$$P(X_* = x) = \frac{\binom{m}{x}^{\nu} p^x (1-p)^{m-x} \left[ \sum_{a_1, \ldots, a_{n_1}}^{x} \binom{x}{a_1, \ldots, a_{n_1}}^{\nu} \right] \left[ \sum_{b_1, \ldots, b_{n_2}}^{m-x} \binom{m-x}{b_1, \ldots, b_{n_2}}^{\nu} \right]}{G(p, \nu, m, n_1, n_2)}, \tag{3.16}$$

where

$$G(p,\nu,m,n_1,n_2) = \sum_{x=0}^{m} \binom{m}{x}^{\nu} p^x (1-p)^{m-x} \left[ \sum_{\substack{a_1,\ldots,a_{n_1}=0 \\ a_1+\cdots+a_{n_1}=x}}^{x} \binom{x}{a_1, \ldots, a_{n_1}}^{\nu} \right] \left[ \sum_{\substack{b_1,\ldots,b_{n_2}=0 \\ b_1+\cdots+b_{n_2}=m-x}}^{m-x} \binom{m-x}{b_1, \ldots, b_{n_2}}^{\nu} \right]$$

is the normalizing constant. Analogous to the manner in which a CMP variable conditioned on the sum of two CMP variables results in a CMB distribution (Equation (3.8)), a sCMP variable conditioned on the sum of two sCMP variables results in a gCMB distribution, i.e. for two independent sCMP($\lambda_i, \nu, n_i$) distributed random variables, $Y_i$, for $i = 1, 2$, $(Y_1 \mid (Y_1 + Y_2 = m))$ has a gCMB$\left(\frac{\lambda_1}{\lambda_1+\lambda_2}, \nu, m, n_1, n_2\right)$ distribution (Sellers et al., 2017b). Special cases of the gCMB distribution include the CMB($m, p, \nu$) (when $n_1 = n_2 = 1$), binomial($m, p$) (for $n_1 = n_2 = 1$ and $\nu = 1$), a NH ($\nu = 0$ and $\lambda < 1$), and hypergeometric ($\nu \to \infty$) distribution.

The gCMB distribution has the probability generating, moment generating, and characteristic functions

$$\Pi_{X_*}(t) = \frac{H\left(\frac{tp}{1-p}, \nu, m, n_1, n_2\right)}{H\left(\frac{p}{1-p}, \nu, m, n_1, n_2\right)}, \quad M_{X_*}(t) = \frac{H\left(\frac{pe^t}{1-p}, \nu, m, n_1, n_2\right)}{H\left(\frac{p}{1-p}, \nu, m, n_1, n_2\right)},$$

$$\phi_{X_*}(t) = \frac{H\left(\frac{pe^{it}}{1-p}, \nu, m, n_1, n_2\right)}{H\left(\frac{p}{1-p}, \nu, m, n_1, n_2\right)}, \quad (3.17)$$

respectively, where

$$H(\theta,\nu,m,n_1,n_2) = \sum_{x=0}^{m} \binom{m}{x}^{\nu} \theta^x \left[ \sum_{a_1,\ldots,a_{n_1}}^{x} \binom{x}{a_1, \ldots, a_{n_1}}^{\nu} \right] \left[ \sum_{b_1,\ldots,b_{n_2}}^{m-x} \binom{m-x}{b_1, \ldots, b_{n_2}}^{\nu} \right].$$

### 3.3.3 The Conway–Maxwell–multinomial (CMM) Distribution

The multinomial distribution can likewise be considered more broadly in a manner to accommodate associations between trials. Where the multinomial distribution effectively describes categorical data, this model assumes that the underlying trials are independent and identically distributed. The CMM distribution extends the multinomial distribution as a distribution that can consider trials with positive or negative associations (Kadane and Wang, 2018; Morris et al., 2020). Like the CMB distribution, the CMM distribution includes a dispersion parameter that allows for this broader range of associations between trials, thus making it a more flexible alternative to other generalized multinomial distributions such as the Dirichlet-multinomial or random-clumped multinomial models. A $\text{CMM}_k(m, \boldsymbol{p}, \nu)$ distributed random variable $\boldsymbol{Y} = (Y_1, \ldots, Y_k)$ has the probability mass function

$$P(\boldsymbol{Y} = \boldsymbol{y}) = \frac{1}{C(\boldsymbol{p}, \nu; m)} \binom{m}{y_1 \cdots y_k}^{\nu} \prod_{j=1}^{k} p_j^{y_j}, \tag{3.18}$$

for $\nu \in \mathbb{R}$, $\boldsymbol{y} \in \Omega_{m,k} = \{(y_1, \ldots, y_k) \in \{0, 1, 2, \ldots, m\}^k$ such that $\sum_{j=1}^{k} y_j = m\}$, where $\boldsymbol{p} = (p_1, \ldots, p_k)$ satisfying $\sum_{j=1}^{k} p_j = 1$, $\binom{m}{y_1 \cdots y_k} = \frac{m!}{y_1! y_2! \cdots y_k!}$, and $C(\boldsymbol{p}, \nu; m) = \sum_{\boldsymbol{y} \in \Omega_{m,k}} \binom{m}{y_1 \cdots y_k}^{\nu} \prod_{j=1}^{k} p_j^{y_j}$ is the normalizing constant. Analogous to the CMB distribution, the CMM distribution also relates to the CMP as the conditional distribution of one CMP random variable conditioned on the outcome from the sum of the independent CMP random variables. For an independent sample of $\text{CMP}(\lambda_j, \nu)$ random variables $Y_j, j = 1, \ldots, k$, the distribution of $\boldsymbol{Y} = (Y_1, \ldots, Y_k)$ given that $\sum_{j=1}^{k} Y_j = m$ is $\text{CMM}_k(m, \boldsymbol{p}, \nu)$, where $\boldsymbol{p} = (\lambda_1 / \sum_{j=1}^{k} \lambda_j, \ldots, \lambda_k / \sum_{j=1}^{k} \lambda_j)$. For the special case where $Y_1, \ldots, Y_k$ are a random sample of $\text{CMP}(\lambda, \nu)$ random variables, this conditional form simplifies to a $\text{CMM}_k(m, \boldsymbol{p}, \nu)$, where $\boldsymbol{p} = (1/k, \ldots, 1/k)$. Both derivations assume $\nu \geq 0$, where $\lambda < 1$ if $\nu = 0$, given the CMP assumptions; however, this constraint is not generally required for the CMM distribution. Other special cases of the $\text{CMM}_k(m, \boldsymbol{p}, \nu)$ include the binomial$(m, p)$ (when $k = 2$) and multinomial $\text{Mult}_k(m, \boldsymbol{p})$ distributions (for $k > 2$) when $\nu = 1$, discrete uniform with probabilities $\binom{m+k-1}{m}^{-1}$ on $\Omega_{m,k}$ when $\nu = 0$ and $\boldsymbol{p} = (1/k, \ldots, 1/k)$, and $\text{CMB}(m, p, \nu)$ for $k = 2$ and $\nu \neq 1$ (Kadane and Wang, 2018; Morris et al., 2020). In the extreme limiting cases of $\nu$ (thus representing the utmost cases of positive and negative association), the CMM distribution accumulates at either the vertex points $(\nu \to -\infty)$ or the center points $(\nu \to \infty)$. For a random sample $\boldsymbol{Y}_1, \ldots, \boldsymbol{Y}_n$, the CMM distribution can be represented in an exponential family form as

$$P(\boldsymbol{Y}_1 = \boldsymbol{y}_1, \ldots, \boldsymbol{Y}_n = \boldsymbol{y}_n \mid \boldsymbol{p}, \nu) \propto \exp\left(\sum_{i=1}^{k-1} \ln\left(\frac{p_i}{p_k}\right) \sum_{j=1}^{n} y_{ij} - \nu \sum_{j=1}^{n} \ln\left(\prod_{i=1}^{k} y_{ij}!\right)\right),$$

with respective sufficient statistics, $S_0 = \sum_{j=1}^{n} \ln(y_{1j}! \ldots y_{kj}!)$ for $\nu$ and $S_i = \sum_{j=1}^{n} y_{ij}$ for $p_i$, $i = 1, \ldots, k-1$. The CMM distribution likewise can be a member of an exponential dispersion family if $\nu$ is known.

The $\text{CMM}_k(m, \boldsymbol{p}, \nu)$ probability mass function (Equation (3.18)) can be reparametrized as

$$P(\boldsymbol{Y} = \boldsymbol{y}) = \frac{1}{S(\boldsymbol{\pi}, \nu; m)} \binom{m}{y_1 \ldots y_k}^{\nu} \prod_{j=1}^{k-1} \pi_j^{y_j} \tag{3.19}$$

$$= \exp\left(-\ln(S(\boldsymbol{\pi}, \nu; m)) + \nu \ln(m!) - \nu \sum_{j=1}^{k} \ln(y_j!) + \sum_{j=1}^{k-1} y_j \ln(\pi_j)\right), \tag{3.20}$$

where $\boldsymbol{\pi} = (\pi_1, \ldots, \pi_{k-1}) = (p_1/p_k, \ldots, p_{k-1}/p_k)$ denotes the baseline odds, and $S(\boldsymbol{\pi}, \nu; m) = \sum_{\boldsymbol{y} \in \Omega_{m,k}} \binom{m}{y_1 \ldots y_k}^{\nu} \prod_{j=1}^{k} p_j^{y_j} = C(\boldsymbol{p}, \nu; m)/p_k^m$ denotes the transformed normalization constant (Morris et al., 2020). The exponential family form provided in Equation (3.20) implies that there exists a conjugate family of the form

$$h((\boldsymbol{\pi}, \nu) \mid \boldsymbol{a}, b, c) \propto \exp(\boldsymbol{\pi} \cdot \boldsymbol{a} + b\nu - c \ln(S(\boldsymbol{\pi}, \nu; m))), \tag{3.21}$$

with hyperparameters $\boldsymbol{a} = (a_1, \ldots, a_k)$, $b$, and $c$, where Equation (3.21) is proper if and only if $c > 0$, $a_i > 0$, for all $i$ such that $\sum_{i=1}^{k} \frac{a_i}{c} = m$, and

$$-\ln(m!) < \frac{b}{c} < -\sum_{i=1}^{k} \left\{ \left( \frac{a_i}{c} - \left\lfloor \frac{a_i}{c} \right\rfloor \right) \ln \left\lceil \frac{a_i}{c} \right\rceil + \ln \left\lfloor \frac{a_i}{c} \right\rfloor ! \right\}$$

(Kadane and Wang, 2018). Further, for a probability function with a finite moment generating function, a proper prior distribution can be attained. To that end, the CMM probability generating, moment generating, and characteristic functions are

$$\Pi_{\boldsymbol{Y}}(\boldsymbol{t}) = C((t_1 p_1, \ldots, t_k p_k), \nu; m)/C(\boldsymbol{p}, \nu; m)$$
$$M_{\boldsymbol{Y}}(\boldsymbol{t}) = C((p_1 e^{t_1}, \ldots, p_k e^{t_k}), \nu; m)/C(\boldsymbol{p}, \nu; m) \text{ and}$$
$$\phi_{\boldsymbol{Y}}(\boldsymbol{t}) = C((p_1 e^{it_1}, \ldots, p_k e^{it_k}), \nu; m)/C(\boldsymbol{p}, \nu; m), \text{ respectively,}$$

from which the reparameterized forms can likewise be determined. The resulting CMM means and covariances are

$$E(Y_j) = mp_j + p_j \frac{\partial \ln C(\boldsymbol{p}, \nu; m)}{\partial p_j} - p_j \sum_{i=1}^{k-1} p_i \frac{\partial \ln C(\boldsymbol{p}, \nu; m)}{\partial p_i}, \quad j = 1, \ldots, k-1,$$

$$= \pi_j \frac{\partial \ln S(\boldsymbol{\pi}, \nu; m)}{\partial \pi_j}, \quad j = 1, \ldots, k-1, \text{under the reparametrized form}$$

$$E(Y_k) = m - \sum_{j=1}^{k-1} E(Y_j)$$

$$\text{Cov}(Y_i, Y_j) = \pi_i \frac{\partial E(Y_j)}{\partial \pi_i}, \quad i,j \in \{1, \ldots, k-1\}.$$

Finally, while the multinomial distribution is closed under summations, marginals, conditionals, and coordinate groupings for a given partition of the index set $\{1, \ldots, k\}$, the CMM is only closed under conditionals; see Morris et al. (2020) for details.

*Maximum Likelihood Estimation and Statistical Computing*

For a random sample $\boldsymbol{Y}_1, \ldots, \boldsymbol{Y}_n$ of $\text{CMM}_k(m, \boldsymbol{p}, \nu)$ distributed random variables with baseline odds $\boldsymbol{\pi} = (p_1/p_k, \ldots, p_{k-1}/p_k)$, the log-likelihood is

$$\ln L(\boldsymbol{\pi}, \nu) = -n \ln S(\boldsymbol{\pi}, \nu; m) + \nu \sum_{i=1}^{n} \ln \binom{m}{y_{i1} \cdots y_{ik}} + \sum_{i=1}^{n} \sum_{j=1}^{k-1} y_{ij} \ln(\pi_j). \tag{3.22}$$

While it does not produce score equations that allow for a closed-form solution for the maximum likelihood estimators, optimization techniques can be performed computationally in `R` (e.g. via the `optim` function). Limitations exist, however, with such an approach because the optimization scheme can quickly become intractable. Instead, Morris et al. (2020) transform the exponential family form of the log-likelihood (via Equation (3.22)) to the form of the Poisson log-likelihood (as performed by Lindsey and Mersch (1992)) in order to reduce the computational complexity and time to obtain estimates and associated standard errors. An extension of the Lindsey and Mersch (1992) method can likewise be considered for the regression setting to estimate varying values for $m, \boldsymbol{p}$, and/or $\nu$. Morris et al. (2020) consider a multinomial logit link $\ln(\boldsymbol{\pi}_i) = \boldsymbol{x}_i'\boldsymbol{\beta}$ to associate the baseline odds with predictors, while the varying dispersion $\boldsymbol{\nu}$ is associated with predictors through the identity link, $\boldsymbol{\nu} = \boldsymbol{z}'\boldsymbol{\gamma}$.

### 3.3.4 CMB and CMM as Sums of Dependent Bernoulli Random Variables

While Borges et al. (2014) refer to the CMB$(m, p, \nu)$ as being over- or under-dispersed relative to the binomial distribution, such terminology is generally avoided in the literature; instead, the CMB distribution is motivated as the sum of exchangeable Bernoulli random variables, i.e. for $m$ Bernoulli random variables $B_1, \ldots, B_m$, we have

$$P(B_1 = b_1, \ldots, B_m = b_m) \propto \binom{m}{x}^{\nu-1} p^x(1-p)^{m-x} \tag{3.23}$$

and $\sum_{i=1}^{m} b_i = x$, where the $b$s are positively (negatively) correlated when $\nu < (>)1$ (Kadane, 2016; Morris et al., 2020; Shmueli et al., 2005). In fact, the correlation between two such Bernoullis $B_i$ and $B_j$ ($i \neq j$, $i, j = 1, \ldots, m$) equals

$$\rho = \frac{p(1-p)(1-4^{\nu-1})}{(p+(1-p)2^{\nu-1})(1-p(1-2^{\nu-1}))} \tag{3.24}$$

(Borges et al., 2014). This result demonstrates that the $b$s are positively (negatively) correlated for $\nu < (>)1$. The CMB simplifies to a binomial$(m, p)$ distribution when $\nu = 1$. Accordingly, a CMB$(m, p, \nu)$ distributed random variable $X$ and a Bernoulli random variable (say, $B_1$) satisfy the relation $\frac{1}{n}E(X) = E(B_1) = p$ when $\nu = 1$; however, this is not generally true (Daly and Gaunt, 2016).

CMM$_k(m, \boldsymbol{p}, \nu)$ can analogously be derived as a sum of dependent unit-multinomial random variables, $\boldsymbol{B}_1, \ldots, \boldsymbol{B}_m \in \Omega_{1,k}$ with joint distribution

$$P(\boldsymbol{B}_1 = \boldsymbol{b}_1, \ldots, \boldsymbol{B}_m = \boldsymbol{b}_m) \propto \binom{m}{y_1 \cdots y_k}^{\nu-1} p_1^{y_1} \cdots p_k^{y_k},$$

where $\boldsymbol{y} = \sum_{i=1}^{m} \boldsymbol{b}_i$, $\nu \in \mathbb{R}$ (Morris et al., 2020). Kadane and Wang (2018) show that there exists a unique distribution on $\boldsymbol{B}_1, \ldots, \boldsymbol{B}_m$ such that they are order $m$ exchangeable and $\sum_{i=1}^{m} \boldsymbol{B}_i$ has the same distribution as $P(\boldsymbol{Y} = \boldsymbol{y}) = p_{\boldsymbol{y}} \geq 0$, where $\sum_{\boldsymbol{y} \in \Omega_{m,k}} p_{\boldsymbol{y}} = 1$. Other properties of these unit-multinomials include the following for $i, i' \in \{1, \ldots, m\}$, where $i \neq i'$ and $\boldsymbol{e}_j$ denotes the $j$th column of a $k \times k$ identity matrix:

1. $P(\boldsymbol{B}_i = \boldsymbol{e}_j) = E(Y_j/m)$, for $j = 1, \ldots, k$
2. $P(\boldsymbol{B}_i = \boldsymbol{e}_j, \boldsymbol{B}_{i'} = \boldsymbol{e}_j) = [m(m-1)]^{-1}E[Y_j(Y_j-1)]$, for $j = 1, \ldots, k$
3. $P(\boldsymbol{B}_i = \boldsymbol{e}_j, \boldsymbol{B}_{i'} = \boldsymbol{e}_l) = [m(m-1)]^{-1}E[Y_jY_l]$, for $j, l \in \{1, \ldots, k\}$ and $j \neq l$
4. $E(\boldsymbol{B}_i) = E(\boldsymbol{Y}/m)$

5. $V(\boldsymbol{B}_i) = \text{Diag}[E(\boldsymbol{Y}/m)] - E(\boldsymbol{Y}/m)E(\boldsymbol{Y}/m)'$
6. $\text{Cov}(\boldsymbol{B}_i, \boldsymbol{B}_{i'}) = [m(m-1)]^{-1}[E(\boldsymbol{Y}\boldsymbol{Y}') - \text{Diag}[E(\boldsymbol{Y})]] - m^{-2}E(\boldsymbol{Y})E(\boldsymbol{Y}')$.

As with the CMB case, $\boldsymbol{B}_1, \ldots, \boldsymbol{B}_m$ are positively (negatively) correlated when $\nu < (>) 1$.

### 3.3.5 R Computing

The `COMMultReg` package (Raim and Morris, 2020) in `R` produces density functions of interest relating to the CMB and CMM distributions and also conducts maximum likelihood estimation associated with CMM regression. For given inputs $m, p, \nu$, the `d_cmb` function computes the probability mass at the value $P(X = x)$ (on the raw or log scale) at a value of interest $x$, where `normconst_cmb` computes the associated CMB normalizing constant (likewise on either scale). Along with these functions, `e_cmb` and `v_cmb` compute the associated mean and variance of the CMB$(m, p, \nu)$ distribution. Finally, `r_cmb` is a random number generator that produces $n$ terms from the CMB$(m, p, \nu)$ distribution. Similarly, the `d_cmm`, `normconst_cmm`, `e_cmm`, `v_cmm`, and `r_cmm` functions compute the probability, normalizing constant, expectation, variance, and randomly generated data associated with the CMM$(m, p, \nu)$ distribution.

More broadly, `COMMultReg` provides the `d_cmb_sample` and `d_cmm_sample` functions, respectively, to compute the density contributions attributed to each of the random variables $X_i$, $i = 1, \ldots, n$, where $X_i$ is distributed as CMB$(m_i, p_i, \nu_i)$ or CMM$_k(m_i, p_i, \nu_i)$, respectively. `d_cmb_sample` contains the inputs `x` of outcomes along with the parameters `m`, `p`, and `nu`. Other `d_cmb_sample` inputs include `take_log`, a logical vector that allows analysts to obtain the density contributions on a log scale if interested, and `normalize`, another logical vector that determines whether or not to include the normalizing constant in the contribution computations. Meanwhile, `d_cmm_sample` includes the $n \times k$ matrix of outcomes `X` and probabilities `P`, respectively, and the $n$-dimensional vector of dispersions `nu`. This function likewise includes the logical inputs, `take_log` and `normalize` as described.

Code 3.1 illustrates several `cmb` codes, including computing the density $P(X = 8)$ and normalizing constant (on the log and raw scales, respectively), and expectation and variance of the CMB$(m = 10, p = 0.75, \nu = 1)$ distribution. As expected, the resulting density, expected value, and variance calculations are precisely those of the binomial$(m = 10, p = 0.75)$ distribution as this is the special case of the associated CMB distribution

Code 3.1 Illustrative R code and output to produce the density and normalizing constant (on the log or raw scale) respectively, expected value, and variance of a Conway-Maxwell-binomial (CMB($m, p, \nu$)) distribution. This example assumes $\nu = 1$ such that CMB($m, p, \nu = 1$)=binomial($m, p$).

```
> m = 10
> p = 0.75
> nu = 1

> d_cmb(8, m, p, nu, take_log = TRUE)
[1] -1.267383
> d_cmb(8, m, p, nu)
[1] 0.2815676
> normconst_cmb(m, p, nu, take_log = TRUE)
[1] 0
> normconst_cmb(m, p, nu)
[1] 1
> e_cmb(m, p, nu)
[1] 7.5
> v_cmb(m, p, nu)
[1] 1.875
```

when $\nu = 1$. In particular, given the form of the normalizing constant $S(m, p, \nu)$ as outlined in Equation (3.6), it makes sense that it would equal 1 when $\nu = 1$ because, in this case, the value is determined by summing the binomial probability mass function $P(X = x)$ over the entire space of values for $x$.

Table 3.1 more broadly supplies illustrative results (on the raw scale) for $\nu \in \{0, 1, 2\}$, demonstrating the effect of over-dispersion (under-dispersion) relative to the binomial distribution. The normalizing constant increases with $\nu$ such that, for $\nu < 1$ (i.e. over-dispersion), the normalizing constant $S(m, p, \nu)$ is likewise less than 1; meanwhile, for $\nu > 1$ (i.e. under-dispersion), $S(m, p, \nu)$ is greater than 1. Meanwhile, the expected value decreases as $\nu$ increases such that $E(X) > (<)\ mp$ when $\nu < (>)\ 1$. The density and variance results, however, demonstrate that the respective outcomes are maximized when $\nu = 1$ (i.e. under the binomial model). The resulting dispersion indexes ($\text{DI}(X) = V(X)/E(X)$) are thus likewise maximized when $\nu = 1$. This supports discussion urging analysts to move from describing relationships as over- or under-dispersed relative to the binomial to instead noting them as sums of positively or negatively correlated Bernoulli random variables; see Section 3.3.4.

The cmm functions analogously produce the density and normalizing constant (on the raw or log scales, respectively), expected value and variance

Table 3.1 *Probability, normalizing constant, expected value, and variance calculations (rounded to three decimal places) for CMB(m =10, p =0.75, $\nu$) distributions where $\nu$ = {0, 1, 2}.*

| | $\nu = 0$ | $\nu = 1$ | $\nu = 2$ |
|---|---|---|---|
| $P(X = 8)$ | 0.074 | 0.282 | 0.135 |
| $S(m, p, \nu)$ | 0.084 | 1.000 | 93.809 |
| $E(X)$ | 9.500 | 7.500 | 6.411 |
| $V(X)$ | 0.749 | 1.875 | 1.222 |

for the $\text{CMM}_k(m, p, \nu)$ distribution. Here, however, one needs to define the number of categories $k$ to ensure that the probability and outcome vectors, respectively $\boldsymbol{p} = (p_1, \ldots, p_k)$ and $\boldsymbol{y} = (y_1, \ldots, y_k)$, satisfy $\sum_{j=1}^{k} p_j = 1$ and $\sum_{j=1}^{k} y_j = m$. The expectation output likewise has length $k$ such that the sum of the expectations equals $m$, while the variance output produces a $k \times k$ matrix representing the variance–covariance matrix associated with the different categories. Code 3.2 demonstrates the `cmm` codes for a $\text{CMM}_3(m = 10; \boldsymbol{p} = (0.1, 0.35, 0.55); \nu = 1)$ distributed random variable $\boldsymbol{Y}$, computing the density $P(\boldsymbol{Y} = (0, 2, 8))$, normalizing constant, expected value, and variance. The resulting density, expected value, and variance–covariance calculations demonstrate that the $\text{CMM}_3(m = 10; \boldsymbol{p} = (0.1, 0.35, 0.55); \nu = 1)$ distribution is the special case of the $\text{Mult}_3(m = 10; \boldsymbol{p} = (0.1, 0.35, 0.55))$ distribution. The normalizing constant again equals 1 because the value is determined by summing the multinomial probability mass function $P(\boldsymbol{Y} = \boldsymbol{y})$ over all values satisfying $\sum_{j=1}^{3} y_j = m = 10$; see Equation (3.18).

Table 3.2 illustrates the effect of positive and negative correlated Bernoulli random variables, relative to the multinomial distribution. The provided table illustrates these correlations via varying $\nu \in \{0, 1, 2\}$. As was the case for Table 3.1, the probability values decrease and the normalizing constants increase as $\nu$ increases. The mean and variance–covariance relationships are more difficult to recognize, however, given the dimensionality associated with the number of categories and constraint $m$.

Similarly, `d_cmb_sample` and `d_cmm_sample` allow for density computations for multiple observations with respective probabilities and dispersion values; see Codes 3.3 and 3.4. In Code 3.3, `d_cmb_sample(y,m,p,nu)` considers three density computations where, for each scenario, $m = 10$, and the density computations are determined

Table 3.2 *Probability, normalizing constant, expected value, and variance calculations (rounded to three decimal places) for* $CMM_3(\mathrm{m} = 10;\ \mathbf{p} = (0.1, 0.35, 0.55);\ \nu)$ *distributions where* $\nu = \{0, 1, 2\}$.

| | $\nu = 0$ | $\nu = 1$ | $\nu = 2$ |
|---|---|---|---|
| $P(\boldsymbol{Y} = (0, 2, 8))$ | 0.121 | 0.046 | 0.002 |
| $C(m, \boldsymbol{p}, \nu)$ | 0.008 | 1.000 | 990.340 |
| $E(\boldsymbol{Y})$ | $(0.221\ \ 1.665\ \ 8.114)$ | $(1.000\ \ 3.500\ \ 5.500)$ | $(1.798\ \ 3.610\ \ 4.592)$ |
| $V(\boldsymbol{Y})$ | $\begin{pmatrix} 0.269 & -0.011 & -0.258 \\ -0.011 & 3.902 & -3.891 \\ -0.258 & -3.891 & 4.149 \end{pmatrix}$ | $\begin{pmatrix} 0.900 & -0.350 & -0.550 \\ -0.350 & 2.275 & -1.925 \\ -0.550 & -1.925 & 2.475 \end{pmatrix}$ | $\begin{pmatrix} 0.846 & -0.375 & -0.470 \\ -0.375 & 1.245 & -0.870 \\ -0.470 & -0.870 & 1.340 \end{pmatrix}$ |

Code 3.2 Illustrative R code and output to produce density and normalizing constant (on the raw scale) respectively, expected value, and variance of a Conway-Maxwell-multinomial (CMM($m, \boldsymbol{p}, \nu$)) distribution. This example assumes three groups (i.e. $k = 3$) and $\nu = 1$, i.e. the special case of the $\text{CMM}_3(m, \boldsymbol{p}, \nu = 1)$ distribution equals the $\text{Mult}_3(m, \boldsymbol{p})$ distribution.

```
> m = 10
> p = c(0.1, 0.35, 0.55)
> nu = 1

> d_cmm(c(0,2,8), p, nu)
[1] 0.04615833
> normconst_cmm(m, p, nu)
[1] 1
> e_cmm(m, p, nu)
[1] 1.0 3.5 5.5
> v_cmm(m, p, nu)
      [,1]   [,2]   [,3]
[1,]  0.90 -0.350 -0.550
[2,] -0.35  2.275 -1.925
[3,] -0.55 -1.925  2.475
```

elementwise given inputs

$$p = \begin{pmatrix} 0.25 \\ 0.50 \\ 0.75 \end{pmatrix}, \quad \nu = \begin{pmatrix} 0.0 \\ 0.5 \\ 1.0 \end{pmatrix}, \text{ and } y = \begin{pmatrix} 6 \\ 7 \\ 8 \end{pmatrix}.$$

The third element of the resulting output matches the `d_cmb` output provided in Code 3.1. Code 3.4 analogously considers three $\text{CMM}_3$ density computations via the `d_cmm_sample(X,P,nu)` command where, for each scenario, $m = 10$, and the density computations are determined row-wise given inputs

$$P = \begin{pmatrix} 0.10 & 0.35 & 0.55 \\ 0.20 & 0.30 & 0.50 \\ 0.40 & 0.40 & 0.20 \end{pmatrix}, \quad \nu = \begin{pmatrix} 1 \\ 0 \\ -1 \end{pmatrix}, \text{ and } X = \begin{pmatrix} 0 & 2 & 8 \\ 4 & 5 & 1 \\ 3 & 4 & 3 \end{pmatrix}.$$

The first element of the resulting output matches the `d_cmm` output provided in Code 3.2.

The `cmm_reg` function in `COMMultReg` (Raim and Morris, 2020) performs a CMM regression as described in Morris et al. (2020). Accordingly,

Code 3.3 *Illustrative example of* `d_cmb_sample`.

```
> m = rep(10, 3)
> p = c(.25,.5,.75)
> nu= c(0,.5,1)
> y=c(6,7,8)
> d_cmb_sample(y,m,p,nu)
             [,1]
[1,] 0.0009144999
[2,] 0.1237691420
[3,] 0.2815675735
```

Code 3.4 Illustrative example of `d_cmm_sample`.

```
> p.mat <- matrix(c(0.1,0.35,0.55,0.2,0.3,0.5,0.4,0.4,
                    0.2),byrow=TRUE,ncol=3)
> x.mat <- matrix(c(0,2,8,4,5,1,3,4,3),byrow=TRUE,
                  ncol=3)
> nu <- c(1,0,-1)
> d_cmm_sample(X=x.mat,P=p.mat,nu)
             [,1]
[1,] 4.615833e-02
[2,] 4.808816e-04
[3,] 1.235335e-05
```

this function assumes a sample $\boldsymbol{Y}_i$ of CMM$_k(m_i, \boldsymbol{p}_i, \nu_i)$ random variables with a multinomial logit link $\ln\left(\frac{p_{ij}}{p_{i1}}\right) = \boldsymbol{x}_i\boldsymbol{\beta}_{j-1}$ and identity link $\nu_i = \boldsymbol{g}_i\boldsymbol{\gamma}$ for the dispersion, $i = 1, \ldots, n$ and $j = 2, \ldots, k$. As noted in the multinomial logit link, it is assumed by default that the first category serves as the baseline; however, the analyst can change this (say) to the $k$th category by specifying `base = k`. Parameter estimation is performed via the Newton–Raphson method, where the algorithm converges when the sum of absolute differences among consecutive estimates is sufficiently small; otherwise, the algorithm fails to converge if it reaches the maximum number of iterations without satisfying the convergence tolerance requirement. The default settings for the tolerance level (`tol`) and maximum number of iterations (`max_iter`), respectively, are $1 \times 10^{-8}$ and 200. If issues occur where convergence is not achieved, then the analyst can update the tolerance level and/or the maximum number of iterations in order to provide more flexibility. R output includes the usual coefficient table containing the estimates and associated standard errors, along with the maximum log-likelihood and AIC; see Raim and Morris (2020) for details.

## 3.4 CMP-Motivated Generalizations of the Negative Binomial Distribution

The NB distribution is a well-studied discrete distribution popular for its ability to handle data over-dispersion; see Section 1.2. While it is recognized for this accommodation, extensive study surrounds it with the goal being to develop even more flexible count distributions that broaden the scope of the NB model. This section describes four such models where at least one extra parameter serves to reflect and account for added variation – the GCMP (Section 3.4.1), the COMNB (Section 3.4.2), the COMtNB (Section 3.4.3), and the ECMP (Section 3.4.4) distributions. Table 3.3 lists these CMP-inspired generalizations of the NB, noting what models each contains as special cases. Naturally, because these distributions all contain the CMP, the classical models it subsumes (i.e. the Poisson, geometric, and Bernoulli) are likewise included here under the appropriate conditions.

### *3.4.1 The Generalized COM–Poisson (GCMP) Distribution*

The GCMP (Imoto, 2014) is a three-parameter distribution denoted as GCMP$(r, \gamma, \theta)$ with a flexible form that can model both dispersion and tail behavior via the probability mass function

$$P(X = x) = \frac{\Gamma(\gamma + x)^r \theta^x}{x! C(r, \gamma, \theta)}, \quad x = 0, 1, 2, \ldots, \tag{3.25}$$

with parameters $r$, $\gamma$, and $\theta$, where the normalizing constant $C(r, \gamma, \theta) = \sum_{k=0}^{\infty} \frac{\Gamma(\gamma+k)^r \theta^k}{k!}$ is defined for $\{(r, \gamma, \theta){:}r < 1,\ \gamma > 0,\ \theta > 0;$ or $r =$

Table 3.3 *Summary of CMP-motivated generalizations of the negative binomial (NB) distribution – the generalized COM–Poisson (GCMP), the Conway-Maxwell-negative binomial (COMNB), the Conway-and-Maxwell-type negative binomial (COMtNB), and the extended Conway–Maxwell–Poisson (ECMP) – and the special cases contained by each of them (noted with √), namely any of the following: NB, CMP, Conway-Maxwell-binomial (CMB), COMNB, COMtNB, exponentially weighted Poisson (EWP), and GCMP.*

| General | Special cases | | | | | | |
|---|---|---|---|---|---|---|---|
| distribution | NB | CMP | COM–Pascal | COMNB | COMtNB | EWP | GCMP |
| GCMP | √ | √ | | | | | |
| COMNB | √ | √ | √ | | | | |
| COMtNB | √ | √ | | | √ | | |
| ECMP | √ | √ | √ | √ | √ | √ | √ |

$1,\ \gamma > 0,\ 0 < \theta < 1\}$. This distribution contains the CMP($\theta, \nu$) model when $\gamma = 1$, and $r = 1 - \nu$ (and accordingly contains the Poisson($\theta$) when $\nu = 1$, geometric with success probability $1 - \theta$ when $\nu = 0$ and $\theta < 1$, and Bernoulli with success probability $\theta/(1+\theta)$ when $\nu \to \infty$); the tail distribution is longer (shorter) than CMP when $\gamma > (<)1$. The GCMP distribution also reduces to the NB($\gamma, \theta$) model when $r = 1$. More broadly, $0 < r < 1$ models data over-dispersion and $r < 0$ captures under-dispersion (Imoto, 2014).

The GCMP($r, \gamma, \theta$) distribution can represent a unimodal or a bimodal structure where one of the modes is at zero. It is unimodal for $r \leq 0$ or $\{0 < r < 1,\ \ \gamma \geq 1\}$ with the mode occurring at $M = \frac{x}{x+1}\left(\frac{\gamma+x}{\gamma+x-1}\right)^r < 1$. For $0 < r < 1, 0 < \gamma < 1$, and $\theta\gamma^r < 1$, the GCMP distribution is bimodal with one of those modes occurring at zero; thus, it accounts for excess zeroes without the need to incorporate zero-inflation. Equation (3.25) can be represented as

$$P(X = x) = \exp\left(r \ln \Gamma(\gamma + x) - \ln(x!) + x \ln \theta - \ln C(r, \gamma, \theta)\right),$$

thus has the form of an exponential family for $r$ and $\ln\theta$ with $\gamma$ viewed as a nuisance parameter. It likewise belongs to the power series family of distributions and can be represented as a weighted Poisson distribution (see Section 1.4) where $w(x) = [\Gamma(\gamma + x)]^r$. Its ratio of successive probabilities

$$\frac{P(X = x)}{P(X = x - 1)} = \frac{\theta(\gamma - 1 + x)^r}{x} \tag{3.26}$$

shows that the GCMP distribution has longer tails than the CMP distribution when $\gamma > 1$ and $0 < r < 1$ and shorter tails than the CMP distribution when $\gamma > 1$ and $r < 0$.

The $k$th factorial moment for the GCMP distribution is

$$E[(X)_k] \doteq E[X(X-1)\cdots(X-k+1)] = \frac{C(r, \gamma + k, \theta)\theta^k}{C(r, \gamma, \theta)}$$

from which one can determine the mean $\mu = E(X) = E[(X)_1]$ and variance $V(X) = E[(X)_2] + E[(X)_1] - [E[(X)_1]]^2$. Membership in the power series family of distributions meanwhile provides the recursion

$$\mu_{k+1} = \theta\frac{\partial \mu_k}{\partial \theta} + k\mu_2\mu_{k-1}, \quad k = 1, 2, \ldots,$$

where $\mu_k = E(X - \mu)^k$ denotes the $k$th central moment of the GCMP distribution (e.g. $\mu_2 = V(X)$). Either result produces the moments

$$\mu = \frac{C(r,\gamma+1,\theta)\theta}{C(r,\gamma,\theta)} = \theta\frac{\partial C(r,\gamma,\theta)}{\partial\theta} \approx \theta^{1/(1-r)} + \frac{(2\gamma-1)r}{2(1-r)} \tag{3.27}$$

$$\begin{aligned}\mu_2 &= \frac{C(r,\gamma+2,\theta)\theta^2}{C(r,\gamma,\theta)} + \frac{C(r,\gamma+1,\theta)\theta}{C(r,\gamma,\theta)} - \frac{[C(r,\gamma+1,\theta)]^2\theta^2}{[C(r,\gamma,\theta)]^2}\\ &= \theta\frac{\partial\mu_1}{\partial\theta} + \mu_2\mu_1 \approx \frac{\theta^{1/(1-r)}}{1-r},\end{aligned} \tag{3.28}$$

where the approximations are determined based on an approximation to $C(r,\gamma,\theta)$ (see the Approximating the Normalizing Constant section below for details); for example, $\mu_3 \approx \frac{\theta^{1/(1-r)}}{(1-r)^2}$ (Imoto, 2014). The approximations for $\mu_1$, $\mu_2$, and $\mu_3$ equal the CMP($\theta$, $\nu = 1 - r$) moment approximations (e.g. Equations (2.14) and (2.15)) when $\gamma = 1$, and (in particular) their analogous Poisson moments when $r = 0$.

### *Approximating the Normalizing Constant, $C(r,\gamma,\theta)$*

The GCMP distribution likewise involves a normalizing constant $C(r,\gamma,\theta) = \sum_{k=0}^{\infty}\frac{[\Gamma(\gamma+k)]^r\theta^k}{k!}$ that is an infinite sum that converges for $r < 1$ or $|\theta| < 1$ when $r = 1$. Thus, analogous to the discussion regarding the CMP normalizing constant (Section 2.8), two methods exist to approximate $C(r,\gamma,\theta)$. The first approach is to truncate the series to some $m$ such that

$$C_m(r,\gamma,\theta) = \sum_{k=0}^{m}\frac{[\Gamma(\gamma+k)]^r\theta^k}{k!}, \tag{3.29}$$

where $\frac{(\gamma-1+m)^r\theta}{m} < 1$. Accordingly, the absolute truncation error is

$$R_m(r,\gamma,\theta) = C(r,\gamma,\theta) - C_m(r,\gamma,\theta) < \frac{[\Gamma(\gamma+m+1)]^r\theta^{m+1}}{(m+1)!\left(1-\frac{(\gamma-1+m)^r\theta}{m}\right)},$$

while the relative truncation error is

$$RR_m(r,\gamma,\theta) = \frac{R_m(r,\gamma,\theta)}{C_m(r,\gamma,\theta)} < \frac{[\Gamma(\gamma+m+1)]^r\theta^{m+1}}{(m+1)!\left(1-\frac{(\gamma-1+m)^r\theta}{m}\right)C_m(r,\gamma,\theta)}.$$

Computational complexities exist when $r < 1$ large and $\theta > 0$; analysts are advised to restrict $\theta$ so that $0 < \theta < 1$ when $r$ converges to 1 (e.g. $\theta < 10^{1-r}$) to maintain a small relative error (Imoto, 2014). The second approach approximates $C(r, \gamma, \theta)$ via the Laplace method; the asymptotic formula for the GCMP normalizing constant is

$$\tilde{C}(r, \gamma, \theta) = \frac{\theta^{(2\gamma-1)r/[2(1-r)]}(2\pi)^{r/2}\exp\left[(1-r)\theta^{1/(1-r)}\right]}{\sqrt{1-r}}. \tag{3.30}$$

While derived for integer-valued $r < 0$, Imoto (2014) utilizes numerical studies to illustrate that this result holds more broadly for real-valued $r < 1$. Accordingly, $C(r, \gamma, \theta)$ can be approximated via the truncation method (Equation (3.29)) when $\gamma$ is small and $r < 0$ or $\theta < 10^{1-r}$, while the $C(r, \gamma, \theta)$ approximation (Equation (3.30)) can be used when $\gamma$ is large, $\theta > 10^{1-r}$, and $0 < r < 1$.

*Parameter Estimation*

Imoto (2014) conducts parameter estimation via the method of moments, estimation by four consecutive probabilities, and the method of maximum likelihood. Under the method of moments procedure, analysts equate the first three sample central moments to their true counterparts (i.e. $\mu_i$, $i = 1, 2, 3$, as provided in Equations (3.27)–(3.28) for $\mu_1$ and $\mu_2$, with subsequent discussion containing $\mu_3$). Using the respective central moment approximations produces the system of equations,

$$\begin{cases} \tilde{\theta}\frac{\partial \ln C(\tilde{r},\tilde{\gamma},\tilde{\theta})}{\partial \tilde{\theta}} & \approx \tilde{\theta}^{1/(1-\tilde{r})} + \frac{(2\tilde{\upsilon}-1)\tilde{r}}{2(1-\tilde{r})} \\ \tilde{\theta}\frac{\partial E(X)}{\partial \tilde{\theta}} & \approx \frac{\tilde{\theta}^{1/(1-\tilde{r})}}{1-\tilde{r}} \\ \tilde{\theta}\frac{\partial V(X)}{\partial \tilde{\theta}} & \approx \frac{\tilde{\theta}^{1/(1-\tilde{r})}}{(1-\tilde{r})^2}; \end{cases}$$

thus the method of moments estimators are $\tilde{r} \approx 1 - \frac{\mu_2}{\mu_3}$, $\tilde{\gamma} \approx \frac{\mu_2(\mu_1\mu_3-\mu_2^2)}{\mu_3(\mu_3-\mu_2)} + \frac{1}{2}$, and $\tilde{\theta} \approx \left(\frac{\mu_2^2}{\mu_3}\right)^{\mu_2/\mu_3}$. While this estimation method is easily computable, situations exist where estimation via the method of moments cannot be utilized because the resulting estimates may not satisfy the prerequisite parameter constraints. The second approach approximates the first four true consecutive probabilities by their respective observed frequencies, re-representing Equation (3.26) as $\frac{P_{x+1}}{P_x} = \frac{\theta(\gamma+x)^r}{x+1}$, along with the equations

$$\ln\left(\frac{x+2}{x+1}\frac{P_{x+2}P_x}{P_{x+1}^2}\right) = r\ln\left(\frac{\gamma+x+1}{\gamma+x}\right) \tag{3.31}$$

$$\frac{\ln\left(\frac{x+3}{x+2}\frac{P_{x+3}P_{x+1}}{P_{x+2}^2}\right)}{\ln\left(\frac{x+2}{x+1}\frac{P_{x+2}P_x}{P_{x+1}^2}\right)} = \frac{\ln\left(\frac{\gamma+x+2}{\gamma+x+1}\right)}{\ln\left(\frac{\gamma+x+1}{\gamma+x}\right)}. \tag{3.32}$$

This approach is analogous to the method of moments where, by substituting the observed (absolute or relative) frequency that $x$ events occurred for $P_x$ into the above three equations, analysts can obtain rough estimates $\breve{r}$, $\breve{\gamma}$, and $\breve{\theta}$ that lie in the parameter space for a chosen $x$ value. Choosing $x = 0$ as the basis for the consecutive probabilities approach provides a good estimation for a dataset with excess zeroes (Imoto, 2014).

The method of maximum likelihood is the third approach for parameter estimation where, given the GCMP log-likelihood

$$\ln L(r,\gamma,\theta) = r\sum_{j=0}^{J} f_j \ln\Gamma(\gamma+j) + \ln\theta\sum_{j=0}^{J} jf_j - n\ln C(r,\gamma,\theta) - \sum_{j=0}^{J} f_j \ln(j!) \tag{3.33}$$

based on a sample of $n$ observations and observed frequencies $f_j$, $j = 0,\ldots,J$, the maximum likelihood estimators $\hat{r}$, $\hat{\gamma}$, and $\hat{\theta}$ can be determined by solving the system of score equations

$$\begin{cases} E[\ln\Gamma(\gamma+X)] &= \sum_{j=0}^{J} \ln\Gamma(\gamma+j)\frac{f_j}{n} \\ E[\psi(\gamma+X)] &= \sum_{j=0}^{J} \psi(\gamma+j)\frac{f_j}{n} \\ E(X) &= \sum_{j=0}^{J} j\frac{f_j}{n}, \end{cases} \tag{3.34}$$

where $\psi(y) = \partial_y \ln\Gamma(y) = \frac{\Gamma'(y)}{\Gamma(y)}$. The iterative scheme

$$\begin{pmatrix} r_{k+1} \\ \gamma_{k+1} \\ \theta_{k+1} \end{pmatrix} = \begin{pmatrix} r_k \\ \gamma_k \\ \theta_k \end{pmatrix} + [I(r_k,\gamma_k,\theta_k)]^{-1} \begin{pmatrix} \sum_{j=0}^{J} \ln\Gamma(\gamma+j)\frac{f_j}{n} - E(\ln\Gamma(\gamma+X)) \\ \sum_{j=0}^{J} \psi(\gamma+j)\frac{f_j}{n} - E(\psi(\gamma+X)) \\ \sum_{j=0}^{J} j\frac{f_j}{n} - E(X) \end{pmatrix} \Bigg|_{r=r_k,\gamma=\gamma_k,\theta=\theta_k}$$

can solve the system of equations contained in Equation (3.34), thus determining a local maximum. Meanwhile, the global maximum can be determined via a profile maximum likelihood estimation procedure because, for a given $\gamma$, the GCMP distribution belongs to the exponential family. Under this scenario, the profile MLEs can be determined by solving the system of equations comprised in Equation (3.34), again via the scoring method. The Fisher information matrix can be derived from the variance–covariance matrix of $(\ln(\Gamma(\gamma+X)), r\psi(\gamma+X), X)$, thus aiding in determining associated standard errors for the parameter estimates.

*The GCMP Queueing Process*

Imoto (2014) derives the GCMP distribution from a queueing model with arrival rate $\lambda(\gamma + x)^r$ and service rate $\mu x$ for a queue of length $x$; the arrival rate thus increases (decreases) with $r > 0$ ($r < 0$), and the service rate is directly proportional to the length $x$. This distribution satisfies the difference equation

$$P(x, t + h) = [1 - \lambda(\gamma + x)^r h - \mu x h]P(x, t) + \lambda(\gamma + x - 1)^r hP(x - 1, t) + \mu(x + 1)hP(x + 1, t) \tag{3.35}$$

for small $h$ which, setting $\theta = \lambda/\mu$ and for $h \to 0$, becomes

$$\frac{\partial P(x, t)}{\partial t} = (\gamma + x - 1)^r \theta P(x - 1, t) + [(\gamma + x)^r \theta + x]P(x, t) + (x + 1)P(x + 1, t).$$

Under the stationary state assumption and letting $P(x, t) = P(x)$, the difference equation is

$$P(x + 1) = \frac{x}{x + 1}P(x) + \frac{(\gamma + x)^r}{x + 1}P(x) - \frac{(\gamma + x - 1)^r \theta}{x + 1}P(x - 1).$$

The queuing scenario considered by Conway and Maxwell (1962) is a special case of Equation (3.35) where $\gamma = 1$; see Section 2.1 for discussion.

### 3.4.2 The COM–Negative Binomial (COMNB) Distribution

The COMNB distribution (Zhang et al., 2018) is another three-parameter generalized count distribution that includes the NB as a special case. Given the NB distribution probability function described in Equation (1.6), the coefficient $\frac{\Gamma(\gamma+x)}{x!\Gamma(\gamma)}$ is updated as $\left(\frac{\Gamma(\gamma+x)}{x!\Gamma(\gamma)}\right)^{\nu}$ to inspire the probability mass function

$$P(X = x) = \frac{1}{D(\gamma, \nu, p)} \left(\frac{\Gamma(\gamma + x)}{x!\Gamma(\gamma)}\right)^{\nu} p^x (1 - p)^{\gamma}, \quad x = 0, 1, 2, \ldots, \tag{3.36}$$

where $D(\gamma, \nu, p) = \sum_{j=0}^{\infty} \left( \frac{\Gamma(\gamma+j)}{j!\Gamma(\gamma)} \right)^{\nu} p^j (1-p)^{\gamma}$ is the normalizing constant.[2] The COMNB$(\gamma, \nu, p)$ distribution contains several well-known models as special cases and relationships with other flexible distributions. When $\nu = 1$, COMNB$(\gamma, \nu, p)$ reduces to the NB$(\gamma, p)$ distribution; in particular, COMNB$(\gamma, \nu = 1, p)$ further reduces to the geometric distribution with success probability $1 - p$ when $\gamma = 1$ or $\nu = 0$. For $\gamma \in \mathbb{Z}^+$, the COMNB simplifies to what can be referred to as the COM–Pascal distribution. The COMNB$(\gamma, \nu, p)$ further converges to a CMP$(\lambda, \nu)$, where $\lambda = \gamma^{\nu} p/(1-p)$ as $\gamma \to \infty$ (hence includes Poisson$(\lambda)$ for $\nu = 1$, geometric with success probability $1 - \lambda$ for $\nu = 0$ and $\lambda < 1$, and Bernoulli with success probability $\lambda/(1+\lambda)$ for $\nu \to \infty$). For two independent COMNB$(\gamma_i, \nu, p)$ random variables, $X_i$, $i = 1, 2$, the conditional distribution of $X_1$ given the sum $X_1 + X_2 = x_1 + x_2$ has a Conway–Maxwell negative hypergeometric COMNH$(z = x_1 + x_2, \nu, \gamma_1, \gamma_2)$ for all $z$, where the COMNH distribution is described in more detail in Section 3.7.1.

The COMNB not only contains the many special-case distributions and relationships with other flexible models as described earlier in this section, it further is a special case itself of broader distributional forms and presents additional interesting distributional properties. The COMNB$(\gamma, \nu, p)$ distribution is log-concave and strongly unimodal when $\gamma \geq 1$, and it is log-convex and discrete infinitely divisible when $\gamma \leq 1$. It is a discrete compound Poisson distribution for $\gamma \leq 1$ and is a discrete pseudo compound Poisson distribution[3] when $\gamma > 1$ and $p\gamma^{\nu} < 1$, or $\gamma \leq 1$. The COMNB is another example of a weighted Poisson distribution – here, the weight function is $w(x) = [\Gamma(1+x)]^{1-\nu}[\Gamma(\gamma+x)]^{\nu}$; see Section 1.4 for details regarding weighted Poisson distributions. The COMNB can accommodate over-dispersed (under-dispersed) data if

[2] An alternate parametrization is achieved by raising the entire NB probability to the $\nu$th power and dividing the result by its appropriate normalizing constant; thus producing the resulting distribution

$$P(X = x) = \frac{1}{D(\gamma, \nu, p^{1/\nu})} \left( \frac{\Gamma(\gamma + x)}{x!\Gamma(\gamma)} \right)^{\nu} \left(p^{1/\nu}\right)^x \left(1 - p^{1/\nu}\right)^{\gamma}, \quad x = 0, 1, 2, \ldots. \tag{3.37}$$

[3] A discrete compound Poisson distribution is a distribution whose probability generating function (pgf) is $\Pi(s) = \sum_{j=0}^{\infty} p_j s^j = \exp\left(\sum_{i=1}^{\infty} \alpha_i \lambda(s^i) - 1\right)$, $|s| \leq 1$, where $\alpha_i \geq 0$ such that $\sum_{i=1}^{\infty} \alpha_i = 1$, and $\lambda > 0$. A discrete pseudo compound Poisson distribution has the same pgf, however the constraints for $\alpha_i$, $i = 1, \ldots, \infty$ are modified to $\alpha_i \in \mathbb{R}$ such that $\sum_{i=1}^{\infty} |\alpha_i| < \infty$ and $\sum_{i=1}^{\infty} \alpha_i = 1$ (Zhang et al., 2018).

$$\sum_{i=0}^{\infty}\left(\frac{1-\nu}{(\mathrm{i}+x+1)^2}+\frac{\nu}{(\mathrm{i}+x+\gamma)^2}\right) > (<)0$$

for all $x \in \mathbb{N}$, implying that it is over-dispersed when $\nu > 0$ and $\gamma < 1$, or $\nu < 1$ and $\gamma > 0$. This distribution's recursive formula is

$$\frac{P(X=x-1)}{P(X=x)} = \frac{1}{p}\left(\frac{x}{x-1+\gamma}\right)^{\nu}. \tag{3.38}$$

Thus, for $\gamma < 1$ and $\nu > 1$, the COMNB is more flexible than NB at accommodating discrete distributions with a considerably high number of excess zeroes.

Other interesting properties hold for this flexible model. A random variable $X$ has a COMNB$(\gamma, \nu, p)$ distribution if and only if

$$E[X^{\nu}g(X) - (X+\gamma)^{\nu}g(X+1)p] = 0$$

holds for any bounded function $g:\mathbb{Z}^{+} \rightarrow \mathbb{R}$. Analogous to considering the CMB distribution as a sum of equicorrelated Bernoulli random variables (see Section 3.3.4), the COMNB can be viewed as the sum of equicorrelated geometric random variables $Z_i$, $i = 1, \ldots, m$, with joint distribution

$$P(Z_1 = z_1, \ldots, Z_m = z_m) \propto \binom{m+x-1}{x}^{\nu-1} p^{x}(1-p)^{m},$$

where $\sum_{i=1}^{m} z_i = x$. Finally, for two COMNB$(\gamma, \nu, p_i)$ random variables, $X_i$, $i = 1, 2$, $X_1$ is less than $X_2$ with respect to stochastic order, likelihood ratio order, hazard rate order, and mean residual life order for $p_1 \leq p_2$, if $\nu \leq 1$ or $\gamma \geq 1$. Similarly, for two COMNB$(\gamma, \nu_i, p)$ random variables, $X_i$, $i = 1, 2$, $X_1$ is less than $X_2$ with respect to each of these respective orders if $\nu_1 \leq \nu_2$ and $\gamma \geq 1$. In particular, for the case where $X_2$ is an NB random variable, $X_1$ is stochastically less than $X_2$ when $\nu_2 > \nu_1 = 1$.

*Parameter Estimation*

Two approaches can be considered for estimating $\gamma, \nu, p$: estimation via recursive formulae, and maximum likelihood estimation. For the recursive formulae, one solves the system of equations

$$\begin{cases} \frac{f_{x+1}}{f_x} = p\left(\frac{x+\gamma}{x+1}\right)^{\nu} \\ \ln\left(\frac{f_x f_{x+2}}{f_{x+1}^2}\right) = \nu \ln\left(\frac{x+\gamma+1}{x+2}\cdot\frac{x+1}{\gamma+x}\right) \\ \frac{\ln\left(\frac{f_x f_{x+2}}{f_{x+1}^2}\right)}{\ln\left(\frac{f_{x+1} f_{x+3}}{f_{x+2}^2}\right)} = \frac{\ln\left(\frac{x+\gamma+1}{x+2}\cdot\frac{x+1}{\gamma+x}\right)}{\ln\left(\frac{x+\gamma+2}{x+3}\cdot\frac{x+2}{\gamma+x+1}\right)} \end{cases} \tag{3.39}$$

to obtain the estimators $\breve{\gamma}, \breve{\nu}, \breve{p}$, where $f_x, x = 0, 1, 2, \ldots$, denotes the sample relative frequencies. Alternatively, maximum likelihood estimation finds those values $\hat{\gamma}, \hat{\nu}, \hat{p}$ that maximize the log-likelihood

$$\ln L(\gamma, \nu, p; \boldsymbol{x}) = \nu \sum_{i=1}^{n} \ln \left( \frac{\Gamma(\gamma + x_i)}{x_i!} \right) + \ln(p) \sum_{i=1}^{n} x_i + n\gamma \ln(1-p) \\ - n\nu \ln(\Gamma(\gamma)) - n \ln(D(\gamma, \nu, p)) \tag{3.40}$$

via the corresponding score equations

$$\begin{cases} \nu \sum_{i=1}^{n} \psi(\gamma + x_i) + n \ln(1-p) - n\nu\psi(t) - n\frac{\partial \ln(D(\gamma,\nu,p))}{\partial \gamma} = 0 \\ \sum_{i=1}^{n} \ln \left( \frac{\Gamma(\gamma+x_i)}{x_i!} \right) - n \ln(\Gamma(\gamma)) - n\frac{\partial \ln(D(\gamma,\nu,p))}{\partial \nu} = 0 \\ \frac{1}{p} \sum_{i=1}^{n} x_i + \frac{n\gamma}{p-1} - n\frac{\partial \ln(D(\gamma,\nu,p))}{\partial p} = 0, \end{cases} \tag{3.41}$$

where $\psi(a) = \frac{\partial \ln(\Gamma(a))}{\partial a}$ is the digamma function. Equation (3.41) does not have a closed form; thus, analysts must utilize a scoring method to solve the system of equations or a Newton-type procedure that optimizes Equation (3.40) such as `optim` in R; under either scenario, the recursive formula estimates can serve as starting values for determining the MLEs (Zhang et al., 2018).

### *3.4.3 The COM-type Negative Binomial (COMtNB) Distribution*

Chakraborty and Ong (2016) introduce what they refer to as a COMtNB distribution. This model with parameters $(\xi, p, \eta)$ has a respective probability mass function and cumulative distribution function of the form

$$P(X = x) = \frac{p^x \xi^{(x)}}{{}_1H_{\eta-1}(\xi; 1; p)(x!)^{\eta}}, \qquad x = 0, 1, 2, \ldots \tag{3.42}$$

$$P(X \leq x) = 1 - \frac{\xi^{(x+1)} p^{x+1}}{[(x+1)!]^{\eta-1}} \frac{{}_2H_{\eta}(\xi + x + 1; 1; x + 2; p)}{{}_1H_{\eta-1}(\xi; 1; p)}, \tag{3.43}$$

where ${}_1H_{\eta-1}(\xi; 1; p) = \sum_{k=0}^{\infty} \frac{\xi^{(k)} p^k}{(k!)^{\eta}}$ is a special case of the general hypergeometric series form ${}_mH_{\eta}(a_1, a_2, \ldots, a_m; b; p) = \sum_{k=0}^{\infty} \frac{a_1^{(k)} a_2^{(k)} \cdots a_m^{(k)}}{\{b^{(k)}\}^{\eta}} \frac{p^k}{k!}$. Given the parameters $(m, \eta, \boldsymbol{a}, b, p)$,

$$a^{(k)} \doteq a(a+1) \cdots (a+k-1) = \frac{\Gamma(a+k)}{\Gamma(a)} \tag{3.44}$$

denotes the rising factorial and $\Gamma(a) = \int_0^{\infty} x^{a-1} e^{-x} dx$ is the usual gamma function. The parameter space of interest is $\{(\xi, p, \eta): \{\xi > 0, p > 0, \eta > 1\} \cup$

$\{\xi > 0, 0 < p < 1, \eta = 1\}\}$; the probability mass function is undefined for $\eta < 1$ because ${}_1H_{\eta-1}(\xi; 1; p)$ does not converge under this condition.

The COMtNB$(\xi, p, \eta)$ distribution contains several special cases. When $\eta = 1$, the COMtNB reduces to the NB$(\xi, p)$ distribution, while it simplifies to the CMP$(p, \eta - 1)$ distribution when $\xi = 1$ (which contains the Poisson$(p)$ when $\eta = 2$, geometric with success probability $1 - p$ when $\eta = 1$ and $p < 1$, and Bernoulli with success probability $p/(1+p)$ when $\eta \to \infty$). The CMP is also attainable as a limiting distribution of the COMtNB; for a fixed $\xi p = \lambda < \infty$, the COMtNB$(\xi, p, \eta)$ distribution converges to the CMP$(\lambda, \eta)$ distribution as $\xi \to \infty$. Given this result, one can again see that the associated special cases hold for the CMP distribution; in particular, the COMtNB$(\xi, p, \eta)$ distribution further converges to the Bernoulli distribution with success probability, $\frac{\lambda}{1+\lambda}$, as $\eta \to \infty$. Meanwhile, the COMtNB$(\xi = 1, p = \eta e^{\phi}, \eta = 2)$ is the exponentially weighted Poisson distribution with weight function $e^{\phi x}$, i.e. a Poisson distribution with rate parameter $\eta e^{\phi}$. More broadly, for integer-valued $\eta \geq 1$, Equations (3.42)–(3.43) simplify to

$$P(X = x) = \frac{p^x \xi^{(x)}}{{}_1F_{\eta-1}(\xi; 1, 1, \ldots, 1; p)(x!)^{\eta}}, \qquad x = 0, 1, 2, \ldots,$$

$$P(X \leq x) = 1 - \frac{\xi^{(x+1)} p^{x+1}}{[(x+1)!]^{\eta-1}} \frac{{}_2F_{\eta}(\xi + x + 1; x+2, x+2, \ldots, x+2; p)}{{}_1F_{\eta-1}(\xi; 1, 1, \ldots, 1; p)},$$

where ${}_1F_{\eta-1}(\xi; 1, 1, \ldots, 1; p) = \sum_{j=0}^{\infty} \frac{\xi^{(j)} p^j}{(j!)^{\eta}}$, $\xi \geq 0$, $0 < p < 1$.

When $\xi > 1$, the COMtNB$(\xi, p, \eta)$ probability distribution is log-concave, has an increasing failure rate function, and is strongly unimodal. In fact, that log-concavity implies that (1) the truncated COMtNB distribution is likewise log-concave, (2) a convolution with any other discrete distribution will produce a new log-concave distribution, (3) the COMtNB has at most an exponential tail, and (4) for any integer $i$, the ratio $\frac{P(X = x+i)}{P(X = x)}$ is nonincreasing in $x$. In particular, the COMtNB distribution satisfies the recurrence equation

$$\frac{P(X = x+1)}{P(X = x)} = \frac{(\xi + x)p}{(x+1)^{\eta}}, \tag{3.45}$$

where $P(X = 0) = \frac{1}{{}_1H_{\eta-1}(\xi; 1; p)}$. The COMtNB$(\xi, p, \eta)$ further has a unique mode at $X = r$ if $\frac{r^{\eta}}{\xi + r - 1} < p < \frac{(r+1)^{\eta}}{\xi + r}$, $r = 1, 2, 3, \ldots$, where the probability function is nondecreasing with mode 0 if $\xi p < 1$. Alternatively, the COMtNB$(\xi, p, \eta)$ has two modes at $X = r$ and $r - 1$ if $(\xi + r - 1)p = r^{\eta}$. This more general form implies that the modes are at 0 and 1 if $\xi p = 1$ (i.e. setting $r = 1$).

The COMtNB is a member of several families of discrete distributions. It is a member of the generalized hypergeometric family, and the modified power series distribution family when $\xi$ and $\eta$ are known (Gupta, 1974; Kemp and Kemp, 1956). The COMtNB($\xi, p, \eta$) can likewise be recognized as either a weighted NB($\xi, p$) distribution with weight function $1/(x!)^{\eta-1}$ and NB probability function provided in Equation (1.6) or as a weighted CMP($p, \eta$) distribution where the CMP probability is described in Equation (2.8) with $0 < p < 1$ and weight function $\xi^{(x)}$. The COMtNB($\xi, p, \eta$) distribution belongs to the exponential family when $\xi$ is fixed and has joint sufficient statistics $S_1 = \sum_{i=1}^{n} x_i$ and $S_2 = \sum_{i=1}^{n} \ln x!$ for $p$ and $\eta$. For varying $\xi$, however, the COMtNB distribution is not a member of the exponential family (analogous to its special-case NB distribution).

The COMtNB probability and moment generating functions can be expressed as

$$\Pi_X(s) = \frac{{}_1H_{\eta-1}(\xi; 1; ps)}{{}_1H_{\eta-1}(\xi; 1; p)} = \frac{{}_1F_{\eta-1}(\xi; 1, 1, \ldots, 1; ps)}{{}_1F_{\eta-1}(\xi; 1, 1, \ldots, 1; p)}, \qquad 0 < ps \leq 1, \tag{3.46}$$

$$M_X(t) = \frac{{}_1H_{\eta-1}(\xi; 1; pe^t)}{{}_1H_{\eta-1}(\xi; 1; p)} = \frac{{}_1F_{\eta-1}(\xi; 1, 1, \ldots, 1; pe^t)}{{}_1F_{\eta-1}(\xi; 1, 1, \ldots, 1; p)}, \tag{3.47}$$

and the $k$th factorial moment $\mu_{(k)} \doteq E(X_{(k)}) = E[X(X-1)\cdots(X-k+1)]$ equals

$$\begin{aligned}
\mu_{(k)} &= \frac{\xi^{(k)} p^k}{(k!)^{\eta-1}} \frac{{}_1H_{\eta-1}(\xi + k; k+1; p)}{{}_1H_{\eta-1}(\xi; 1; p)} \\
&= \frac{\xi^{(k)} p^k}{(k!)^{\eta-1}} \frac{{}_1F_{\eta-1}(\xi + k; k+1, k+1, \ldots, k+1; p)}{{}_1F_{\eta-1}(\xi; 1, 1, \ldots, 1; p)},
\end{aligned}$$

where the respective latter expressions hold for integer-valued $\eta \geq 1$. More broadly, letting $\mu'_k \doteq E(X^k)$ and $\mu_k \doteq E[(X-\mu)^k]$ denote the $k$th moment and central moment, respectively, the following recursions hold:

$$\mu_{(k+1)} = p\frac{d}{dp}\mu_{(k)} + \mu_{(k)}\mu_{(1)} - k\mu_{(k)} \tag{3.48}$$

$$\mu'_{k+1} = p\frac{d}{dp}\mu'_k + \mu'_k\mu'_1 \tag{3.49}$$

$$\mu_{k+1} = p\frac{d}{dp}\mu_k + k\mu_{k-1}\mu_2. \tag{3.50}$$

The dispersion index (see Chapter 1) can be used to describe the association of parameter constraints with data dispersion, stemming from the COMtNB distribution relative to various baseline models. Comparing the COMtNB distribution with the Poisson model demonstrates the greatest extent of model flexibility with regard to dispersion. When $\eta = 1$ (i.e. the special case of the NB($\xi, p$) distribution), the COMtNB is over-dispersed, while when $\xi = 1$ (i.e. the CMP($p, \eta - 1$) special case), the COMtNB($\xi, p, \eta$) model is equi-dispersed for $\eta = 2$ and over-dispersed (under-dispersed) for $\eta < (>)2$. For fixed $\xi$ and $\eta$ with respect to $p$, the COMtNB($\xi, p, \eta$) distribution is equi-dispersed if $E(X) = cp$ and over-dispersed (under-dispersed) if $E(X) > (<)cp$, where $c$ is a constant. Meanwhile, for constants $c > 0$ and $d$ with respect to $p$, equi-dispersion exists if ${}_1H_{\eta-1}(\xi; 1; p) = \exp(c\lambda + d)$, and COMtNB data are over-dispersed (under-dispersed) if ${}_1H_{\eta-1}(\xi; 1; p) = \exp(c\lambda + d + g(\lambda))$, where $g(\lambda) > (<)0$ is an increasing (decreasing) function of $p$. More broadly, the COMtNB distribution is over-dispersed (under-dispersed) relative to the CMP when $\xi > (<)1$, and it is under-dispersed relative to the NB distribution when $\eta > 1$.

As was demonstrated with the CMP distribution (see Sections 3.2 and 3.3.1), one can consider the sum of COMtNB random variables, and the conditional distribution of a COMtNB random variable on the sum of COMtNB random variables. The sum of $n$ independent and identically distributed COMtNB($\xi, p, \eta$) random variables produces a continuous bridge between the NB($n\xi, p$) (for $\eta = 1$ and $p < 1$), the Poisson($np$) (for $\eta = 2$ and $\xi = 1$), and the binomial$\left(n, \frac{\lambda}{1+\lambda}\right)$ distributions (for fixed $\xi$ or $p$ such that $\xi p = \lambda$ and $\eta \to \infty$). Meanwhile, for two independent COMtNB($\xi_i, p, \eta$) random variables $X_i$, $i = 1, 2$, the conditional distribution of $X_1 = x \mid (X_1 + X_2 = s)$ is proportional to $\frac{\xi_1^{(x)} \xi_2^{(s-x)}}{(\xi_1+\xi_2)^{(s)}} \binom{s}{x}^\eta$. Chakraborty and Ong (2016) refer to the resulting probability function as the CMPtNH distribution; see Section 3.7.2 for further details.

## *Survival Analysis and Stochastic Ordering*

The survival and failure rate functions are

$$\begin{aligned}
S(t) = P(X \geq t) &= \frac{\xi^{(t)} p^t}{(t!)^\eta} \frac{{}_2H_{\eta-1}(\xi + t; t + 1; p)}{{}_1H_{\eta-1}(\xi; 1; p)} \\
&= \frac{\xi^{(t)} p^t}{(t!)^\eta} \frac{{}_2F_\eta(\xi + t, 1; t + 1, t + 1, \ldots, t + 1; p)}{{}_1F_{\eta-1}(\xi; 1, 1, \ldots, 1; p)} \\
r(t) = \frac{P(X = t)}{P(X \geq t)} &= \frac{1}{{}_2H_\eta(\xi + t, 1; t + 1; p)} \\
&= \frac{1}{{}_2F_\eta(\xi + t, 1; t + 1, t + 1, \ldots, t + 1; p)},
\end{aligned}$$

where the latter respective expressions hold for integer-valued $\eta \geq 1$. A COMtNB-distributed random variable is meanwhile smaller than its corresponding NB-distributed random variable with respect to the likelihood ratio order. This inequality implies that the COMtNB random variable is smaller than the NB random variable with regard to hazard rate order and mean residual life order, respectively.

### *Parameter Estimation and Hypothesis Testing*

The maximum likelihood estimators for $\xi, p$, and $\eta$ are determined via the log-likelihood

$$\ln L(\xi, p, \eta; \boldsymbol{x}) = \sum_{i=1}^{k} f_i \ln(\xi^{(x_i)}) + \left(\sum_{i=1}^{k} f_i x_i\right) \ln p - \eta \sum_{i=1}^{k} f_i \ln x_i! - n \ln \sum_{j=0}^{\infty} \frac{\xi^{(j)} p^j}{(j!)^\eta}$$

and corresponding score equations, where the $n$ COMtNB outcomes are grouped into $k$ classes $\boldsymbol{x} = (x_1, \ldots, x_k)$, with corresponding frequencies $\boldsymbol{f} = (f_1, \ldots, f_k)$ such that $n = \sum_{i=1}^{k} f_i$. As usual, these estimates cannot be derived in a closed form; however, numerical optimization procedures can determine them.

Recognizing that the COMtNB$(\xi, p, \eta)$ simplifies to the NB$(\xi, p)$ when $\eta = 1$, a natural inquiry is to test whether a significant amount of data dispersion exists relative to the NB such that one should instead consider the flexible COMtNB distribution. Accordingly, one can consider the hypotheses, $H_0\colon \eta = 1$ versus $H_1 : \eta \neq 1$, and conduct a likelihood ratio test with the statistic, $\Lambda = \frac{L(\hat{\xi}_0, \hat{p}_0, \eta=1; \boldsymbol{x})}{L(\hat{\xi}, \hat{p}, \hat{\eta}; \boldsymbol{x})}$, where $\hat{\xi}_0$ and $\hat{p}_0$ are the MLEs under the null hypothesis (i.e. the estimates under an assumed NB$(\xi, p)$ model) and $\hat{\xi}$, $\hat{p}$ and $\hat{\eta}$ are the MLEs assuming a COMtNB$(\xi, p, \eta)$ model. Analogous to other works (e.g. Sellers and Shmueli, 2010), Chakraborty and Ong (2016) consider a two-sided test for $\eta$ because the direction of data dispersion (i.e. over- or under-dispersion) is not important here; rather, the test infers whether statistically significant data dispersion exists (in either direction) such that the NB model is inappropriate. As usual, under $H_0$, $-2 \ln \Lambda$ converges to a chi-squared distribution with one degree of freedom.

### *The COMtNB Normalizing Constant*

The COMtNB normalizing function ${}_1H_{\eta-1}(\xi; 1; p) = \sum_{k=0}^{\infty} \frac{\xi^{(k)} p^k}{(k!)^\eta}$ is a special case of the generalized form of a hypergeometric series, ${}_mH_n(a_1, a_2, \ldots, a_m; b; p) = \sum_{k=0}^{\infty} \frac{a_1^{(k)} a_2^{(k)} \cdots a_m^{(k)}}{\{b^{(k)}\}^n} \frac{p^k}{k!}$ given parameters

$(m, n, \boldsymbol{a}, b, p)$, and ${}_1H_{\eta-1}(\xi; 1; p)$ itself contains several special-case functions. When $\eta \in \mathbb{Z}$, $\eta \geq 1$, ${}_1H_{\eta-1}(\xi; 1; p) = {}_1F_{\eta-1}(\xi; 1; p)$, a particular generalized hypergeometric function. Additional special cases regarding ${}_1H_{\eta-1}(\xi; 1; p)$ include

1. ${}_1H_0(\xi; 1; p) = \frac{1}{(1-p)^{\xi}}$,
2. ${}_1H_{\eta-1}(1; 1; p) = Z(p, \eta - 1)$, i.e. the CMP normalizing function described in Chapter 2,
3. $\lim_{\eta \to \infty} {}_1H_{\eta-1}(\xi; 1; p) = 1 + \xi p$, and
4. $\lim_{\xi \to \infty} {}_1H_{\eta-1}(\xi; 1; p) = Z(\lambda, \eta)$, when $0 < \xi p = \lambda < \infty$.

Beyond these special cases, ${}_1H_{\eta-1}(\xi; 1; p)$ can be determined via a truncated approximation at some $M$ so that ${}_1H_{\eta-1}(\xi; 1; p) \approx \sum_{k=0}^{M} \frac{\xi^{(k)} p^k}{(k!)^{\eta}}$, where $\frac{(\xi+M)p}{(M+1)^{\eta}}$ is sufficiently small.

### 3.4.4 Extended CMP (ECMP) Distribution

The ECMP (Chakraborty and Imoto, 2016) is a four-parameter distribution that contains the COMtNB and GCMP distributions along with their respective special cases. The ECMP$(\gamma, p, \beta, \alpha)$ distribution has the probability

$$P(X = x) = \frac{\left[\gamma^{(x)}\right]^{\alpha} p^x}{{}_1G^{\alpha}_{\beta-1}(\gamma; 1; p)(x!)^{\beta}} = \left(\frac{[\Gamma(\gamma + x)]}{[\Gamma(\gamma)]}\right)^{\alpha} \frac{p^x}{{}_1G^{\alpha}_{\beta-1}(\gamma; 1; p)(x!)^{\beta}}, \tag{3.51}$$

where ${}_1G^{\alpha}_{\beta-1}(\gamma; 1; p) = \sum_{k=0}^{\infty} \frac{[\gamma^{(k)}]^{\alpha} p^k}{(k!)^{\beta}}$ is the normalizing constant, and $\gamma^{(k)}$ is a rising factorial as defined in Equation (3.44); a detailed discussion regarding ${}_1G^{\alpha}_{\beta-1}(\gamma; 1; p)$ is provided later in this section. This distribution is defined for $\gamma \geq 0$, $p > 0$, and $0 \leq \alpha < \beta$ or where $\gamma > 0$, $0 < p < 1$, and $\alpha = \beta$. In particular, while the ECMP requires $\alpha \leq \beta$, these parameters are not otherwise constrained (i.e. $\alpha, \beta \in \mathbb{R}$) as they are for the GCMP ($\alpha \leq 1$) or the COMtNB ($\beta \geq 1$) distributions. This distribution can take various shapes, including a nonincreasing probability distribution with mode at 0, a unimodal shape with a nonzero mode, or a bimodal shape with one zero and one nonzero mode. Further, the survival and failure rate functions are

$$\begin{aligned} S(t) &= \frac{\gamma^{(t)} p^t}{(t!)^{\beta}} \frac{{}_2G^{\alpha}_{\beta-1(\gamma+t,1;\, t+1;\, p)}}{{}_1G^{\alpha}_{\beta-1}(\gamma; 1; p)} \\ &= \frac{\gamma^{(t)} p^t}{(t!)^{\beta}} \frac{{}_{\alpha+1}F_{\beta-1}(\gamma + t, 1; t+1, t+1, \ldots, t+1; p)}{{}_{\alpha}F_{\beta-1}(\gamma; 1, 1, \cdots, 1; p)} \end{aligned} \tag{3.52}$$

$$r(t) = \frac{1}{{}_2G^{\alpha}_{\beta-1}(\gamma+t, 1; t+1; p)}$$
$$= \frac{1}{{}_{\alpha+1}F_{\beta}(\gamma+t, 1; t+1, t+1, \ldots, t+1; p)}, \quad (3.53)$$

where the respective latter representations in Equations (3.52)–(3.53) hold when $\alpha, \beta \in \mathbb{Z}$. The ECMP satisfies the recurrence

$$\frac{P(X = x+1)}{P(X = x)} = \frac{(\gamma + x)^{\alpha} p}{(x+1)^{\beta}}, \quad (3.54)$$

with $P(X = 0) = \frac{1}{{}_1G^{\alpha}_{\beta-1}(\gamma; 1; p)}$; this relationship generalizes Equation (3.45) by Chakraborty and Ong (2016). Equation (3.54) implies that, for $\alpha > (<)1$, the ECMP($\gamma, p, \beta, \alpha$) distribution has a longer (shorter) tail than the GCMP($\gamma, p, \alpha$) distribution, while, for $\beta < (>)1$, the ECMP has a longer (shorter) tail than the COMtNB($\gamma, p, \beta$) model. Finally, the ECMP($\gamma, p, \beta, \alpha$) can be represented as a member of an exponential family in $(\ln(p), \beta, \alpha)$ for a given $\gamma$ and as a weighted Poisson($p$) distribution as described in Section 1.4 with weight $w(x) = \frac{[\Gamma(\gamma+x)]^{\alpha}}{[\Gamma(1+x)]^{\beta-1}}$.

The ECMP($\gamma, p, \beta, \alpha$) distribution is log-concave when all of the constraints $\{\gamma > 1,\ p > 0,\ \alpha \leq \beta\}$ hold; this result confirms that the GCMP($\gamma, p, \beta$) is log-concave for $\{\gamma \geq 1,\ p > 0,\ \beta < 1\}$. The general constrained space of log-concavity for the ECMP distribution implies that the distribution is "strongly unimodal" with an increasing failure rate function (Chakraborty and Imoto, 2016). The ECMP has a mode of 0 when $p\gamma^{\alpha} < 1$, and the mode occurs at $X = r$, $r = 1, 2, 3, \ldots$, when $\frac{r^{\beta}}{(\gamma+r-1)^{\alpha}} < p < \frac{(r+1)^{\beta}}{(\gamma+r)^{\alpha}}$. Meanwhile, the ECMP is bimodal with modes at $r$ and $r-1$ when $p(\gamma+r-1)^{\alpha} = r^{\beta}$; in particular, the modes are 0 and 1 when $p\gamma^{\alpha} = 1$ (i.e. $r = 1$). Conversely, the ECMP($\gamma, p, \beta, \alpha$) distribution is log-convex when $0 < \gamma \leq 1$, $p > 0$, and $\alpha = \beta$. Under such circumstances, ECMP is infinitely divisible, has a DFR function, an increasing mean residual life function, and the variance is bounded above by $p\gamma^{\alpha}$. Finally, the $k$th factorial moment is

$$(\mu)_k = \frac{\{\gamma^{(k)}\}^{\alpha} p^k}{(k!)^{\beta-1}} \frac{{}_1G^{\alpha}_{\beta-1}(\gamma+k, k+1, p)}{{}_1G^{\alpha}_{\beta-1}(\gamma, 1, p)} \quad (3.55)$$
$$= \frac{\{\gamma^{(k)}\}^{\alpha} p^k}{(k!)^{\beta-1}} \frac{{}_{\alpha}F_{\beta-1}(\gamma+k; k+1, k+1, \ldots, k+1; p)}{{}_{\alpha}F_{\beta-1}(\gamma; 1, 1, \cdots, 1; p)}, \quad (3.56)$$

where Equation (3.56) holds for $\alpha, \beta \in \mathbb{Z}$. In particular, the mean of the ECMP($\gamma, p, \beta, \alpha$) distribution is approximately $p^{1/(\beta-\alpha)} + \frac{1-\beta+(2\gamma-1)\alpha}{2(\beta-\alpha)}$, where this approximation holds when $p$ is large and $|\beta - \alpha|$ is small.

The ECMP distribution contains several special cases, including the NB, CMP, GCMP, COMNB, and COMtNB distributions. The NB($\gamma, p$) whose probability function is as defined in Equation (1.5) is a special case of the ECMP($\gamma, p, \beta, \alpha$) when $\alpha = 1$ and $\beta = 0$, while the COMNB($\gamma, \delta, p$) distribution (discussed in Section 3.4.2) is achieved when $\alpha = \beta = \delta$. The CMP distribution (as described in Chapter 2) can meanwhile be attained in various ways. The ECMP distribution contains the CMP($p, \beta - \alpha$) (in particular, the CMP($p, \beta$) when $\alpha = 0$) when $\gamma = 1$ and converges to the CMP($\lambda, \beta$) distribution as $\gamma \rightarrow \infty$, with $0 < \lambda \doteq \gamma^{\alpha} p < \infty$ fixed. Accordingly, the ECMP distribution contains the Poisson, geometric, and Bernoulli distributions (i.e. the special CMP cases) under the appropriate conditions. The ECMP further contains the GCMP($\gamma, p, \alpha$) when $\beta = 1$ and the COMtNB($\gamma, p, \beta$) when $\alpha = 1$; refer to Sections 3.4.1 and 3.4.3, respectively, for discussion regarding the GCMP and COMtNB distributions. With respect to likelihood ratio order, the ECMP($\gamma, p, \beta, \alpha$) random variable is smaller than the COMtNB($\gamma, p, \beta$) random variable when $\alpha < 1$, and is smaller than the GCMP($\gamma, p, \alpha$) random variable when $\beta > 1$. Both cases imply the same progression with regard to hazard rate ordering and thus mean residual life ordering under the respective constraints.

Data dispersion for the ECMP distribution can be reflected relative to several other distributions. Considering the ECMP distribution as a weighted Poisson($p$) distribution (see Section 1.4 for details) with weight $w(x) = \frac{[\Gamma(\gamma+x)]^{\alpha}}{[\Gamma(1+x)]^{\beta-1}}$ implies that the ECMP($\gamma, p, \beta, \alpha$) addresses over-dispersion relative to the Poisson distribution (1) for all $\gamma$ when $\alpha \geq 0$ and $\beta < 1$, or (2) when $\{0 < \gamma \leq 1, \alpha \leq \beta \leq \alpha + 1\}$ or $\{\gamma > 1, \beta \leq 1\}$ is true, and $\{\alpha > 0, \beta \geq 1\}$ or $\{\alpha < 0, \beta < 1\}$ also holds. Meanwhile, ECMP($\gamma, p, \beta, \alpha$) accommodates under-dispersion relative to the Poisson when (1) $\alpha < 0$ and $\beta \geq 1$ for all $\gamma$, or (2) for $\{\alpha > 0, \beta \geq 1\}$ or $\{\alpha < 0, \beta < 1\}$ when $\{0 < \gamma \leq 1, \beta \geq \alpha + 1\}$ or $\{\gamma > 1, \beta \geq 1\}$ hold true. More broadly, the ECMP($\gamma, p, \beta, \alpha$) is over-dispersed (under-dispersed) relative to the COMtNB($\gamma, p, \beta$) when $0 < \alpha \leq \beta$ ($\beta \geq \alpha \leq 0$), while ECMP is over-dispersed (under-dispersed) relative to the GCMP($\gamma, p, \alpha$) when $\alpha \leq \beta < 1$ ($1 < \alpha \leq \beta$).

### *Parameter Estimation*

Maximum likelihood estimation is a popular procedure for parameter estimation; however, the ECMP log-likelihood,

$$\ln L(\gamma, p, \beta, \alpha; \boldsymbol{f}) = \alpha \sum_{i=0}^{k} f_i \ln(\gamma^{(i)}) - \beta \sum_{i=0}^{k} f_i \ln(i!)$$
$$+ \ln(p) \sum_{i=0}^{k} \text{i} f_i - n \ln\left({}_1G^{\alpha}_{\beta-1}(\gamma, 1, p)\right), \tag{3.57}$$

can have multiple local maxima, where $\boldsymbol{f} = (f_1, \ldots, f_k)$ denotes the observed frequencies at $0, \ldots, k$ such that $k$ denotes the largest observation, and $n = \sum_{i=0}^{k} f_i$, thus the log-likelihood may not contain a unique solution. Chakraborty and Imoto (2016) suggest conducting a profile likelihood approach to determine the estimators where, for a given $\gamma$, analysts maximize Equation (3.57) with respect to $p, \alpha$, and $\beta$. Given its exponential family form for a fixed $\gamma$, profile MLEs can be attained from which the overall MLEs $(\hat{\gamma}, \hat{p}, \hat{\alpha}, \hat{\beta})$ can be identified. The authors further propose the starting values

$$\alpha_k^{(0)}(\gamma) = s_{2,k}(\gamma), \tag{3.58}$$
$$\beta_k^{(0)}(\gamma) = \begin{cases} s_{1,k}(\gamma) \text{ for } s_{1,k}(\gamma) > s_{2,k}(\gamma) \\ s_{2,k}(\gamma) \text{ otherwise} \end{cases} \tag{3.59}$$
$$p_l^{(0)}(\gamma) = \frac{(l+1)\beta_k^{(0)}(\gamma)}{(\gamma + l)\alpha_k^{(0)}(\gamma)} \tag{3.60}$$

for a given $\gamma$ from an ECMP$(\gamma, p, \beta, \alpha)$ distribution, where $l$ is the smallest observation and $s_{1,k}(\gamma), s_{2,k}(\gamma)$ satisfy

$$\begin{pmatrix} s_{1,k}(\gamma) \\ s_{2,k}(\gamma) \end{pmatrix} = \begin{pmatrix} \ln\left(\frac{k+1}{k+2}\right) & \ln\left(\frac{\gamma+k+1}{\gamma+k}\right) \\ \ln\left(\frac{k+2}{k+3}\right) & \ln\left(\frac{\gamma+k+2}{\gamma+k+1}\right) \end{pmatrix}^{-1} \begin{pmatrix} \ln\left(\frac{f_{k+2}}{f_{k+1}}\right) - \ln\left(\frac{f_{k+1}}{f_k}\right) \\ \ln\left(\frac{f_{k+3}}{f_{k+2}}\right) - \ln\left(\frac{f_{k+2}}{f_{k+1}}\right) \end{pmatrix} \tag{3.61}$$

for chosen $k$ such that the inverse matrix exists.

### *The ECMP Normalizing Constant, ${}_1G^{\alpha}_{\beta-1}(\gamma; 1; p)$, and Its Approximation*

The normalizing constant ${}_1G^{\alpha}_{\beta-1}(\gamma; 1; p) = \sum_{k=0}^{\infty} \frac{[\gamma^{(k)}]^{\alpha} p^k}{(k!)^{\beta}}$ is a special case of the generalized form of the Chakraborty and Ong (2016) hypergeometric-type series,

$$_mG^c_d(a_1, a_2, \ldots, a_m; b; p) = \sum_{k=0}^{\infty} \frac{\left[a_1^{(k)}\right]^c a_2^{(k)} \cdots a_m^{(k)}}{[b^{(k)}]^d} \frac{p^k}{k!}. \tag{3.62}$$

Equation (3.62) can likewise be represented as

$${}_mG^c_d(a_1,a_2,\ldots,a_m;b;p)={}_{c+m-1}F_d(\underbrace{a_1,a_1,\ldots,a_1}_{c\text{ times}},a_2,\ldots,a_m;\underbrace{b,b,\ldots,b}_{d\text{ times}};p)$$

for $\alpha, \beta$, and $m \in \mathbb{Z}^+$. The following properties hold as special cases of ${}_1G^{\alpha}_{\beta-1}(\gamma;1;p)$, thus defining the normalizing constant for certain special-case distributions:

1. ${}_1G^{\alpha}_{\beta-1}(\gamma;1;p) = {}_{\alpha}F_{\beta-1}(\gamma;1,1,\ldots,1;p)$ for integer-valued $\alpha, \beta$,
2. ${}_1G^{1}_{\beta-1}(\gamma;1;p) = {}_1H_{\beta-1}(\gamma;1;p)$ as defined in Section 3.4.3,
3. $\frac{{}_1G^{\alpha}_{0}(\gamma;1;p)}{(\Gamma(\gamma))^{\alpha}} = C(\alpha,\gamma,p)$ as defined in Section 3.4.1,
4. ${}_1G^{\alpha}_{\beta-1}(1;1;p) = Z(p,\beta-\alpha)$ as described in Chapter 2,
5. ${}_1G^{1}_{0}(\gamma;1;p) = \frac{1}{(1-p)^{\gamma}}$,
6. ${}_1G^{x}_{x}(1;1;p) = \exp(p)$,
7. $\lim_{\beta\to\infty} {}_1G^{\alpha}_{\beta-1}(\gamma;1;p) = 1+\gamma^{\alpha}p$, and
8. $\lim_{\gamma\to\infty} {}_1G^{\alpha}_{\beta-1}(\gamma;1;p) = Z(\lambda,\beta)$, where $0<\lambda \doteq \gamma^{\alpha}p<\infty$.

Special cases (2)–(5) above define the respective normalizing constants associated with the following distributions: COMtNB$(\gamma,p,\beta)$, GCMP$(\gamma,p,\alpha)$, COMtNB$(\gamma=1,p,\beta-\alpha)=$CMP$(p,\beta-\alpha)$, and NB$(\gamma,p)$. Other special-case distributions that can be obtained from the ECMP distribution include the CMP$(p,\beta)$ distribution when $\alpha=0$ (as shown via Case 4), and the Poisson$(p)$ distribution when $\gamma=1$ and $\alpha=\beta$ (i.e. Case 6), or when $\alpha=0$ and $\beta=1$. The CMP$(\lambda,\beta)$ distribution is likewise asymptotically attainable as $\gamma\to\infty$ such that $0<\lambda\doteq\gamma^{\alpha}p<\infty$ remains fixed (Case 8). Another special case, namely a new generalization of the NB distribution, is defined as

$$P(X=k) = \frac{1}{{}_1G^{\delta}_{\delta-1}(\gamma;1;p)}\left(\binom{\gamma+k-1}{k}\right)^{\delta} p^k; \tag{3.63}$$

this distribution is obtained as a special case of ECMP, with $\alpha=\beta=\delta$, and is log-convex for $0<\gamma\leq 1$ (Chakraborty and Imoto, 2016).

As is the case with other CMP-motivated distributions, these extensions rely on a normalizing constant whose form involves an infinite summation. Because there does not exist a closed form for this summation except for special cases, Chakraborty and Imoto (2016) suggest two approaches toward approximating the normalizing constant: truncating the infinite sum or determining an asymptotic approximation via Laplace's method. By truncating the summation, we approximate ${}_1G^{\alpha}_{\beta-1}(\gamma;1;p)$ with

${}_1G^{\alpha}_{\beta-1,M}(\gamma;1;p) = \sum_{k=0}^{M} \frac{[\gamma^{(k)}]^{\alpha} p^k}{(k!)^{\beta}}$ for some $M \in \mathbb{Z}^+$ such that $\epsilon_M = \frac{(\gamma-M+1)^{\alpha} p}{M^{\beta}} < 1$. This requirement produces a relative truncation error

$$R_M(\gamma,p,\beta,\alpha) < \frac{\left[\gamma^{(M+1)}\right]^{\alpha} p^{M+1}}{[(1-\epsilon_M)(M+1)!]^{\beta}\, {}_1G^{\alpha}_{\beta-1,M}(\gamma;1;p)}.$$

This approximation is good for $\beta - \alpha \geq 1$, producing a truncation value $M$ that is not large. For $0 < \beta - \alpha < 1$ and $p > 1$; however, $M$ needs to be considerably large. Under such situations, analysts are urged to restrict $p$ so that $p < 1$ when $\beta - \alpha \to 0$. Alternatively, ${}_1G^{\alpha}_{\beta-1}(\gamma;1;p)$ can be approximated via the Laplace method to obtain

$$ {}_1G^{\alpha}_{\beta-1}(\gamma;1;p) \approx \frac{p^{[1-\beta+(2\gamma-1)\alpha]/2(\beta-\alpha)} \exp\left((\beta-\alpha)p^{1/(\beta-\alpha)}\right)}{(2\pi)^{(\beta-\alpha-1)/2}\sqrt{\beta-\alpha}[\Gamma(\gamma)]^{\alpha}}. \qquad (3.64)$$

This reduces to the asymptotic approximation for the COMtNB normalizing constant (Equation (3.30)) when $\beta = 1$ (Imoto, 2014) and to the analogous approximation for the CMP normalizing constant (Equation (2.12)) when $\gamma = 1$ or $\alpha = 0$ (Minka et al., 2003).

*Formulating the ECMP($\gamma, p, \beta, \alpha$) Distribution*

The ECMP model can be formulated either via a flexible queuing process or by an exponential combination. Akin to the derivation of the CMP distribution, here one can consider a queuing system with arrival and service rates, $\lambda_x = (\gamma + x)^{\alpha}\lambda$ and $\mu_x = x^{\beta}\mu$, respectively, such that the recursive relationship $\frac{P(X=x+1)}{P(X=x)} = \frac{\lambda_x}{\mu_{x+1}}$ holds; let $1/\mu$ and $1/\lambda$ denote the respective normal mean service and arrival times for a solo unit in a system, and $\beta$ and $\gamma$, respectively, denote pressure coefficients. Given this framework, one can follow the Conway and Maxwell (1962) queuing derivation as discussed in Section 2.1, now letting the arrival and service rates increase exponentially as $x$ increases to obtain the ECMP($\gamma, p, \beta, \alpha$) probability function with $p = \lambda/\mu$ for the probability of the system being in state $x$ (Chakraborty and Imoto, 2016; Zhang, 2015). Alternatively, the ECMP($\gamma, p, \beta, \alpha$) can be viewed as an exponential combination of the NB($\gamma, \lambda$) and CMP($\mu, \theta$) distributions, where $p = \lambda^{\alpha}\mu^{1-\alpha}$ and $\beta = \theta(1-\alpha) + \alpha$. Accordingly, $\alpha$ converging to 0 implies more weight toward the NB model, while $\beta$ converging to 1 tends closer to the CMP distribution (Chakraborty and Imoto, 2016).

*Alternative Parametrization and Associated Results*

An alternative ECMP parametrization (denoted as ECMPa$(\gamma, p, \beta, \alpha)$) has the probability

$$P(X = x) = \frac{\frac{\Gamma(\gamma+x)^{\alpha}}{(x!)^{\beta}} p^x}{\sum_{i=0}^{\infty} \frac{\Gamma(\gamma+i)^{\alpha}}{(i!)^{\beta}} p^i}, \quad x = 0, 1, 2, \ldots$$

(Zhang, 2015). This formulation is infinitely divisible for $\gamma, \alpha$, and $\beta$ that satisfy the constraint

$$\frac{1}{2^{\beta}} \left(1 + \frac{1}{\gamma}\right)^{\alpha} \geq 1.$$

Further, the Stein identity holds under either parametrization, i.e. letting $\mathbb{N} = \{0, 1, 2, \ldots\}$, a random variable $X$ has the ECMP/ECMPa$(\gamma, p, \beta, \alpha)$ distribution if and only if, for any bounded function $g \colon \mathbb{N} \to \mathbb{R}$,

$$E[X^{\beta} g(X) - p(X + \gamma)^{\alpha} g(X + 1)] = 0$$

holds.

An extended negative hypergeometric distribution (ENHG) distribution is meanwhile derived via the conditional distribution involving ECMP distributions. Denoted as ENHG$(s, \gamma_1, \gamma_2, \alpha, \beta)$, its probability distribution has the form

$$P(Y = y) = \frac{[\Gamma(\gamma_1 + y)\Gamma(\gamma_2 + s - y)]^{\alpha}}{[y!(s - y)!]^{\beta} H(s, \gamma_1, \gamma_2, \alpha, \beta)},$$

where $H(s, \gamma_1, \gamma_2, \alpha, \beta) = \sum_{i=0}^{s} \frac{[\Gamma(\gamma_1+i)\Gamma(\gamma_2+s-i)]^{\alpha}}{[i!(s-i)!]^{\beta}}$ is the normalizing constant. This form can be obtained by considering the conditional distribution of $X_1$ given $X_1 + X_2 = s$ for two independent ECMP$(\gamma_i, p, \alpha, \beta)$ random variables, $X_i$, $i = 1, 2$ (Zhang, 2015).

## 3.5 Conway–Maxwell Katz (COM–Katz) Class of Distributions

The Katz class (also referred to as the $(a, b, 0)$ class) of distributions is that family of distributions satisfying the recursion

$$P(X = k) = \left(a + \frac{b}{k}\right) P(X = k - 1), \quad k = 1, 2, \ldots, \tag{3.65}$$

where $a, b \in \mathbb{R}$. It contains several well-known distributions as special cases, including the Poisson, binomial, and NB distributions. The family has been applied in diverse applications such as minefield clearance (Washburn, 1996) and control chart theory (Fang, 2003). Zhang et al. (2018)

introduce a "COM-type extension" of this class where Equation (3.65) is modified to

$$P(X = k) = \left(a + \frac{b}{k}\right)^{\nu} P(X = k-1), \quad k = 1, 2, \ldots, \tag{3.66}$$

for some constants $a, b \in \mathbb{R}$ and $\nu \in \mathbb{R}^{+}$; we will refer to this generalized family of distributions as the COM–Katz[4] class of distributions.

The degenerate, CMP($\lambda$, $\nu$), CMB, and COMNB distributions all belong to the COM–Katz class. The CMP($\lambda$, $\nu$) satisfies this form by letting $a = 0$ and $b = \lambda^{1/\nu}$; the CMP recursion is provided in Equation (2.44). The CMB$(m, p, \nu)$ recursion property (Equation (3.7)) can likewise be represented as Equation (3.66); in this case, $a = -1$ and $b = m + 1$. Finally, the COMNB distribution is another special case of this form, where $a = p^{1/\nu}$ and $b = (\gamma - 1)p^{1/\nu}$; see Equation (3.38). The Brown and Xia (2001) birth–death processes with arrival rate $\lambda_x = c[a(x+1) + b]^{\nu}$ and service rate $\mu_x = cx^{\nu}$ such that the ratio of consecutive probabilities $\frac{P(X=x)}{P(X=x-1)} = \frac{\lambda_{x-1}}{\mu_x}$, $x = 1, 2, \ldots$, are another example that lies within the COM–Katz class of distributions.

## 3.6 Flexible Series System Life-Length Distributions

Two popular two-parameter distributions have been considered as tools for describing the life length of systems that arise (say) in the biological or engineering sciences. The exponential–geometric (EG) distribution has the probability density function

$$f_{\text{EG}}(y; \kappa, p) = \frac{\kappa(1-p)e^{-\kappa y}}{(1 - pe^{-\kappa y})^2}, \quad y \geq 0, \tag{3.67}$$

for a random variable $Y$, where $\kappa > 0$ and $p \in (0, 1)$ (Adamidis and Loukas, 1998). The exponential–Poisson (EP) distribution meanwhile has the density

$$f_{\text{EP}}(y; \kappa, \lambda) = \frac{\kappa\lambda}{1 - e^{-\lambda}} e^{-\lambda - \kappa y + \lambda \exp(-\kappa y)}, \quad y \geq 0 \tag{3.68}$$

(Kuş, 2007). Both model pdfs are decreasing functions with a mode at 0 and have a DFR; other respective distributional properties, including their respective modal values, are provided in Table 3.4. In both cases, these distributions are derived by considering $Y = \min(\{X_i\}_{i=1}^{C})$ for a random sample of exponentially distributed random variables $\{X_i\}_{i=1}^{C}$, with density

[4] While one could consider referring to the COM–Katz class as the $(a, b, \nu, 0)$ (to maintain consistency with the Katz and $(a, b, 0)$ names), we refrain from doing so here.

Table 3.4 *Distributional properties of the exponential-geometric (EG) and exponential-Poisson (EP) distributions*

| | EG | EP |
|---|---|---|
| Probability density function, $f(y)$ | $\kappa(1-p)e^{-\kappa y}(1-pe^{-\kappa y})^{-2}$ | $\frac{\kappa\lambda}{1-e^{-\lambda}}e^{-\lambda-\kappa y+\lambda\exp(-\kappa y)}$ |
| Cumulative dist. function, $F(y)$ | $(1-e^{-\kappa y})(1-pe^{-\kappa y})^{-1}$ | $\left(e^{\lambda\exp(-\kappa y)}-e^{\lambda}\right)\left(1-e^{\lambda}\right)^{-1}$ |
| Survival function, $S(y)$ | $(1-p)e^{-\kappa y}(1-pe^{-\kappa y})^{-1}$ | $\left(1-e^{\lambda\exp(-\kappa y)}\right)\left(1-e^{\lambda}\right)^{-1}$ |
| Hazard function, $h(y)$ | $\kappa(1-pe^{-\kappa x})^{-1}$ | $\frac{\kappa\lambda e^{-\lambda-\beta y+\lambda\exp(-\kappa y)}\left(1-e^{\lambda}\right)}{\left(1-e^{-\lambda}\right)\left(1-e^{\lambda\exp(-\kappa y)}\right)}$ |
| Moments, $E(Y^r)$ | $(1-p)r!(p\kappa^r)^{-1}\sum_{j=1}^{\infty}(p^j/j^r)$ | $\frac{\lambda\Gamma(r+1)}{(e^{\lambda}-1)\kappa^r}F_{r+1,r+1}$ $([1,1,\ldots,1],[2,2,\ldots,2],\lambda)$ |
| Modal value | $\kappa(1-p)^{-1}$ | $\kappa\lambda(1-e^{-\lambda})^{-1}$ |
| Median | $\kappa^{-1}\ln(2-p)$ | $\kappa^{-1}\ln\left[\ln\left(\frac{e^{\lambda}+1}{2}\right)^{-1}\lambda\right]$ |

$f(x;\kappa)=\kappa e^{-\kappa x}, x>0$, where $C$ denotes either a geometric random variable with probability $P(C=j;p)=(1-p)p^{j-1}$, for $j=1,2,3,\ldots$ (Adamidis and Loukas, 1998) or a zero-truncated Poisson random variable, where $P(C=j;\lambda)=\frac{e^{-\lambda}\lambda^j}{(1-e^{-\lambda})j!}, j=1,2,3,\ldots$ (Kuş, 2007).

The following sections consider generalizations of the EG and EP distributions in two ways. Section 3.6.1 introduces the flexible ExpCMP distribution as an extension of EG and EP. Here, $C$ is instead assumed to be a zero-truncated CMP distribution with probability mass function

$$P(C=j;\lambda,\nu)=\frac{\lambda^j}{[Z(\lambda,\nu)-1](j!)^{\nu}},\quad j=1,2,3,\ldots; \tag{3.69}$$

this choice of distribution for $C$ contains the geometric distribution with success probability $1-\lambda$ (when $\nu=0$ and $\lambda<1$) and zero-truncated Poisson with rate parameter $\lambda$ ($\nu=1$) as special cases. Section 3.6.2 develops a further generalized distribution by considering both the zero-truncated CMP random variable $C$ with instead a Weibull distribution for $X$. This results in a marginal distribution that is WCMP.

### *3.6.1 The Exponential-CMP (ExpCMP) Distribution*

The ExpCMP distribution is a flexible model to study the minimum lifetimes of a series of exponentially distributed random variables. A random variable $Y$ that has an ExpCMP($\kappa, \lambda, \nu$) distribution has the density function

$$f(y; \kappa, \lambda, \nu) = \frac{\kappa}{Z(\lambda, \nu) - 1} \sum_{j=1}^{\infty} \frac{j\lambda^j}{(j!)^{\nu}} \exp(-j\kappa y), \quad y > 0, \tag{3.70}$$

where $\kappa, \lambda > 0$ and $\nu \geq 0$ (Cordeiro et al., 2012). This distribution is derived by compounding an exponential distribution with a CMP distribution. Letting $C$ be a zero-truncated CMP($\lambda, \nu$) distributed random variable, consider a random sample $X_1, \ldots, X_j$ given $C = j$ of exponentially distributed random variables (independent of $C$) with mean $1/\kappa$. Letting $Y = \min\{X_i\}_{i=1}^{C}$, its conditional density function given $C = j$ is exponential with mean $(j\kappa)^{-1}$, i.e. $f(y \mid j; \kappa) = j\kappa \exp(-j\kappa y), y \geq 0$.

The ExpCMP distribution is a weighted exponential distribution of the form $f(y; \kappa, \lambda, \nu) = \sum_{j=1}^{\infty} w_j f_e(y; j\kappa)$, where

$$f_e(y;\, j\kappa) = j\kappa \exp(-j\kappa y) \quad \text{and} \quad w_j = \frac{\lambda^j}{[Z(\lambda, \nu) - 1](j!)^{\nu}}. \tag{3.71}$$

Its representation implies that the ExpCMP has the respective cumulative distribution and hazard rate functions

$$F(y; \kappa, \lambda, \nu) = 1 - \sum_{j=1}^{\infty} w_j \exp(-j\kappa y) = 1 - \frac{Z(\lambda e^{-\kappa y}, \nu)}{Z(\lambda, \nu) - 1} \tag{3.72}$$

$$h(y; \kappa, \lambda, \nu) = \frac{\sum_{j=1}^{\infty} w_j f_e(y; j\kappa)}{\sum_{j=1}^{\infty} w_j \exp(-j\kappa y)} = \frac{\kappa\lambda \left(\frac{\partial Z(\lambda e^{-\kappa y}, \nu)}{\partial \lambda}\right)}{Z(\lambda e^{-\kappa y}, \nu)}. \tag{3.73}$$

The ExpCMP distribution has a DFR whose moments and moment generating function are

$$E(Y^r) = \frac{\Gamma(r+1)}{\kappa^r} E(C^{-r}) = \frac{\Gamma(r+1)}{\kappa^r} \sum_{j=1}^{\infty} \frac{w_j}{j^r}$$

$$= \frac{\Gamma(r+1)}{\kappa^r} \sum_{j=1}^{\infty} \frac{\lambda^j}{j^r [Z(\lambda, \nu) - 1](j!)^{\nu}} \tag{3.74}$$

$$M_Y(t) = \sum_{j=1}^{\infty} w_j \left(1 - \frac{t}{j\kappa}\right)^{-1} = \frac{E(1 - t(\kappa C)^{-1})}{Z(\lambda, \nu) - 1}, \tag{3.75}$$

where $C$ is a zero-truncated CMP random variable with probability defined in Equation (3.69). Meanwhile, the ExpCMP quantiles can be determined by solving $Z(\lambda e^{-\kappa y}, \nu) = (1 - p)[Z(\lambda, \nu) - 1]$ for any probability, $p$. The

ExpCMP($\kappa, \lambda, \nu$) distribution contains the EP and EG distributions as special cases; the ExpCMP reduces to the EP distribution when $\nu = 1$ and to the EG distribution when $\nu = 0$ and $\lambda < 1$ (Cordeiro et al., 2012).

*Order Statistics*

Order statistics and their associated properties can prove informative in applications such as reliability theory or quality control. Given a random sample of $n$ ExpCMP random variables, $Y_1, \ldots, Y_n$, the $i$th-order statistic $Y_{(i)}$ has the probability

$$\begin{aligned} P(Y_{(i)} = y) &= \frac{\kappa}{B(\mathrm{i}, n-\mathrm{i}+1)} \left( \sum_{j=0}^{\infty} j w_j e^{-j\kappa y} \right) \\ &\quad \left[ \sum_{\ell=0}^{n-\mathrm{i}} (-1)^{\ell} \binom{n-\mathrm{i}}{\ell} \left( \sum_{m=0}^{\infty} (-w_m e^{-m\kappa y}) \right)^{\mathrm{i}+\ell-1} \right] \\ &= \sum_{j=1}^{\infty} \sum_{m=0}^{\infty} \rho(j,m) f_e(y; (j+m)\kappa), \end{aligned} \tag{3.76}$$

where

$$\rho(j,m) = \frac{j w_j \sum_{\ell=0}^{n-\mathrm{i}} (-1)^{\ell} \binom{n-\mathrm{i}}{\ell} e_{m,\mathrm{i}+\ell-1}}{(j+m) B(\mathrm{i}, n-\mathrm{i}+1)}, \tag{3.77}$$

$B(a,b) = \frac{\Gamma(a)\Gamma(b)}{\Gamma(a+b)}$ is the Beta function, $e_{m,\mathrm{i}+\ell-1} = \frac{1}{m} \sum_{j=1}^{m} [(\mathrm{i}+\ell-1)j - j + m](-w_j) e_{m-j,\mathrm{i}+\ell-1}$, and $w_j$ and $f_e(y;\ \cdot)$ are defined in Equation (3.71). This latter representation of the order statistic density function of the ExpCMP allows for easy determination of the cumulative distribution function, moment generating function, and moments associated with the $i$th-order statistic, namely

$$F_{i:n}(y) = 1 - \sum_{j=1}^{\infty} \sum_{m=0}^{\infty} \rho(j,m) \exp(-(j+m)\kappa y), \tag{3.78}$$

$$M_{Y_{(i)}}(t) = \sum_{j=1}^{\infty} \sum_{m=0}^{\infty} \rho(j,m) \left[ 1 - \frac{t}{(j+m)\kappa} \right]^{-1}, \text{ and} \tag{3.79}$$

$$E(Y_{(i)}^r) = \frac{\Gamma(r+1)}{\kappa^r} \sum_{j=1}^{\infty} \sum_{m=0}^{\infty} \frac{\rho(j,m)}{(j+m)^r}, \tag{3.80}$$

where $\rho(j,m)$ is defined in Equation (3.77).

*Parameter Estimation*

Parameter estimation can be achieved via the method of maximum likelihood, where

$$\ln L(\kappa, \lambda, \nu; \boldsymbol{y}) = n \ln(\kappa\lambda) + \sum_{i=1}^{n} \ln\left(\frac{\partial Z(\lambda e^{-\kappa y_i}, \nu)}{\partial \lambda}\right) - n \ln(Z(\lambda, \nu) - 1) \tag{3.81}$$

is the log-likelihood optimized at the estimates $\{\hat{\kappa}, \hat{\lambda}, \hat{\nu}\}$ given an observed sample $\boldsymbol{y} = (y_1, \ldots, y_n)$. The associated system of score equations is

$$\begin{cases} \frac{n}{\kappa} + \sum_{i=1}^{n} \left(\frac{-y_i Z(\lambda e^{-\kappa y_i}, \nu) E(C_{\dagger i}^2)}{Z(\lambda e^{-\kappa y_i}, \nu) E(C_{\dagger i})}\right) = 0 \\ \frac{n}{\lambda} + \sum_{i=1}^{n} \left(\frac{Z(\lambda e^{-\kappa y_i}, \nu) E(C_{\dagger i}(C_{\dagger i} - 1))}{\lambda Z(\lambda e^{-\kappa y_i}, \nu) E(C_{\dagger i})}\right) - \sum_{i=1}^{n} \left(\frac{Z(\lambda, \nu) E(C_\star)}{\lambda [Z(\lambda, \nu) - 1]}\right) = 0 \\ \sum_{i=1}^{n} \left(\frac{Z(\lambda, \nu) E(\ln(C_\star!))}{Z(\lambda, \nu) - 1}\right) - \sum_{i=1}^{n} \left(\frac{Z(\lambda e^{-\kappa y_i}, \nu) E[C_{\dagger i} \ln(C_{\dagger i}!)]}{Z(\lambda e^{-\kappa y_i}, \nu) E(C_{\dagger i})}\right) = 0 \end{cases} \tag{3.82}$$

which is solved with respect to $\kappa$, $\lambda$, and $\nu$, where $C_\star$ and $C_{\dagger i}$ are CMP($\lambda$, $\nu$) and CMP($\lambda e^{-\kappa y_i}$, $\nu$) random variables, respectively. Cordeiro et al. (2012) utilize the `compoisson` (Dunn, 2012) and `gamlss` (Stasinopoulos and Rigby, 2007) packages in `R` to perform statistical computations, where the RS method (Rigby and Stasinopoulos, 2005) is used in the `gamlss` package to optimize the log-likelihood. The usual distribution theory results hold, i.e. for a collection of parameters $\theta = \{\kappa, \lambda, \nu\}$ with maximum likelihood estimators $\hat{\theta} = \{\hat{\kappa}, \hat{\lambda}, \hat{\nu}\}$, the distribution of $\sqrt{n}(\hat{\theta} - \theta)$ converges to a trivariate normal distribution with zero mean and variance–covariance matrix that is the inverse of the resulting Fisher information matrix. This form allows for analysts to obtain the respective standard errors associated with the estimates and to construct approximate confidence intervals for $\theta$ along with the survival and risk functions.

### 3.6.2 The Weibull–CMP (WCMP) Distribution

The WCMP distribution generalizes the ExpCMP. Let $C \geq 1$ denote the number of components in a series system and assume that $C$ has a zero-truncated CMP($\lambda$, $\nu$) distribution with probability mass function as defined in Equation (3.69). Letting $X_1, \ldots, X_C$ be independent and identically distributed Weibull random variables with scale parameter $1/\kappa$ and shape parameter $\xi$ (density function provided, e.g., in Jiang and Murthy (2011)) and $Y = \min\{X_i\}_{i=1}^{C}$ being the smallest of these random variables, $Y$ has the conditional distribution (given $C = j$)

$$f(y \mid j; \kappa, \xi) = \kappa \xi j (\kappa y)^{\xi - 1} e^{-j(\kappa y)^{\xi}}, \quad y > 0.$$

This combination of distributions compounds to the WCMP distribution and has density, cumulative distribution, and hazard rate functions of the form

$$f(y;\kappa,\lambda,\nu,\xi) = \frac{\kappa\xi(\kappa y)^{\xi-1}}{Z(\lambda,\nu)-1}\sum_{j=1}^{\infty}\frac{j[\lambda e^{-(\kappa y)^{\xi}}]^{j}}{(j!)^{\nu}}$$
$$= \xi(\kappa y)^{\xi-1}\frac{\kappa\lambda\frac{\partial Z(\lambda e^{-(\kappa y)^{\xi}},\nu)}{\partial\lambda}}{Z(\lambda,\nu)-1},\quad y>0, \tag{3.83}$$

$$F(y;\kappa,\lambda,\nu,\xi) = 1 - \frac{Z\left(\lambda e^{-(\kappa y)^{\xi}},\nu\right)-1}{Z(\lambda,\nu)-1} \tag{3.84}$$

$$h(y;\kappa,\lambda,\nu,\xi) = \frac{\xi(\kappa y)^{\xi-1}\kappa Z\left(\lambda e^{-(\kappa y)^{\xi}},\nu\right)E(C_{\ddagger})}{Z\left(\lambda e^{-(\kappa y)^{\xi}},\nu\right)-1} = \frac{\lambda\xi(\kappa y)\frac{\partial Z\left(\lambda e^{-(\kappa y)^{\xi}},\nu\right)}{\partial\lambda}}{Z\left(\lambda e^{-(\kappa y)^{\xi}},\nu\right)-1} \tag{3.85}$$

for a random variable $Y$, where $C_{\ddagger}$ has a CMP$\left(\lambda e^{-(\kappa y)^{\xi}},\nu\right)$ distribution (Gupta and Huang, 2017). The WCMP$(\kappa,\lambda,\nu,\xi)$ distribution allows for a decreasing, increasing, or downward U-shaped distributional form. It further has a strictly decreasing density and hazard rate function when $0<\xi\leq 1$, while the density function is unimodal when $\xi>1$. The WCMP$(\kappa,\lambda,\nu,\xi)$ model generalizes the EG, EP, and ExpCMP distributions; the ExpCMP$(\kappa,\lambda,\nu)$ is achieved when $\xi=1$, while the EG$(\kappa, p = 1-\lambda)$ and EP$(\kappa,\lambda)$ distributions are attained when $\nu=0$ and $\lambda<1$, and $\nu=1$, respectively. Finally, the WCMP distribution has the moment generating function and associated moments

$$M_Y(t) = \sum_{k=0}^{\infty}\left(\frac{t}{\kappa}\right)^{k}\xi\left(1+\frac{k}{\xi}\right)E\left(C_{\star}^{-k/\xi}\right)$$
$$E(Y^r) = \kappa^{-r}\xi\left(\frac{r}{\xi}+1\right)E\left(C_{\star}^{-r/\xi}\right),$$

where $C_{\star}$ has a CMP$(\lambda,\nu)$ distribution.

*Parameter Estimation and Hypothesis Testing*

Maximum likelihood estimation can be performed to estimate $\kappa$, $\lambda$, $\nu$, and $\xi$. The log-likelihood stemming from a random sample $Y_1,\ldots,Y_n$ of WCMP distributed random variables has the form

$$\begin{aligned}\ln L(\kappa, \lambda, \nu, \xi; y) &= n \ln(\lambda\kappa\xi) + (\xi - 1)\sum_{i=1}^{n} \ln(\kappa y_i) \\ &+ \sum_{i=1}^{n} \ln\left(\frac{\partial Z(\lambda e^{-(\kappa y_i)^{\xi}}, \nu)}{\partial \lambda}\right) \\ &- n \ln(Z(\lambda, \nu) - 1). \end{aligned} \tag{3.86}$$

The associated score functions can be obtained and the resulting estimates $\{\hat{\kappa}, \hat{\lambda}, \hat{\nu}, \hat{\xi}\}$ determined such that they solve the system of equations; see Gupta and Huang (2017) for details. The Rao's score test can meanwhile be utilized for hypothesis testing; the corresponding test statistic is $T = S'I^{-1}S$, where $S$ denotes the $4 \times 1$ score vector and $I$ the $4 \times 4$ Fisher information matrix, and each is evaluated at the parameter MLEs under the null hypothesis; $T$ has a chi-squared distribution where the degrees of freedom equal the number of null hypothesis constraints. The respective components that comprise the score vector and information matrix do not have a closed form, thus computations must be conducted via a computing tool, e.g. R. Nonetheless, the usual regularity conditions imply convergence, i.e. $\sqrt{n}(\hat{\boldsymbol{\theta}} - \boldsymbol{\theta}) \rightarrow N_4(\mathbf{0}, \boldsymbol{I}(\boldsymbol{\theta})^{-1})$, where $\boldsymbol{\theta} = \{\kappa, \lambda, \nu, \xi\}$.

Simulation studies show that, given fixed values for $\lambda$ and $\kappa$, $\hat{\nu}$ and $\hat{\xi}$ are close to their true respective values as illustrated through a small absolute bias, and their respective standard errors are close to those obtained via the observed Fisher information matrix. Per the usual, an increased sample size associates with decreased standard errors for the estimates. By fixing one of the two parameters (be it $\nu$ or $\xi$), the coverage probability associated with the confidence interval for the other parameter closely approximates its associated nominal value (Gupta and Huang, 2017).

## 3.7 CMP-Motivated Generalizations of the Negative Hypergeometric Distribution

The negative hypergeometric (NH) distribution is a popular classical distribution recognized as being over-dispersed relative to the Poisson model. As the NB distribution counts the number of successes in a space where draws are done with replacement and assuming a certain number of failures, the NH distribution is derived when those draws are made without replacement.

The NH distribution can be derived from a NB random variable conditioned on the sum of two independent NB variables with the same success probability. As described in Section 3.4, however, the

NB distribution can allow for several generalizations motivated by the COM–Poisson distribution to account for added dispersion. This section accordingly considers some COM–Poisson-inspired NH distributions derived from such conditional probabilities in regard to their corresponding COM–Poisson-inspired NB random variables – a COM-negative hypergeometric (COMNH) (Section 3.7.1), COM–Poisson-type negative hypergeometric (CMPtNH) (Section 3.7.2), and Conway-Maxwell negative hypergeometric (CMNH) (Section 3.7.3) distributions.

### *3.7.1 The COM-negative Hypergeometric (COMNH) Distribution, Type I*

The Type I COM-negative hypergeometric (COMNH) distribution (Zhang et al., 2018) has the probability mass function

$$P(Y=y)=\frac{1}{N(h,\nu,\gamma_1,\gamma_2)}\left(\binom{h}{y}\frac{B(\gamma_1+y,\gamma_2+h-y)}{B(\gamma_1,\gamma_2)}\right)^{\nu},\quad y=0,1,\ldots,h \tag{3.87}$$

where $B(\alpha,\beta)=\frac{\Gamma(\alpha)\Gamma(\beta)}{\Gamma(\alpha+\beta)}$ denotes the Beta coefficient for two terms $\alpha$ and $\beta$, and $N(h,\nu,\gamma_1,\gamma_2)=\sum_{y=0}^{h}\left(\binom{h}{y}\frac{B(\gamma_1+y,\gamma_2+h-y)}{B(\gamma_1,\gamma_2)}\right)^{\nu}$ is the normalizing constant. The COMNH$(h,\nu,\gamma_1,\gamma_2)$ reduces to the NH distribution when $\nu=1$.

Two substantive results connect the COMNB and COMNH distributions. The first result says that two independent random variables, $X_i$, $i=1,2$, are COMNB$(\gamma_i,\nu,p)$ distributed as defined in Section 3.4.2 if and only if the conditional distribution of $X_1$ given the sum $X_1+X_2=x_1+x_2$ has a COMNH$(h=x_1+x_2,\nu,\gamma_1,\gamma_2)$ for all $h=x_1+x_2$. Secondly, a COMNH$(h,\nu,\gamma_1,\gamma_2)$ distributed random variable with finite $\gamma_1$ and $h/(\gamma_2+h)=p^{1/\nu}$ where $p\in(0,1)$ converges to a COMNB$(\gamma_1,\nu,p)$ distribution as $\gamma_2\to\infty$.

### *3.7.2 The COM–Poisson-type Negative Hypergeometric (CMPtNH) Distribution*

The COM–Poisson-type negative hypergeometric (i.e. CMPtNH $(s,\eta,\xi_1,\xi_2)$) distribution (Chakraborty and Ong, 2016) has the probability mass function

$$P(X=x)=\frac{1}{(C(s,\eta,\xi_1,\xi_2))}\frac{\xi_1^{(x)}\xi_2^{(s-x)}}{(\xi_1+\xi_2)^{(s)}}\binom{s}{x}^{\eta},\quad x=0,1,\ldots,s,$$

where $C(s, \eta, \xi_1, \xi_2) = \sum_{x=0}^{s} \frac{\xi_1^{(x)} \xi_2^{(s-x)}}{(\xi_1+\xi_2)^{(s)}} \binom{s}{x}^{\eta}$ denotes the normalizing constant, and $a^{(b)}$ denotes the rising factorial for values $a, b$ as defined in Equation (3.44). For $\eta = 1$, this distribution reduces to the NH distribution. Meanwhile, the CMPtNH distribution converges to a $\text{CMB}\left(s, \frac{\xi_1}{\xi_1+\xi_2}, \eta\right)$ distribution (see Section 3.3.1 for details) when $\xi_1 + \xi_2 \to \infty$ and $0 < \frac{\xi_1}{\xi_1+\xi_2} < \infty$. The CMPtNH distribution likewise converges to the COMtNB($\xi_1, p, \eta$) distribution as described in Section 3.4.3 when $\eta \geq 1$, $s^{\eta} \to \infty$, $\frac{1}{\xi_2} \to 0$, and $\frac{s^{\eta}}{\xi_2} = p$.

Chakraborty and Ong (2016) provide two derivations for the CMPtNH model – either as a conditional distribution, or via a compounding method. For two independent COMtNB($\xi_i, p, \eta$) random variables $X_i$, $i = 1, 2$, the conditional distribution of $X_1 = x \mid X_1 + X_2 = s$ is CMPtNH($s, \eta, \xi_1, \xi_2$) distributed. Alternatively, the CMPtNH distribution can be derived as a compounded CMB($s, p, \eta$) where $p$ is Beta($\xi_1, \xi_2$) distributed.

### 3.7.3 The COM-Negative Hypergeometric (CMNH) Distribution, Type II

Roy et al. (2020) offers the most extensive effort generalizing a NH distribution, introducing the Type II COM-negative hypergeometric (CMNH) distribution that further includes the COMNB and CMP distributions as limiting cases. The CMNH($r, \nu, mp, m$) has the probability function

$$P(X = x) = \frac{\binom{x+r-1}{x}^{\nu} \binom{m-x-r}{m-mp-x}}{H(r, \nu, np, m)}, \quad x = 0, 1, \ldots, m - mp, \tag{3.88}$$

where $H(r, \nu, mp, m) = \sum_{j=0}^{m-mp} \binom{j+r-1}{j}^{\nu} \binom{m-j-r}{m-mp-j}$ is the normalizing constant. This distribution is log-concave with an increasing failure rate when $\nu > 0$ and $1 < r < mp$, and has a U-shaped failure rate when $\nu < 0$. For the special case where $\nu = 1$, the CMNH reduces to the NH model with probability

$$P(X = x) = \frac{\binom{x+r-1}{x} \binom{m-x-r}{m-mp-x}}{\binom{m}{m-mp}}, \quad x = 0, 1, \ldots, m - mp.$$

Meanwhile, as $m \to \infty$, CMNH converges to the COMNB($r, \nu, q$) model as described in Section 3.4.2, which (in turn) converges to CMP($\lambda, \nu$) when $r \to \infty$ and $q \to 0$ simultaneously such that $\lambda = r^{\nu} q$ remains fixed (Roy et al., 2020).

Dispersion is controlled by the combination of CMNH parameters. For $\nu = 1$, the NH distribution is always over-dispersed. Through simulation studies, however, Roy et al. (2020) note that the CMNH distribution can achieve data over- or under-dispersion for certain combinations of its parameters. In particular, the degree of over-dispersion decreases as $r$ increases while, for certain combinations of $r$ and $\nu$, the CMNH can achieve data under-dispersion. Likewise, for certain $r$ fixed, it can achieve data under-dispersion with certain combinations of $p$ and $\nu$.

The CMNH probability generating function is

$$\Pi(s) = \frac{{}_{\nu+1}F_{\nu}(r, r, \ldots, r, mp - m; 1, 1, \ldots, 1, r - m; s)}{{}_{\nu+1}F_{\nu}(r, r, \ldots, r, mp - m; 1, 1, \ldots, 1, r - m; 1)}$$

with mean and variance

$$E(X) = \frac{r^{\nu}(m - mp)}{m - r} \frac{{}_{\nu+1}F_{\nu}(r + 1, r + 1, \ldots, r + 1, mp - m + 1; 2, 2, \ldots, 2, r - m + 1; 1)}{{}_{\nu+1}F_{\nu}(r, r, \ldots, r, mp - m; 1, 1, \ldots, 1, r - m; 1)}$$

$$V(X) = E(X(X - 1)) + E(X) - (E(X))^2,$$

where

$$E(X(X - 1)) = \frac{(r(r + 1))^{\nu}(m - mp)(m - mp - 1)}{(m - r)(m - r - 1)2^{\nu-1}} \cdot \frac{{}_{\nu+1}F_{\nu}(r + 2, r + 2, \ldots, r + 2, mp - m + 2; 3, 3, \ldots, 3, r - m + 2; 1)}{{}_{\nu+1}F_{\nu}(r, r, \ldots, r, mp - m; 1, 1, \ldots, 1, r - m; 1)}.$$

More broadly, the $k$th moment can be expressed as

$$E(X^k) = \frac{r^{\nu}(m - mp)}{mp + 1} E\left(\frac{1}{X}\right)^{\nu-k}.$$

The mean and variance for the CMNH distribution do not have a closed form except for the special case where $\nu = 1$, i.e. the NH distribution.

Parameter estimation can be achieved via maximum likelihood estimation where the log-likelihood associated with observations $\boldsymbol{x} = (x_1, \ldots, x_n)$ from an assumed CMNH random sample is

$$\ln L(\nu, p; \boldsymbol{x}) = \sum_{i=1}^{n} \nu \ln \binom{x_i + r - 1}{x_i} + \sum_{i=1}^{n} \ln \binom{m - x_i - r}{m - mp - x_i} - n \ln \sum_{j=0}^{m-mp} \binom{j + r - 1}{j}^{\nu} \binom{m - j}{m - mp - j + r}, \quad (3.89)$$

where the resulting score equations do not have closed-form solutions, thus the estimates $\hat{\nu}, \hat{p}$ can be attained numerically via the `optim` function in `R`. Roy et al. (2020) show that, for $\nu > 0$, $\hat{\nu}$ has small standard error with a smaller bias and better coverage probability when $p \geq 0.5$. Meanwhile, $\hat{p}$ has a small bias and small standard error when $\nu > 0$. They see, however, that for large, positive $\nu$ and small $p$, the coverage probability of $p$ is less optimal. The test of hypotheses $H_0{:}\nu = 1$ versus $H_1{:}\nu \neq 1$ determines whether an observed sample stem from a NH distribution or is statistically significantly different from NH such that the CMNH should be considered, recognizing some measure of data dispersion. The resulting likelihood ratio test statistic is $\Lambda = \frac{L(\hat{p}_0, \nu=1 \mid \boldsymbol{x})}{L(\hat{p}, \hat{\nu} \mid \boldsymbol{x})}$ where, under $H_0$, $-2\ln(\Lambda)$ converges to a chi-squared distribution with one degree of freedom.

## 3.8 Summary

The COM–Poisson distribution has motivated an array of various distributional extensions, including flexible analogs of the Skellam, binomial, multinomial, NB, and NH distributions, etc. As a result, these distributional extensions can model categorical data with a fixed number of trials (CMB, gCMB, CMM), unbounded (sCMP, GCMP, COMNB, COMtNB, ECMP) or bounded (COMNH, CMPtNH, CMNH) count data, differences in counts (CMS), or life length of systems (ExpCMP, WCMP). The respective distributions maintain distributional relationships that are consistent with their classical counterparts; yet, they allow for added flexibility to account for data dispersion.

Unfortunately, only a select few of the discussed models provide a supporting `R` package. Interested analysts, however, can utilize the associated section discussion to develop `R` functionality for the model(s) of interest.

# 4

# Multivariate Forms of the COM–Poisson Distribution

Multivariate count data analysis is gaining interest throughout society, and such data require multivariate model development that takes into account the discrete nature while likewise accounting for data dispersion. Given the potential dependence structures that can exist in a multivariate setting, research has typically been established first for a bivariate distribution whose definition has then been broadened to satisfy a multivariate framework. A bivariate Poisson (BP) distribution, for example, is a natural choice for modeling count data stemming from two correlated random variables. This construct, however, is limited by the underlying univariate model assumption that the data are equi-dispersed. Alternative bivariate models include a bivariate negative binomial (BNB) and a bivariate generalized Poisson (BGP) distribution, which themselves suffer from analogous limitations as described in Chapter 1. The BNB of Famoye (2010), for example, cannot address data under-dispersion, while the BGP distribution of Famoye and Consul (1995) has limited capacity to address data under-dispersion. Another form of dispersion is measured via a generalized dispersion index (GDI) (Kokonendji and Puig, 2018). For a random vector $\boldsymbol{X} = (X_1, \ldots, X_d)$, the GDI is measured as

$$\mathrm{GDI}(X) = \frac{(\sqrt{E(\boldsymbol{X})})'(\mathrm{Cov}(\boldsymbol{X}))\sqrt{E(\boldsymbol{X})}}{(E(\boldsymbol{X}))'E(\boldsymbol{X})}, \tag{4.1}$$

where $\mathrm{Cov}(\boldsymbol{X})$ denotes the variance–covariance matrix of $\boldsymbol{X}$. The GDI assesses dispersion relative to the uncorrelated multivariate Poisson (MP) distribution, where $\mathrm{GDI} = 1$ implies equi-dispersion, and $\mathrm{GDI} > (<) 1$ infer over-dispersion (under-dispersion) relative to the uncorrelated MP distribution.

While the aforementioned distributions motivate the need to instead consider a multivariate analog of the univariate CMP, such model development (as with those for other discrete distributions) varies in order to take into account (or results in) certain distributional qualities. A variety of approaches have been proposed in the literature to develop multivariate discrete distributions; see Kocherlakota and Kocherlakota (1992) and Johnson et al. (1997) for discussions regarding potential approaches to develop bivariate and multivariate discrete distributions, respectively. This chapter summarizes such efforts where, for each approach, readers will first learn about any bivariate COM–Poisson (BCMP) distribution formulations, followed by any multivariate analogs. Accordingly, because these models are multidimensional generalizations of the univariate COM–Poisson, they each contain their analogous forms of the Poisson, Bernoulli, and geometric distributions as special cases. The methods discussed in this chapter are the trivariate reduction method (Section 4.1), the compounding method (Section 4.2), the Sarmanov family of distributions (Section 4.3), and the use of copulas (Section 4.4). These methods introduce differing attributes associated with their model forms; thus, the following discussion will maintain notation that not only notes the dimensionality (be it bivariate or multivariate) but also the approach by which the distribution is established in order to provide the reader with added clarity regarding the multiple forms (for example) of BCMP distributions. Section 4.5 illustrates and compares model performance for these various model constructs on two real datasets. Finally, Section 4.6 concludes the chapter with a summary and discussion.

While bivariate and multivariate COM–Poisson distributions can be developed motivated by any of the parametrizations discussed in Section 2.6, all of the presented models considered in this chapter assume an underlying CMP model; hence, we maintain the general acronyms BCMP and MultiCMP,[1] respectively, throughout this chapter to denote bivariate and multivariate COM–Poisson distributions.

## 4.1 Trivariate Reduction

The trivariate reduction (also referred to as the "variables in common") method is a popular approach for developing bivariate count distributions. Letting $W_1, W_2, W_{12}$ denote independently distributed random variables

[1] While notation recognizing multivariate distributions will typically begin simply with "M," we avoid doing so in reference to the multivariate CMP distribution to prevent confusion with the general use of MCMP to denote mean-parametrized COM–Poisson models.

(with or without commonly associated parameters), the trivariate reduction approach sets (say)

$$\begin{aligned} X_1 &= W_1 + W_{12} \\ X_2 &= W_2 + W_{12} \end{aligned}$$

to derive the joint probability $P(X_1 = x_1, X_2 = x_2)$ that is mutually dependent on $W_{12}$. Weems et al. (2021) developed a BCMP distribution via the trivariate reduction method, i.e. letting $W_k$ be CMP$(\lambda_k, \nu)$ distributed[2] $(k = 1, 2)$ as described in Chapter 2 and $W_{12}$ likewise be CMP$(\lambda_{12}, \nu)$. Under this construct, the resulting joint probability is

$$\begin{aligned} P(X_1 = x_1, X_2 = x_2) &= \frac{\lambda_1^{x_1}\lambda_2^{x_2}}{Z(\lambda_1, \nu)Z(\lambda_2, \nu)Z(\lambda_{12}, \nu)(x_1!x_2!)^{\nu}} \\ &\quad \times \sum_{j=0}^{\min(x_1, x_2)} \left(\frac{\lambda_{12}}{\lambda_1\lambda_2}\right)^j \left[\binom{x_1}{j}\binom{x_2}{j} j!\right]^{\nu} \end{aligned} \tag{4.2}$$

with the joint probability generating function (pgf)

$$\Pi(t_1, t_2) = \frac{Z(\lambda_1 t_1, \nu)}{Z(\lambda_1, \nu)} \cdot \frac{Z(\lambda_2 t_2, \nu)}{Z(\lambda_2, \nu)} \cdot \frac{Z(\lambda_{12} t_1 t_2, \nu)}{Z(\lambda_{12}, \nu)}.$$

The trivariate-reduced bivariate CMP (BCMPtriv) distribution contains the Holgate (1964) BP model as a special case when $\nu = 1$, a trivariate-reduced bivariate geometric ($\nu = 0$; $\lambda_k < 1$, $k = 1, 2$; $\lambda_{12} < 1$) distribution, and a trivariate-reduced bivariate Bernoulli ($\nu \to \infty$) distribution, respectively.

More broadly, the BCMPtriv has marginal pgfs

$$\Pi_{X_k}(t) = \frac{Z(\lambda_k t, \nu)Z(\lambda_{12} t, \nu)}{Z(\lambda_k, \nu)Z(\lambda_{12}, \nu)}, \quad k = 1, 2, \tag{4.3}$$

which demonstrate that $X_k$ has Poisson$(\lambda_k + \lambda_{12})$ marginals as a special case when $\nu = 1$. Meanwhile, for $\lambda_1 = \lambda_2 = \lambda_{12} = \lambda$, the marginal pgfs simplify to $\left(\frac{Z(\lambda t, \nu)}{Z(\lambda, \nu)}\right)^2$, i.e. $X_1$ and $X_2$ are then sCMP$(\lambda, \nu, 2)$ distributed. While these marginal distributions are not themselves CMP distributed, they contain the Poisson$(2\lambda)$, binomial$\left(2, \frac{\lambda}{1+\lambda}\right)$, and negative binomial (NB)$(2, 1-\lambda)$ distributions as special cases. The respective marginal (for $k = 1, 2$) and joint moment generating functions produce the means, variances, and covariance

[2] A more flexible CMP$(\lambda_k, \nu_k)$ representation can likewise be considered but introduces greater mathematical complexity.

$$E(X_k) = \lambda_k \frac{\partial \ln Z(\lambda_k, \nu)}{\partial \lambda_k} + \lambda_{12} \frac{\partial \ln Z(\lambda_{12}, \nu)}{\partial \lambda_{12}} \tag{4.4}$$

$$\approx \lambda_k^{1/\nu} + \lambda_{12}^{1/\nu} - \frac{\nu - 1}{\nu} \tag{4.5}$$

$$E(X_k^2) = \lambda_k \frac{\partial \ln Z(\lambda_k, \nu)}{\partial \lambda_k} + \frac{\lambda_k^2}{Z(\lambda_k, \nu)} \frac{\partial^2 Z(\lambda_k, \nu)}{\partial \lambda_k^2} + \lambda_{12} \frac{\partial \ln Z(\lambda_{12}, \nu)}{\partial \lambda_{12}}$$

$$+ \frac{\lambda_{12}^2}{Z(\lambda_{12}, \nu)} \frac{\partial^2 Z(\lambda_{12}, \nu)}{\partial \lambda_{12}^2} + \lambda_k \lambda_{12} \frac{\partial \ln Z(\lambda_k, \nu)}{\partial \lambda_k} \frac{\partial \ln Z(\lambda_{12}, \nu)}{\partial \lambda_{12}} \tag{4.6}$$

$$V(X_k) = \frac{\partial E(W_k)}{\partial \ln \lambda_k} + \frac{\partial E(W_{12})}{\partial \ln \lambda_{12}} \approx \frac{1}{\nu} \left( \lambda_k^{1/\nu} + \lambda_{12}^{1/\nu} \right) \tag{4.7}$$

$$E(X_1 X_2) = \lambda_1 \lambda_2 \frac{\partial \ln Z(\lambda_1, \nu)}{\partial \lambda_1} \frac{\partial \ln Z(\lambda_2, \nu)}{\partial \lambda_2} + \lambda_1 \lambda_{12} \frac{\partial \ln Z(\lambda_1, \nu)}{\partial \lambda_1} \frac{\partial \ln Z(\lambda_{12}, \nu)}{\partial \lambda_{12}}$$

$$+ \lambda_2 \lambda_{12} \frac{\partial \ln Z(\lambda_2, \nu)}{\partial \lambda_2} \frac{\partial \ln Z(\lambda_{12}, \nu)}{\partial \lambda_{12}}$$

$$+ \lambda_{12} \frac{\partial \ln Z(\lambda_{12}, \nu)}{\partial \lambda_{12}} + \frac{\lambda_{12}^2}{Z(\lambda_{12}, \nu)} \frac{\partial^2 Z(\lambda_{12}, \nu)}{\partial \lambda_{12}^2} \tag{4.8}$$

$$\mathrm{Cov}(X_1, X_2) = E(X_1 X_2) - E(X_1) E(X_2) \tag{4.9}$$

$$\approx \frac{\nu - 1}{2\nu} \left[ 3 \left( \frac{\nu - 1}{2\nu} \right) - \lambda_1^{1/\nu} - \lambda_2^{1/\nu} - 2\lambda_3^{1/\nu} \right] + \frac{1}{\nu} \lambda_3^{1/\nu} \tag{4.10}$$

from which the correlation and GDI can be obtained. While these measures do not offer easily represented closed forms, one can recognize from the bivariate construction and univariate CMP distributional form that the correlation between $X_1$ and $X_2$ is nonnegative. Further, the mean, variance, and covariance approximations can be used to determine an approximate GDI. For example, when $\nu = 1$, the GDI for the BCMPtriv distribution reduces to the GDI of the Holgate (1964) BP distribution,

$$\mathrm{GDI}_{\mathrm{BCMPtriv}(\nu=1)}(X_1, X_2) = \mathrm{GDI}_{\mathrm{BP}}(X_1, X_2) = 1 + \frac{2\lambda_{12}\sqrt{\lambda_1 + \lambda_{12}}\sqrt{\lambda_2 + \lambda_{12}}}{(\lambda_1 + \lambda_{12})^2 + (\lambda_2 + \lambda_{12})^2} \geq 1,$$

implying that the BCMPtriv when $\nu = 1$ is equi-dispersed (over-dispersed) if and only if $\lambda_{12} = (>)\, 0$.

### 4.1.1 Parameter Estimation

Weems et al. (2021) consider parameter estimation via a modified method of moments approach and the method of maximum likelihood to determine estimates for the BCMPtriv parameters, i.e. $\lambda_1$, $\lambda_2$, $\lambda_{12}$, and $\nu$. While the method of moments estimators are typically obtained by equating the system of true moments (Equations (4.4)–(4.8)) with their corresponding sampling estimators, $\overline{X_k}$, $\overline{X_k^2}$, and $\overline{X_1X_2}$, the lack of a closed-form solution motivates a modified approach that approximates the method of moments estimates by minimizing the squared-error loss function

$$\ell(\lambda_1, \lambda_2, \lambda_{12}, \nu; (\boldsymbol{x}_1, \boldsymbol{x}_2)) = \left(E(X_1) - \overline{X_1}\right)^2 + \left(E(X_2) - \overline{X_2}\right)^2 + \left(E(X_1^2) - \overline{X_1^2}\right)^2 + \left(E(X_2^2) - \overline{X_2^2}\right)^2 + \left(E(X_1X_2) - \overline{X_1X_2}\right)^2. \quad (4.11)$$

The maximum likelihood estimators meanwhile optimize the log-likelihood function,

$$\ln L(\lambda_1, \lambda_2, \lambda_{12}, \nu; (\boldsymbol{x}_1, \boldsymbol{x}_2)) = \sum_{k=1}^{2}\sum_{i=1}^{n} x_{ki} \ln \lambda_k - n \sum_{k=1}^{2} \ln Z(\lambda_k, \nu) - n \ln Z(\lambda_{12}, \nu) - \nu \sum_{k=1}^{2}\sum_{i=1}^{n} \ln (x_{ki}!) + \sum_{i=1}^{n} \ln \left( \sum_{j=0}^{\min(x_{1i}, x_{2i})} \left(\frac{\lambda_{12}}{\lambda_1\lambda_2}\right)^j \left[\binom{x_{1i}}{j}\binom{x_{2i}}{j} j!\right]^{\nu} \right). \quad (4.12)$$

Both functions are optimized via the `optim` function with constraints, $\nu \geq 0$, and positive-valued $\lambda_1$, $\lambda_2$, and $\lambda_{12}$. The BP estimates can serve as starting values in both cases, or the method of moments estimates can likewise serve as starting values for the maximum likelihood approach. The corresponding standard errors associated with the respective estimates can meanwhile be determined with the aid of the approximate Hessian matrix available in the `optim` output. Simulation studies showed good performance by both estimators, with the method of maximum likelihood outperforming the modified method of moments (Weems et al., 2021).

### *4.1.2 Hypothesis Testing*

Respective hypothesis tests can address important questions of interest; in both cases, analysts can use the likelihood ratio test to perform testing. The first test considers the hypotheses $H_0: \nu = 1$ versus $H_0: \nu \neq 1$, investigating whether statistically significant data dispersion exists such that the BCMPtriv distribution is preferred over the BP distribution to model a given dataset. For this scenario, the resulting test statistic is

$$\Lambda_\nu = \frac{\max L(\lambda_1, \lambda_2, \lambda_{12}, \nu = 1)}{\max L(\lambda_1, \lambda_2, \lambda_{12}, \nu)}, \quad (4.13)$$

where the numerator and denominator of Equation (4.13) denote the respective maximized likelihood values under $H_0$ and in general. The null hypothesis represents the BP special case; thus, associated statistical computing can be conducted via the `bivpois` package (Karlis and Ntzoufras, 2008) in R. The `lm.bp` function provides the maximum likelihood estimates (MLEs) (reported outputs `lambda1` = $\hat{\lambda}_1$, `lambda2` = $\hat{\lambda}_2$, `lambda3` = $\hat{\lambda}_{12}$) and the associated log-likelihood value (`loglikelihood`) in the subsequent output. Meanwhile, the MLEs and associated log-likelihood for the general BCMPtriv model can be determined, e.g. via the `optim` function in R where the function to be minimized is the negated form of Equation (4.12).

The second test of interest considers the hypotheses, $H_0\colon \lambda_{12} = 0$ versus $H_1\colon \lambda_{12} \neq 0$, assessing whether there exists a statistically significant relation between $X_1$ and $X_2$. Here, the associated test statistic is

$$\Lambda_{12} = \frac{\max L(\lambda_1, \lambda_2, \lambda_{12} = 0, \nu)}{\max L(\lambda_1, \lambda_2, \lambda_{12}, \nu)}, \tag{4.14}$$

where the denominator remains the likelihood value associated with the BCMPtriv distribution evaluated at the MLEs ($\hat{\lambda}_1$, $\hat{\lambda}_2$, $\hat{\lambda}_{12}$, $\hat{\nu}$), while the numerator is the maximized likelihood evaluated at its estimates given the constraint that $\lambda_{12} = 0$. Both of these results can be attained via the `optim` function. Under each scenario with the usual regularity conditions assumed true, the respective $-2 \ln \Lambda$ variables converge to a chi-squared distribution with one degree of freedom, where $\Lambda_\nu$ and $\Lambda_{12}$ are defined in Equations (4.13) and (4.14). These convergence results aid in determining the associated respective $p$-values to draw inference regarding the hypothesis tests.

As a follow-up from the first test, the extent to which statistically significant dispersion exists can then introduce the question whether a trivariate-reduced bivariate Bernoulli ($H_0\colon \nu \to \infty$) or bivariate geometric ($H_0\colon \nu = 0$) distribution is more appropriate. These tests can be considered via a likelihood ratio test statistic whose numerator is the maximized log-likelihood for the null-hypothesized special-case distribution, and the denominator is the maximized log-likelihood for the general BCMPtriv distribution. These tests converge to a mixture between a point mass and the $\chi^2_1$ distribution to account for the respective boundary cases for $\nu$.

### 4.1.3 Multivariate Generalization

This trivariate reduction approach is easily generalizable to $d$ dimensions by defining

$$X_i = W_i + W, \qquad i = 1, 2, \ldots, d, \tag{4.15}$$

where $W_i$, $i = 1, \ldots, d$, and $W$ are mutually independent CMP random variables with varying intensity values, $\lambda_i$, $i = 1, \ldots, d$, and $\lambda$, respectively, and a common dispersion parameter, $\nu$. This trivariate-reduced multivariate CMP (MultiCMPtriv) construct likewise contains marginal distributions that are Poisson distributed with respective intensity parameters, $\lambda_i + \lambda$, $i = 1, \ldots, d$, when $\nu = 1$. Meanwhile, for the special case where $\lambda_1 = \cdots = \lambda_d = \lambda$, $X_1, \ldots, X_d$ are each sCMP$(\lambda, \nu, 2)$ marginally distributed; refer to Section 3.2 for discussion regarding the sCMP distribution. The MultiCMPtriv marginal and joint first moments are, for $i, j \in \{1, \ldots, d\}$, where $i \neq j$,

$$E(X_i) = \lambda_i \frac{\partial \ln Z(\lambda_i, \nu)}{\partial \lambda_i} + \lambda \frac{\partial \ln Z(\lambda, \nu)}{\partial \lambda} \tag{4.16}$$

$$\begin{aligned} E(X_i^2) = {} & \lambda_i \frac{\partial \ln Z(\lambda_i, \nu)}{\partial \lambda_i} + \frac{\lambda_i^2}{Z(\lambda_i, \nu)} \frac{\partial^2 Z(\lambda_i, \nu)}{\partial \lambda_i^2} + \lambda \frac{\partial \ln Z(\lambda, \nu)}{\partial \lambda} \\ & + \frac{\lambda^2}{Z(\lambda, \nu)} \frac{\partial^2 Z(\lambda, \nu)}{\partial \lambda^2} + \lambda_i \lambda \frac{\partial \ln Z(\lambda_i, \nu)}{\partial \lambda_i} \frac{\partial \ln Z(\lambda, \nu)}{\partial \lambda} \end{aligned} \tag{4.17}$$

$$\begin{aligned} E(X_i X_j) = {} & \lambda_i \lambda_j \frac{\partial \ln Z(\lambda_i, \nu)}{\partial \lambda_i} \frac{\partial \ln Z(\lambda_j, \nu)}{\partial \lambda_j} + \lambda_i \lambda \frac{\partial \ln Z(\lambda_i, \nu)}{\partial \lambda_i} \frac{\partial \ln Z(\lambda, \nu)}{\partial \lambda} \\ & + \lambda_j \lambda \frac{\partial \ln Z(\lambda_j, \nu)}{\partial \lambda_j} \frac{\partial \ln Z(\lambda, \nu)}{\partial \lambda} + \lambda \frac{\partial \ln Z(\lambda, \nu)}{\partial \lambda} + \frac{\lambda^2}{Z(\lambda, \nu)} \frac{\partial^2 Z(\lambda, \nu)}{\partial \lambda^2} \end{aligned} \tag{4.18}$$

which again aid in determining the marginal variances, covariances, and correlations. Under the $d$-dimensional form, the respective correlations remain nonnegative.

The method of moments and maximum likelihood estimations as described above are generalizable for the $d$-dimensional analog. Similarly, the hypothesis tests for data dispersion and dependence still hold in concept for the multivariate distribution. The likelihood ratio test statistic $-2 \ln \Lambda$ associated with the hypotheses $H_0: \nu = 1$ versus $H_1: \nu \neq 1$ now equals twice the difference between the respective maximized log-likelihood values associated with the MultiCMPtriv and multivariate Poisson (MP) distributions. Similarly, testing $H_0: \lambda = 0$ versus $H_1: \lambda \neq 0$ produces the test statistic $-2 \ln \Lambda$ that equals twice the difference between the maximized log-likelihood values stemming from the MultiCMPtriv distributions that are structured as described in Equation (4.15) versus its independent construct, respectively.

## 4.2 Compounding Method

While the BCMPtriv model when $\nu = 1$ simplifies to the Holgate (1964) BP, it does not likewise contain the other bivariate special cases (i.e. bivariate Bernoulli and geometric distributions as described in Marshall and Olkin (1985)). Alternatively, the BCMP distribution obtained via the compounding method (BCMPcomp) has the joint probability

$$p(x_1, x_2) = \frac{1}{Z(\lambda, \nu)} \sum_{n=0}^{\infty} \frac{\lambda^n}{(n!)^\nu} \times \sum_{a=n-x_1-x_2}^{n} \binom{n}{a, n-a-x_2, n-a-x_1, x_1+x_2+a-n} p_{00}^{a} p_{10}^{n-a-x_2} p_{01}^{n-a-x_1} p_{11}^{x_1+x_2+a-n}, \tag{4.19}$$

where $p(x_1, x_2) \doteq P(X_1 = x_1, X_2 = x_2)$ and $p_{00} + p_{01} + p_{10} + p_{11} = 1$, and it introduces a flexible bivariate model that contains all three special bivariate cases. Let $(X_1, X_2 \mid n)$ denote a joint conditional bivariate binomial distribution with pgf

$$\Pi_{X_1,X_2}(t_1, t_2 \mid n) = (p_{00} + p_{10}t_1 + p_{01}t_2 + p_{11}t_1t_2)^n,$$

where $n$ denotes the number of trials and assumes a CMP($\lambda$, $\nu$) distribution. These definitions produce the joint unconditional pgf

$$\Pi_{X_1,X_2}(t_1, t_2) = \frac{Z\left(\lambda[1 + p_{1+}(t_1 - 1) + p_{+1}(t_2 - 1) + p_{11}(t_1 - 1)(t_2 - 1)], \nu\right)}{Z(\lambda, \nu)} \tag{4.20}$$

for $(X_1, X_2)$, where $Z(\psi, \nu) = \sum_{s=0}^{\infty} \frac{\psi^s}{(s!)^\nu}$ is the usual notation for the CMP($\psi$, $\nu$) normalizing constant for some intensity $\psi > 0$ and dispersion $\nu \geq 0$. For the special case where $p_{ij} = p_{i+}p_{+j}$, for $i, j = 0, 1$, Equation (4.20) simplifies to

$$\Pi_{X_1,X_2}(t_1, t_2) = \frac{Z\left(\lambda[1 + p_{1+}(t_1 - 1)][1 + p_{+1}(t_2 - 1)], \nu\right)}{Z(\lambda, \nu)}. \tag{4.21}$$

The BCMPcomp($\lambda$, $\nu$, $\boldsymbol{p} = (p_{00}, p_{01}, p_{10}, p_{11})$) distribution contains three special-case bivariate discrete distributions. This construct contains the Holgate (1964) BP[3] with parameters $\lambda$, $p_{00}$, $p_{01}$, $p_{10}$, $p_{11}$, where

[3] The Holgate (1964) BP distribution has the same form whether obtained via trivariate reduction or via the compounding method: $\lambda_1 + \lambda_3 = \lambda p_{1+}$, $\lambda_2 + \lambda_3 = \lambda p_{+1}$, and $\lambda_3 = \lambda p_{11}$.

$p_{00} + p_{01} + p_{10} + p_{11} = 1$, $p_{j+} = p_{j0} + p_{j1}$, and $p_{+j} = p_{0j} + p_{1j}$, for $j = 0, 1$, when $\nu = 1$; the bivariate Bernoulli distribution described in Marshall and Olkin (1985) with $p_{00}^* = \frac{1+\lambda p_{00}}{1+\lambda}$, $p_{10}^* = \frac{\lambda p_{10}}{1+\lambda}$, $p_{01}^* = \frac{\lambda p_{01}}{1+\lambda}$, and $p_{11}^* = \frac{\lambda p_{11}}{1+\lambda}$ as $\nu \to \infty$; and a bivariate geometric distribution when $\nu = 0$, $\lambda < 1$ and $\lambda \{p_{1+}(t_1 - 1) + p_{+1}(t_2 - 1) + p_{11}(t_1 - 1)(t_2 - 1)\} < 1$. While $X_1$ and $X_2$ do not have CMP marginal distributions, they contain respective Poisson distributions as special cases when $\nu = 1$; $X_1$ and $X_2$ are, respectively, Poisson($\lambda p_{1+}$) and Poisson($\lambda p_{+1}$) distributed when $\nu = 1$ with the associated marginal pgfs

$$\Pi_{X_1}(t) = \frac{Z[\lambda\{1 + p_{1+}(t-1)\}, \nu]}{Z(\lambda, \nu)} \quad \text{and}$$
$$\Pi_{X_2}(t) = \frac{Z[\lambda\{1 + p_{+1}(t-1)\}, \nu]}{Z(\lambda, \nu)}. \tag{4.22}$$

While Equations (4.21) and (4.22) demonstrate some similarity when $p_{ij} = p_{i+}p_{+j}$, for $i, j = 0, 1$, it verifies that this constraint does not imply an independence result for $X_1$ and $X_2$.

The joint, marginal, and factorial moment generating functions and cumulant generating function can analogously be attained to aid in determining the distributional probabilities, moments, factorial moments, and cumulants of interest. Defining $Z^{(k)}(\psi, \nu) \doteq \frac{\partial^k}{\partial \psi^k} Z(\psi, \nu)$ for a general intensity $\psi > 0$, the marginal and product means for the BCMPcomp distribution, along with the variances and covariance, are

$$E(X_1) = \lambda p_{1+} \left\{ \frac{\partial \ln Z(\lambda, \nu)}{\partial \lambda} \right\}, \tag{4.23}$$

$$E(X_2) = \lambda p_{+1} \left\{ \frac{\partial \ln Z(\lambda, \nu)}{\partial \lambda} \right\}, \tag{4.24}$$

$$E(X_1 X_2) = \frac{Z''(\lambda, \nu)}{Z(\lambda, \nu)} \lambda^2 p_{1+} p_{+1} + \frac{Z'(\lambda, \nu)}{Z(\lambda, \nu)} \lambda p_{11}, \tag{4.25}$$

$$\text{Cov}(X_1, X_2) = \lambda^2 p_{1+} p_{+1} \left\{ \frac{\partial^2 \ln Z(\lambda, \nu)}{\partial \lambda^2} \right\} + \lambda p_{11} \left\{ \frac{\partial \ln Z(\lambda, \nu)}{\partial \lambda} \right\}, \tag{4.26}$$

$$V(X_1) = \lambda^2 p_{1+}^2 \left\{ \frac{\partial^2 \ln Z(\lambda, \nu)}{\partial \lambda^2} \right\} + \lambda p_{1+} \left\{ \frac{\partial \ln Z(\lambda, \nu)}{\partial \lambda} \right\}, \tag{4.27}$$

$$V(X_2) = \lambda^2 p_{+1}^2 \left\{ \frac{\partial^2 \ln Z(\lambda, \nu)}{\partial \lambda^2} \right\} + \lambda p_{+1} \left\{ \frac{\partial \ln Z(\lambda, \nu)}{\partial \lambda} \right\}, \tag{4.28}$$

where Equations (4.26)-(4.28) determine the correlation $\rho \geq 0$ whose range varies with $\nu$. This construction only allows for positive association with regard to dispersion, i.e., either $X_1$ and $X_2$ are either both over- or under-dispersed. The conditional pgf of $X_2$ given $X_1 = x_1$ is

$$\Pi_{X_2}(t \mid X_1 = x_1) = \left\{ \frac{\lambda p_{1+} + \lambda p_{11}(t-1)}{\lambda p_{1+}} \right\}^{x_1} \frac{Z^{(x_1)}\{\lambda p_{0+} + \lambda p_{01}(t-1), \nu\}}{Z^{(x_1)}(\lambda p_{0+}, \nu)},$$

and the regression of $X_2$ on $X_1$ is $E(X_2 \mid X_1 = x_1) = \frac{p_{11}}{p_{1+}} x_1 + \frac{\lambda p_{01} Z^{(x_1+1)}(\lambda p_{0+},\nu)}{Z^{(x_1)}(\lambda p_{0+},\nu)}$.

The compounding method can likewise be used to develop the broader bivariate sCMP distribution (Sellers et al., 2016); here, the joint conditional pgf has a bivariate Bernoulli distribution conditioned on the number of trials, which itself has a univariate sCMP distribution (see Section 3.2). The resulting unconditional pgf for the bivariate sCMP distribution is

$$\Pi(t_1, t_2) = \left( \frac{Z\left[\lambda\{1 + p_{1+}(t_1-1) + p_{+1}(t_2-1) + p_{11}(t_1-1)(t_2-1)\}, \nu\right]}{Z(\lambda, \nu)} \right)^n, \tag{4.29}$$

and it contains the following special-case bivariate distributions described in Johnson et al. (1997): BP when $\nu = 1$, BNB when $\nu = 0$, and the bivariate binomial when $\nu \to \infty$.

### 4.2.1 Parameter Estimation

Two approaches to estimate the BCMPcomp parameters $\lambda, \nu, \boldsymbol{p} = (p_{00}, p_{01}, p_{10}, p_{11})$ are the method of moments or the method of maximum likelihood. The method of moments estimators $\tilde{\lambda}, \tilde{\nu}, \tilde{\boldsymbol{p}} = (\tilde{p}_{00}, \tilde{p}_{01}, \tilde{p}_{10}, \tilde{p}_{11})$ can be determined via the system of equations,

$$\tilde{p}_{11} = \frac{m_{11} - \tilde{\lambda}^2 \tilde{p}_{1+} \tilde{p}_{+1} \left( \frac{\partial^2 \ln Z(\tilde{\lambda},\tilde{\nu})}{\partial \lambda^2} \right)}{\tilde{\lambda} \left( \frac{\partial \ln Z(\tilde{\lambda},\tilde{\nu})}{\partial \lambda} \right)},$$

$$\tilde{p}_{01} = \tilde{p}_{+1} \left( 1 - \left[ \frac{m_{11} - \tilde{\lambda}^2 \tilde{p}_{1+} \left\{ \frac{\partial^2 \ln Z(\tilde{\lambda},\tilde{\nu})}{\partial \lambda^2} \right\}}{\tilde{\lambda} \left\{ \frac{\partial \ln Z(\tilde{\lambda},\tilde{\nu})}{\partial \lambda} \right\}} \right] \right),$$

$$\tilde{p}_{10} = \tilde{p}_{1+} \left( 1 - \left[ \frac{m_{11} - \tilde{\lambda}^2 \tilde{p}_{+1} \left\{ \frac{\partial^2 \ln Z(\tilde{\lambda},\tilde{\nu})}{\partial \lambda^2} \right\}}{\tilde{\lambda} \left\{ \frac{\partial \ln Z(\tilde{\lambda},\tilde{\nu})}{\partial \lambda} \right\}} \right] \right),$$

$$\tilde{p}_{00} = 1 - \tilde{p}_{10} - \tilde{p}_{01} - \tilde{p}_{11},$$

where $\tilde{p}_{1+} = \bar{x}_1 \left[ \tilde{\lambda} \left\{ \frac{\partial \ln Z(\tilde{\lambda},\tilde{\nu})}{\partial \lambda} \right\} \right]^{-1}$ and $\tilde{p}_{+1} = \bar{x}_2 \left[ \tilde{\lambda} \left\{ \frac{\partial \ln Z(\tilde{\lambda},\tilde{\nu})}{\partial \lambda} \right\} \right]^{-1}$, and the sample covariance is $m_{11} = \frac{1}{n} \sum_{x_1} \sum_{x_2} (x_1 - \bar{x}_1)(x_2 - \bar{x}_2) n_{x_1 x_2}$; $n_{x_1 x_2}$ is the observed frequency in the $(x_1, x_2)$ cell and $n$ is the total sample size. The MLEs $(\hat{\lambda};\ \hat{\nu};\ \hat{\boldsymbol{p}})$ meanwhile maximize the log-likelihood $\ln L(\lambda, \nu, \boldsymbol{p}; \boldsymbol{x}) = \sum_{x_1} \sum_{x_2} n_{x_1 x_2} \ln P(X_1 = x_1, X_2 = x_2)$, where $P(X_1 = x_1, X_2 = x_2)$ is defined in Equation (4.19).

### *4.2.2 Hypothesis Testing*

Hypothesis tests can be conducted to assess whether statistically significant data dispersion exists (via $H_0\colon \nu = 1$ versus $H_1\colon \nu \neq 1$), thus arguing against a BP distribution in favor of the BCMPcomp model. This two-sided test does not address the direction of data dispersion but merely assesses whether statistically significant dispersion (of either type) exists. Utilizing a likelihood ratio test approach, the resulting test statistic is

$$\Lambda = \frac{\max L(\lambda, \nu = 1, \boldsymbol{p})}{\max L(\lambda, \nu, \boldsymbol{p})}. \tag{4.30}$$

The null hypothesis represents the BP special case; thus, associated statistical computing can be conducted via the `bivpois` package (Karlis and Ntzoufras, 2008) by substituting the appropriate definitions for $\lambda_j$, $j = 1, 2, 3$. Given that the Holgate (1964) BP distribution can be derived either via the trivariate reduction or confounding methods with $\lambda_1 + \lambda_3 = \lambda_0 p_{1+}$, $\lambda_2 + \lambda_3 = \lambda_0 p_{+1}$, and $\lambda_3 = \lambda_0 p_{11}$, analysts can utilize the `bivpois` package with `lambda1` $= \lambda_0 p_{10}$, `lambda2` $= \lambda_0 p_{01}$, and `lambda3` $= \lambda_0 p_{11}$, where $\lambda_0$ denotes the value of $\lambda$ under $H_0$. In the context of parameter estimation via the `lm.bp` function, however, this approach faces an identifiability problem in that `lm.bp` can produce MLEs $\hat{\lambda}_j$, $j = 1, 2, 3$; yet, analysts cannot back-solve to uniquely determine $\lambda_0$, $p_{10}$, $p_{01}$ and $p_{11}$. Instead, analysts can use `optim` to determine the log-likelihoods, both in the general parameter space and given the BP constraint $(\nu = 1)$, in order to determine the likelihood ratio test statistic $-2 \ln \Lambda$, where $\Lambda$ is defined in Equation (4.30).

For data that express statistically significant dispersion, analysts can consider whether a bivariate Bernoulli $(H_0\colon \nu \to \infty)$ or bivariate geometric $(H_0\colon \nu = 0)$ distribution attained via the compounding method is an appropriate model of choice. Here, the likelihood ratio test statistic has the maximized log-likelihood under $H_0$ and under the general parameter space for the numerator and denominator, respectively. Given the usual regularity conditions, the resulting test statistic for the $H_0\colon \nu = 1$ versus $H_1\colon \nu \neq 1$ test converges to a $\chi^2_1$ distribution, while the latter tests (i.e. whether considering $H_0\colon \nu = 0$ or $H_0\colon \nu \to \infty$ versus $H_1$: otherwise) converge to a mixture between a point mass and the $\chi_1^2$ distribution to account for the respective boundary cases for $\nu$.

### 4.2.3 R Computing

The `dbivCMP` function contained in the `R` package `multicmp` (Sellers et al., 2017a) computes the joint probability of the BCMPcomp distribution $P(X_1 = x_1, X_2 = x_2)$ with parameters, $\lambda$, $\nu$, $\boldsymbol{p}$. The analyst supplies the intensity and dispersion parameters of interest $\lambda$ and $\nu$, along with bivariate probabilities (`bivprob` = `c(`$p_{00}, p_{01}, p_{10}, p_{11}$`)`) such that $\sum_{i=0}^{1}\sum_{j=0}^{1} p_{ij} = 1$, and the number of terms `maxit` used to estimate the infinite summation of the associated normalizing constants. Setting `nu = 1`, this function produces probabilities equal to those using the `pbivpois` function in the `bivpois` (Karlis and Ntzoufras, 2005) package with `lambda1` $= \lambda p_{10}$, `lambda2` $= \lambda p_{01}$, and `lambda3` $= \lambda p_{11}$. Meanwhile letting `nu = 0`, `dbivCMP` can obtain probabilities equal to those from `dbinom2.or`, the bivariate Bernoulli distribution function contained in the `VGAM` (Yee, 2010) package.

To illustrate `dbivCMP`, suppose that one is interested in determining the probability $P(X_1 = 1, X_2 = 2)$ from a BCMPcomp($\lambda = 10$, $\nu = 1$, $\boldsymbol{p} = (0.4, 0.3, 0.2, 0.1)$) distribution. This result can be determined explicitly via `dbivCMP` by defining all of the relevant inputs (`lambda`, `nu`, `bivprob=`$\boldsymbol{p}$, `x=`$x_1$, `y=`$x_2$) accordingly. For illustrative purposes, `maxit` is set here to 100, but analysts are encouraged to try several values for `maxit` to assess approximation accuracy; in this example, 100 (indeed) suffices.

```
> dbivCMP(lambda=10, nu=1, bivprob=c(0.4, 0.3, 0.2, 0.1),
x=1, y=2, maxit = 100) [1] 0.02974503
```

Note that $\nu = 1$ implies that the BCMPcomp equates to a BP distribution where $\lambda_1 = \lambda p_{10} = 2$, $\lambda_2 = \lambda p_{01} = 3$, and $\lambda_3 = \lambda p_{11} = 1$.

The `multicmp` package also contains the function `multicmpests` to conduct maximum likelihood estimation for the BCMPcomp($\lambda, \nu, \boldsymbol{p}$) distribution. The only required input is `data`, a two-column entity of observed $(x_1, x_2)$ locations used to tabulate the total number of observations at each location. Additional optional inputs, however, include `startvalues` used to begin the Newton-type optimization iterative process, and `max` that serves as the upper bound used to approximate the infinite summation in the normalizing constant. The values ($\lambda = \nu = 1, \boldsymbol{p} = (0.25, 0.25, 0.25, 0.25)$) serve as the default starting values, but analysts have the ability to supply their own starting values. Meanwhile, the input `max` is defaulted to 100, but analysts are encouraged to try different values for `max` to gauge estimate accuracy and robustness. Function outputs from the `multicmpests` call include the resulting parameter estimates for $\lambda$, $\nu, \boldsymbol{p}$, the associated negated log-likelihood, and a likelihood ratio test statistic and associated $p$-value from conducting a dispersion test for the hypotheses $H_0: \nu = 1$ versus $H_1: \nu \neq 1$. Prior to reporting the parameter estimates, analysts can monitor the iterative scheme progress as it works toward optimizing the negated log-likelihood via unconstrained optimization. Because the latter four estimates are intended to represent probabilities, those resulting estimates are first scaled to ensure that their sum equals 1; the scaled estimates are thus presented for $\boldsymbol{p}$.

Code 4.1 provides an illustrative code example, including the function call and associated output; this example maintains the default settings for `max` and `startvalues`. The illustration contains 11 observations: four at (0,0), two at (0,1), two at (1,0), and three at (1,1). Upon running the `multicmpests` command, we see the iteration updates where the reports are offered for every 10 steps until convergence is achieved, followed by the resulting outputs: the parameter estimates, the negated log-likelihood, and the likelihood ratio test statistic and $p$-value. The resulting parameter estimates for this example are $\hat{\lambda} \approx 12.1956$, $\hat{\nu} \approx 33.7667$, and $\hat{\boldsymbol{p}} \approx (0.3115, 0.1967, 0.1967, 0.2951)$, which produced an associated log-likelihood of approximately $-14.763$. Meanwhile, the corresponding dispersion test found these data to be statistically significantly dispersed such that they cannot be estimated via a BP model ($-2 \ln \Lambda \approx 4.9789$ with 1 df; $p$-value = 0.0257). In fact, the estimated dispersion value implies that the data are approximately bivariate Bernoulli. This makes sense because all of the outcomes are contained within the bivariate Bernoulli space.

Code 4.1 Illustrative codes and output using `multicmpests` to conduct maximum likelihood estimation for the BCMPcomp parameters, $\lambda$, $\nu, \boldsymbol{p}$.

```
> x1=c(0,0,0,1,0,1,1,0,0,1,1)
> x2=c(0,0,0,1,1,0,1,0,1,0,1)
> ex <- cbind(x1,x2)
> multicmpests(ex)
Iterating...
  0:       17.284630:   1.00000   1.00000 0.250000
           0.250000 0.250000 0.250000
 10:       14.788795:   3.15429   9.58827 0.721030
           1.17692 1.17692  1.83347
 20:       14.763292:   6.90457  19.5713   3.29352
           2.53269 2.53269  3.78791
```

```
30:      14.763252:   7.96702  22.4503  4.00982
         2.89264 2.89264  4.34215
40:      14.763245:   11.3693  31.5584  6.32790
         4.06862 4.06862  6.10312

The parameter estimates ($par) are as follows:
Parameter         MLE
   lambda 12.1955518
       nu 33.7666649
      p00  0.3114534
      p10  0.1967291
      p01  0.1967291
      p11  0.2950884

Log-likelihood ($negll): 14.76324

Dispersion hypothesis test statistic ($LRTbpd)
and p-value ($pbpd): Likelihood.ratio.test   p.value
             4.978879 0.0256586
```

### 4.2.4 Multivariate Generalization

The compounding method can likewise be applied to create a MultiCMP distribution. Let $X_1, X_2, \ldots, X_d$ denote $d$ random variables that have a joint conditional multivariate binomial distribution with pgf

$$\Pi_{X_1,\ldots,X_d}(t_1, \ldots, t_d \mid n) = \left( \sum_{x_1=0}^{1} \sum_{x_2=0}^{1} \cdots \sum_{x_d=0}^{1} p_{x_1x_2\cdots x_d} \prod_{i=1}^{d} (t_i)^{x_i} \right)^n$$

given the CMP($\lambda$, $\nu$) distributed number of trials, $n$. The unconditional joint pgf for $X_1, X_2, \ldots, X_d$ now becomes

$$\Pi_{X_1,\ldots,X_d}(t_1, \ldots, t_d) = \frac{Z\left[\lambda \left( \sum_{x_1=0}^{1} \cdots \sum_{x_d=0}^{1} p_{x_1x_2\cdots x_d} \prod_{i=1}^{d} (t_i)^{x_i} \right), \nu\right]}{Z(\lambda, \nu)}. \tag{4.31}$$

The multivariate CMP distribution obtained via the compounding method (MultiCMPcomp) likewise contains the MP described in Johnson et al. (1997) when $\nu = 1$, the multivariate Bernoulli (i.e. the Krishnamoorthy (1951) multivariate binomial with one trial) when $\nu \to \infty$, and the multivariate geometric (i.e. the Doss (1979) multivariate negative binomial (MNB) with one success) when $\nu = 0$. While $X_1, \ldots, X_d$ do not have CMP marginal distributions, they contain the Poisson distribution as special cases when $\nu = 1$. Further, as with the BCMPcomp, the correlation between any two of the $d$ random variables is likewise nonnegative. Meanwhile, because the MultiCMPcomp involves only one dispersion parameter ($\nu$), it is only suitable for data with similar dispersion levels in every dimension (Sellers et al., 2021b).

Method of moments or maximum likelihood estimation approaches can again be pursued to conduct parameter estimation, and the proposed hypothesis tests are likewise relevant for consideration in the broader multivariate space. The test for statistically significant data dispersion ($H_0\colon \nu = 1$ versus $H_1\colon \nu \neq 1$) assesses the appropriateness of the MP distribution versus the MultiCMPcomp distribution to model the data. Analogous tests can likewise consider the appropriateness of the multivariate Bernoulli ($H_0\colon \nu \rightarrow \infty$) or multivariate geometric ($H_0\colon \nu = 0$) distribution (both attained via the compounding method). Again, the resulting likelihood ratio test statistic $-2 \ln \Lambda$ for $H_0\colon \nu = 1$ versus $H_1\colon \nu \neq 1$ converges to a $\chi_1^2$ distribution, while the latter tests (whether $H_0\colon \nu = 0$ or $H_0\colon \nu \rightarrow \infty$ versus otherwise) converge to a mixture between a point mass and the $\chi_1^2$ distribution to account for the respective boundary cases for $\nu$ (Balakrishnan and Pal, 2013). As noted for the BCMPcomp distribution, a test for independence is not easily represented mathematically.

## 4.3 The Sarmanov Construction

While the trivariate reduction and compounding methods offer easy BCMP and MultiCMP constructs, a significant drawback to their construction is their inability to allow for negatively correlated data. In contrast, Sarmanov (1966) constructed a bivariate family of distributions that maintains univariate marginal distributions and allows for a broad (−1,1) correlation structure. Given two random variables $X_1$ and $X_2$ with respective probabilities $P(X_j = x_j)$ and mixing functions $\phi_j(X_j)$ such that $E[\phi_j(X_j)] = 0$, $j = 1, 2$, and $1 + \phi_1(x_1)\phi_2(x_2) \geq 0$, a bivariate distribution can be attained of the form

$$P(X_1 = x_1, X_2 = x_2) = P(X_1 = x_1)P(X_2 = x_2)[1 + \gamma\phi_1(x_1)\phi_2(x_2)],$$
$$x_j \in \mathbb{R}, \quad j = 1, 2, \tag{4.32}$$

where $\gamma \in [-1, 1]$. Inspired by this family of distributions, Ong et al. (2021) used it to develop two versions of a BCMP distribution that maintain CMP marginal forms as described in Chapter 2; for ease in discussion, we refer to these models as BCMPsar1 and BCMPsar2, respectively. The first approach (producing BCMPsar1) is motivated by a larger weighted Poisson distributional framework (Kokonendji et al., 2008) such that Equation (4.32) with

$$\phi_j(x_j) = p^\alpha(x_j) - E(p^\alpha(X_j)), \quad j = 1, 2, \tag{4.33}$$

produces the joint probability mass function

$$\begin{aligned} P(X_1 = x_1, X_2 = x_2) &= P(X_1 = x_1)P(X_2 = x_2) \\ &\quad \{1 + \gamma[p^\alpha(x_1) - E(p^\alpha(X_1))][p^\alpha(x_2) - E(p^\alpha(X_2))]\}, \end{aligned}$$

where $P(X_j = x_j)$, $j = 1, 2$, denotes the CMP($\lambda_j$, $\nu_j$) marginal probability as defined in Equation (2.8), $p(x_j)$ is the Poisson($\lambda_j$) probability provided in Equation (1.1), $\alpha > 0$, and

$$E[p^\alpha(X_j)] = \frac{e^{-\lambda_j\alpha} Z(\lambda_j^2, \nu_j + 1)}{Z(\lambda_j, \nu_j)} \sum_{x=0}^{\infty} \frac{\lambda_j^{x_j(\alpha+1)}}{(x_j!)^{\nu_j+\alpha} Z(\lambda_j^2, \nu_j + 1)} \leq 1,$$

where $Z(\phi, \psi) = \sum_{j=0}^{\infty} \frac{\phi^j}{(j!)^\psi}$ for two parameters $\phi$ and $\psi$; see Section 2.8 for detailed discussion regarding the $Z$ function. The BCMPsar1 distribution has the correlation

$$\rho = \frac{\gamma(\lambda_1 - \mu_1)(\lambda_2 - \mu_2)}{\sigma_1 \sigma_2}, \tag{4.34}$$

where $\mu_j$ and $\sigma_j$, respectively, denote the marginal mean and standard deviation for $X_j$, $j = 1, 2$.

The second approach (which produces BCMPsar2) considers Equation (4.32) with

$$\phi_j(x_j) = \theta^{x_j} - \frac{Z(\lambda_j\theta, \nu_j)}{Z(\lambda_j, \nu_j)}, \quad j = 1, 2, \tag{4.35}$$

where $0 < \theta < 1$, thus producing the joint probability mass function

$$\begin{aligned} P(X_1 = x_1, X_2 = x_2) &= P(X_1 = x_1)P(X_2 = x_2) \\ &\quad \left[1 + \gamma\left(\theta^{x_1} - \frac{Z(\lambda_1\theta, \nu_1)}{Z(\lambda_1, \nu_1)}\right)\left(\theta^{x_2} - \frac{Z(\lambda_2\theta, \nu_2)}{Z(\lambda_2, \nu_2)}\right)\right]. \end{aligned}$$

The BCMPsar2 distribution contains the Lee (1996) BP distribution as a special case when $\nu_1 = \nu_2 = 1$ and $\theta = e^{-1}$ and has the correlation coefficient

$$\rho = \frac{\gamma\left(\theta \frac{\partial \frac{Z(\lambda_1\theta,\nu_1)}{Z(\lambda_1,\nu_1)}}{\partial\theta} - \mu_1 \frac{Z(\lambda_1\theta,\nu_1)}{Z(\lambda_1,\nu_1)}\right)\left(\theta \frac{\partial \frac{Z(\lambda_2\theta,\nu_2)}{Z(\lambda_2,\nu_2)}}{\partial\theta} - \mu_2 \frac{Z(\lambda_2\theta,\nu_2)}{Z(\lambda_2,\nu_2)}\right)}{\sigma_1\sigma_2}. \tag{4.36}$$

### 4.3.1 Parameter Estimation and Hypothesis Testing

Ong et al. (2021) introduce the method of moments or maximum likelihood as methods for parameter estimation for either model. For the method of moments approach, the first and second marginal sample moments are equated with their theoretical marginal approximations (Equations (2.14) and (2.15)) to estimate $\lambda_j$ and $\nu_j$, while $\gamma$ is estimated by equating the correlation coefficient with either Equation (4.37) or (4.38), depending on the BCMP formulation of interest (BCMPsar1 or BCMPsar2). Recall, however, that utilizing the approximations for the marginal mean and variance assumes that the necessary constraints hold, i.e. $\nu \leq 1$ or $\lambda > 10^{\nu}$ (Shmueli et al., 2005). Meanwhile, maximum likelihood estimation can be conducted by optimizing the log-likelihood associated with a random sample $\boldsymbol{x} = \{(x_{1i}, x_{2i}) : i = 1, \ldots, n\}$,

$$\ln L(\lambda_1, \lambda_2, \nu_1, \nu_2, \gamma; \boldsymbol{x}) = -n(\ln Z(\lambda_1, \nu_1) + \ln Z(\lambda_2, \nu_2))$$
$$+ \sum_{i=1}^{n} [x_{1i} \ln \lambda_1 + x_{2i} \ln \lambda_2 - \nu_1(\ln x_{1i}!) - \nu_2(\ln x_{2i}!)]$$
$$+ \sum_{i=1}^{n} \ln \left\{ 1 + \gamma \left( \theta^{x_{1i}} - \frac{Z(\lambda_1 \theta, \nu_1)}{Z(\lambda_1, \nu_1)} \right) \left( \theta^{x_{2i}} - \frac{Z(\lambda_2 \theta, \nu_2)}{Z(\lambda_2, \nu_2)} \right) \right\}.$$

Under either approach, Ong et al. (2021) conduct infinite summation computations (namely, $Z(\lambda_j, \nu_j)$, the means $\mu_j$, and variances $\sigma_j^2$, $j = 1, 2$) via recursion with double-precision accuracy; the computations for the means and variances aid in determining the correlation coefficient.

Two hypothesis tests can be considered, namely a test for dispersion ($H_0 \colon \nu_1 = \nu_2 = 1$ versus $H_1 \colon$ otherwise), and a test for independence between the two random variables ($H_0 \colon \gamma = 0$ versus $H_1 \colon \gamma \neq 0$). For both cases, associated inference can be conducted via the likelihood ratio or score tests. The usual regularity conditions imply that these test statistics have an approximate chi-squared distribution with two degrees of freedom for the dispersion test and an approximate chi-squared distribution with one degree of freedom for the independence test.

### 4.3.2 Multivariate Generalization

Lee (1996) extended the Sarmanov family bivariate distributional form to the multivariate case. Given $d$ random variables $X_j$ with probability $P(X_j = x_j)$ and bounded nonconstant functions $\phi_j(t)$ such that $E[\phi_j(X_j)] = 0$, for all $j = 1, \ldots, d$, Equation (4.32) is broadened to produce the multivariate probability mass function

$$P(X_1 = x_1, \ldots, X_d = x_d) = \left(\prod_{i=1}^{d} P(X_i = x_i)\right)\left(1 + \sum_{j_1 <}^{n-1} \sum_{j_2}^{n} \gamma_{j_1 j_2} \phi_{j_1}(x_{j_1}) \phi_{j_2}(x_{j_2})\right.$$
$$\left. + \sum_{j_1 <}^{n-2} \sum_{j_2 <}^{n-1} \sum_{j_3}^{n} \gamma_{j_1 j_2 j_3} \phi_{j_1}(x_{j_1}) \phi_{j_2}(x_{j_2}) \phi_{j_3}(x_{j_3}) + \cdots + \gamma_{1,2,\ldots,d} \prod_{i=1}^{d} \phi_i(x_i)\right).$$

While a published work developing a MultiCMP distribution based on this method does not yet exist, the definitions provided in Ong et al. (2021) seem like natural choices for model development, i.e. let $\phi_j(x_j), j = 1, \ldots, d$, be as defined in Equations (4.33) or (4.35) to, respectively, produce a MultiCMPsar1 or MultiCMPsar2 distribution. The respective two-way correlations are thus

$$\rho = \frac{\gamma(\lambda_j - \mu_j)(\lambda_k - \mu_k)}{\sigma_j \sigma_k}, \quad j \neq k, \tag{4.37}$$

for the MultiCMPsar1 distribution, and

$$\rho = \frac{\gamma \left(\theta \frac{\partial \frac{Z(\lambda_j \theta, \nu_j)}{Z(\lambda_j, \nu_j)}}{\partial \theta} - \mu_j \frac{Z(\lambda_j \theta, \nu_j)}{Z(\lambda_j, \nu_j)}\right)\left(\theta \frac{\partial \frac{Z(\lambda_k \theta, \nu_k)}{Z(\lambda_k, \nu_k)}}{\partial \theta} - \mu_k \frac{Z(\lambda_k \theta, \nu_k)}{Z(\lambda_k, \nu_k)}\right)}{\sigma_k \sigma_k}, \quad j \neq k, \tag{4.38}$$

for the MultiCMPsar2 distribution, where $\mu_j$ and $\sigma_j$, respectively, denote the marginal mean and standard deviation for $X_j, j = 1, 2, \ldots, d$.

## 4.4 Construction with Copulas

Copulas are a popular tool for multivariate distributional development. The foundation underlying copula function consideration is Sklar's theorem that states that, for any collection of random variables $X_1, \ldots, X_d$ with joint and marginal cumulative distribution functions, $F(x_1, \ldots, x_d) = P(X_1 \leq x_1, \ldots, X_d \leq x_d)$ and $F_i(x) = P(X_i \leq x)$, for $i = 1, \ldots, d$, respectively, there exists a copula $C$ such that

$$F(x_1, \ldots, x_d) = C(F_1(x_1), \ldots, F_d(x_d)). \tag{4.39}$$

This result is helpful for determining a joint probability mass function for a multivariate discrete distribution because the multivariate probabilities $P(X_1 = x_1, \ldots, X_d = x_d)$ can be determined by

$$P(X_1 = x_1, \ldots, X_d = x_d) = \sum_{j_1=0}^{1} \cdots \sum_{j_d=0}^{1} (-1)^{j_1+\cdots+j_d} F(x_1 - j_1, \ldots, x_d - j_d) \tag{4.40}$$

$$= \sum_{j_1=0}^{1} \cdots \sum_{j_d=0}^{1} (-1)^{j_1+\cdots+j_d} C(F_1(x_1 - j_1), \ldots, F_d(x_d - j_d)) \tag{4.41}$$

for the copula. For example, a bivariate discrete distribution has probabilities

$$\begin{aligned} P(X_1 = x_1, X_2 = x_2) &= C(F_1(x_1), F_2(x_2)) - C(F_1(x_1 - 1), F_2(x_2)) \\ &\quad - C(F_1(x_1), F_2(x_2 - 1)) + C(F_1(x_1 - 1), F_2(x_2 - 1)). \end{aligned} \tag{4.42}$$

Suggested copulas have been proposed through the literature for MultiCMP model development because of their distinctive qualities (Alqawba and Diawara, 2021; Ötting et al., 2021); see, for example, Table 4.1 that lists various copula functions considered to develop BCMP or MultiCMP distributions. Trivedi and Zimmer (2017), for example, use the Gaussian, Clayton, and Gumbel copulas to develop various BP models; these copulas can likewise be considered for BCMP development. Gaussian copulas are proposed for symmetric distributions, while the Gumbel and reflected Gumbel copulas are suggested for distributions with tail dependence. The Ali–Mikhail–Haq, Clayton, Frank, and Gaussian copulas allow for negative dependence with the Frank and Gaussian copulas likewise being reflection symmetric. More broadly, a max-infinitely-divisible (max-id) copula $C$ satisfies the property that $C\left(u_1^{1/m}, u_2^{1/m}\right)^m$ is a copula for all $m > 0$. Alqawba and Diawara (2021) propose using Gaussian or max-id copulas to determine joint probabilities associated with MultiCMP distributions.

Sklar's theorem not only notes the existence of a copula function such that Equation (4.39) holds; it further notes that, for continuous cumulative distribution functions, the copula is unique. This uniqueness property, however, does not apply for discrete (e.g. the CMP) distributions. Copulas for discrete outcomes are not identifiable, particularly for count distribution families; see Trivedi and Zimmer (2017) and Genest and Nešlehová (2007) for extensive discussion regarding identifiability of copulas and the associated dependence parameter $\zeta$.

Table 4.1 *Bivariate copula functions*

| Copula name | Copula function, $C(u_1,u_2)$ | Range for $\zeta$ |
|---|---|---|
| Ali–Mikhail–Haq | $\frac{u_1 u_2}{1-\zeta(1-u_1)(1-u_2)}$ | $[-1,1)$ |
| Clayton | $\left(\max\left(u_1^{-\zeta}+u_2^{-\zeta}-1,0\right)\right)^{-1/\zeta}$ | $[-1,0)\cup(0,\infty)$ |
| Frank | $-\frac{1}{\zeta}\ln\left(1+\frac{(\exp(-\zeta u_1)-1)(\exp(-\zeta u_2)-1)}{\exp(-\zeta)-1}\right)$ | $(-\infty,0)\cup(0,\infty)$ |
| Gaussian | $\Phi_\zeta(\Phi^{-1}(u_1),\Phi^{-1}(u_2))$ | $[-1,1]$ |
| Gumbel | $\exp[-((-\ln(u_1))^\zeta+(-\ln(u_2))^\zeta)^{1/\zeta}]$ | $[1,\infty)$ |
| Plackett | $\frac{[1+(\zeta-1)(u_1+u_2)]-\sqrt{[1+(\zeta-1)(u_1+u_2)]^2-4u_1u_2\zeta(\zeta-1)}}{2(\zeta-1)}$ | $[0,\infty)$ |
| Reflected Gumbel | $u_1+u_2-1+\exp[-((-\ln(u_1))^\zeta+(-\ln(u_2))^\zeta)^{1/\zeta}]$ | $[1,\infty)$ |

## 4.5 Real Data Examples

This section illustrates the aforementioned models through two real data examples where, for consistency and optimal model comparison, both examples involve bivariate count data. Section 4.5.1 considers a frequently studied over-dispersed dataset regarding the number of shunter accidents that occurred over two consecutive time periods. Section 4.5.2 meanwhile considers the number of professional basketball players from two positions (forward and center, respectively) selected for the 2000–2016 National Basketball Association's (NBA's) All Star games. The latter dataset is an under-dispersed data example that further illustrates the flexibility of the BCMP models.

Both illustrative examples consider and compare (among others) the BCMPtriv, BCMPcomp, BCMPsar1, and BCMPsar2 models. Given the expansive list of potential copulas for consideration in BCMP development (only some of which are listed in Table 4.1) and identifiability issues discussed in Section 4.4, however, the copula approach is not included in subsequent discussion.

### *4.5.1 Over-dispersed Example: Number of Shunter Accidents*

We revisit a dataset first introduced by Arbous and Kerrick (1951) that has received extensive attention in the literature for various bivariate model considerations and developments. This dataset reports the number of accidents incurred by 122 shunters in two consecutive year periods; $X_1$ and $X_2$, respectively, denote the number of accidents observed between 1937–1942 and 1943–1947. While Holgate (1964) models the data via a BP distribution, the data are recognized as being over-dispersed. Other considered models for analysis include the BNB and the BGP distributions, along with the BCMPtriv, BCMPcomp, BCMPsar1, and BCMPsar2 distributions, respectively.

Table 4.2 provides the reported MLEs for the respective models, along with their associated log-likelihood, number of model parameters, the Akaike information criterion (AIC), and difference in AIC ($\Delta_i$) values for model comparison as described in Chapter 1; see Table 1.2 for inferred model support levels based on $\Delta_i$ as described in Burnham and Anderson (2002). The BP model MLEs are determined via the `bivpois` package, while the BNB and BGP estimates are provided from Famoye and Consul (1995), and the respective BCMP estimates are discussed in recent works (Ong et al., 2021; Sellers et al., 2016; Weems et al., 2021). The BNB model has the minimum AIC (691.2) among all of the models considered for comparison, while the BCMPtriv distribution has the smallest AIC among the BCMP models. Further, the BCMPtriv demonstrates substantial support relative to the BNB model ($\Delta_{\text{BCMPtriv}} = 0.8$). Among the other BCMP models considered, the BCMPcomp obtains associated MLEs with a larger log-likelihood than that obtained for the BCMPtriv distribution. This occurs however at the expense of two additional parameters; thus, the AIC for the BCMPcomp model is larger than that obtained via the BCMPtriv distribution.

All of the BCMP models recognize the statistically significant over-dispersion present in the data; for all associated distributions, the estimated dispersion parameter(s) is/are less than 1 and the associated likelihood ratio test statistics produce small, associated $p$-values. The BCMPtriv has an estimated dispersion $\hat{\nu} = 0.438$ with test statistic $-2 \ln \Lambda = 7.25$ and $p$-value $< 0.01$. The BCMPcomp method has $\hat{\nu} = 0.084$ with associated test statistic $-2 \ln \Lambda = 7.862$ and $p$-value equaling 0.0005. Meanwhile, the Sarmanov BCMP distributions each have marginal dispersions that are reportedly less than 1 ($\hat{\nu}_1 = 0.57$ and $\hat{\nu}_2 = 0.53$ for the BCMPsar1; $\hat{\nu}_1 = 0.59$ and $\hat{\nu}_2 = 0.56$ for the BCMPsar2); however, Ong et al. (2021) do not report on their statistical significance.

The observed and estimated shunter accident frequency data stemming from all of the associated model MLEs are provided in Tables 4.3 and 4.4.[4] All of the BCMP models appear to reasonably estimate the observed number of shunter accidents over the combination of respective time periods. The same holds true for the marginal expected frequencies, where

[4] BCMPsar1 and BCMPsar2 estimated frequencies are as reported in Ong et al. (2021)

Table 4.2 *Parameter maximum likelihood estimates (MLEs), log-likelihood (ln L), Akaike information criterion (AIC), difference in AIC ($\Delta_i = AIC_i - AIC_{min}$), goodness-of-fit ($GOF = \sum \frac{(O-E)^2}{E}$, where O and E denote the observed and expected cell frequencies, respectively) measures and associated p-values for various bivariate distributions on the shunters accident dataset: bivariate Poisson (BP); bivariate negative binomial (BNB); bivariate generalized Poisson (BGP); and the BCMP obtained via the trivariate reduction (BCMPtriv), compounding (BCMPcomp), or either Sarmanov family (BCMPsar1 and BCMPsar2, respectively) method.*

| Model | Parameter MLEs | $\ln L$ | No. of param. | AIC | $\Delta_i$ | GOF | $p$-values |
|---|---|---|---|---|---|---|---|
| BP | $\hat{\lambda}_1 = 0.717$, $\hat{\lambda}_2 = 1.012$, $\hat{\lambda}_3 = 0.258$ | −345.635 | 3 | 697.3 | 6.1 | 48.05 | 0.13 |
| BNB | $\hat{m} = 0.891$, $\hat{r} = 3.876$, $\hat{\alpha}_1 = 1.331$, $\hat{\alpha}_2 = 0.095$ | −341.610 | 4 | 691.2 | – | 21.92 | 0.97 |
| BGP | $\hat{\theta}_1 = 0.560$, $\hat{\theta}_2 = 0.837$, $\hat{\theta}_3 = 0.305$, $\hat{\lambda}_1 = 0.151$, $\hat{\lambda}_2 = 0.123$, $\hat{\lambda}_3 = 0.031$ | −341.513 | 6 | 695.0 | 3.8 | 23.59 | 0.93 |
| BCMPtriv | $\hat{\lambda}_1 = 0.517$, $\hat{\lambda}_2 = 0.684$, $\hat{\lambda}_3 = 0.270$, $\hat{\nu} = 0.438$ | −342.009 | 4 | 692.0 | 0.8 | 23.36 | 0.96 |
| BCMPcomp | $\hat{\lambda} = 1.328$, $\hat{\nu} = 0.084$, $\hat{p}_{00} = 0.939$, $\hat{p}_{01} = 0.034$, $\hat{p}_{10} = 0.025$, $\hat{p}_{11} = 0.002$ | −341.704 | 5 | 693.4 | 2.2 | 22.16 | 0.95 |
| BCMPsar1 | $\hat{\lambda}_1 = 0.92$, $\hat{\nu}_1 = 0.57$, $\hat{\lambda}_2 = 0.73$, $\hat{\nu}_2 = 0.53$, $\hat{\alpha} = 0.58$, $\hat{\beta} = 1.00$ | −345.553 | 6 | 703.1 | 11.9 | 37.06 | 0.37 |
| BCMPsar2 | $\hat{\lambda}_1 = 0.94$, $\hat{\nu}_1 = 0.59$, $\hat{\lambda}_2 = 0.75$, $\hat{\nu}_2 = 0.56$, $\hat{\beta} = 1.00$ | −343.503 | 5 | 697.0 | 5.8 | 31.46 | 0.68 |

Table 4.3 *Observed accident data among 122 shunters along with associated count estimates from various bivariate distributions: bivariate negative binomial (BNB), bivariate Poisson (BP), bivariate generalized Poisson (BGP), bivariate geometric (BG), and BCMP obtained via the compounding (BCMPcomp), trivariate reduction (BCMPtriv), or either Sarmanov family (BCMPsar1 and BCMPsar2) methods. Estimated counts determined from MLEs for respective model parameters reported in Table 4.2.*

| $x$ | | $y=0$ | $y=1$ | $y=2$ | $y=3$ | $y=4$ | $y=5$ | $y=6+$ | |
|---|---|---|---|---|---|---|---|---|---|
| 0 | OBS | 21 | 18 | 8 | 2 | 1 | – | – | 50 |
| | BP | 16.72 | 16.92 | 8.56 | 2.89 | 0.73 | 0.15 | 0.02 | 46.00 |
| | BNB | 21.90 | 16.67 | 7.98 | 3.07 | 1.04 | 0.32 | 0.13 | 51.11 |
| | BGP | 22.21 | 16.44 | 7.88 | 3.12 | 1.11 | 0.37 | 0.17 | 51.32 |
| | BCMPtriv | 22.52 | 15.40 | 7.77 | 3.28 | 1.22 | 0.41 | 0.18 | 50.78 |
| | BCMPcomp | 22.48 | 16.10 | 7.88 | 3.14 | 1.09 | 0.34 | 0.13 | 51.16 |
| | BCMPsar1 | 18.08 | 16.60 | 10.02 | 4.82 | 1.99 | 0.73 | 0.24 | 52.47 |
| | BCMPsar2 | 21.09 | 15.44 | 8.63 | 4.05 | 1.65 | 0.59 | 0.19 | 51.65 |
| 1 | OBS | 13 | 14 | 10 | 1 | 4 | 1 | – | 43 |
| | BP | 11.99 | 16.45 | 10.51 | 4.28 | 1.27 | 0.29 | 0.07 | 44.87 |
| | BNB | 12.52 | 13.18 | 8.06 | 3.77 | 1.50 | 0.53 | 0.26 | 39.83 |
| | BGP | 10.70 | 14.51 | 8.67 | 3.84 | 1.46 | 0.51 | 0.25 | 39.93 |
| | BCMPtriv | 11.64 | 14.04 | 8.17 | 3.79 | 1.52 | 0.54 | 0.258 | 39.96 |
| | BCMPcomp | 12.11 | 12.94 | 8.14 | 3.90 | 1.57 | 0.55 | 0.25 | 39.46 |
| | BCMPsar1 | 12.98 | 11.95 | 7.38 | 3.62 | 1.51 | 0.56 | 0.18 | 38.18 |
| | BCMPsar2 | 11.56 | 12.43 | 8.12 | 4.05 | 1.69 | 0.61 | 0.20 | 38.66 |
| 2 | OBS | 4 | 5 | 4 | 2 | 1 | 0 | 1 | 17 |
| | BP | 4.30 | 7.45 | 5.89 | 2.89 | 1.01 | 0.27 | 0.07 | 21.88 |
| | BNB | 4.50 | 6.06 | 4.54 | 2.52 | 1.16 | 0.47 | 0.26 | 19.52 |
| | BGP | 3.97 | 6.11 | 4.93 | 2.55 | 1.06 | 0.39 | 0.19 | 19.20 |
| | BCMPtriv | 4.44 | 6.18 | 4.89 | 2.56 | 1.12 | 0.43 | 0.22 | 19.83 |
| | BCMPcomp | 4.46 | 6.12 | 4.68 | 2.62 | 1.20 | 0.47 | 0.23 | 19.78 |
| | BCMPsar1 | 6.36 | 5.90 | 3.82 | 1.95 | 0.83 | 0.31 | 0.10 | 19.28 |
| | BCMPsar2 | 5.08 | 6.48 | 4.43 | 2.25 | 0.94 | 0.34 | 0.11 | 19.63 |
| 3 | OBS | 2 | 1 | 3 | 2 | 0 | 1 | 0 | 9 |
| | BP | 1.03 | 2.15 | 2.05 | 1.20 | 0.49 | 0.15 | 0.05 | 7.11 |
| | BNB | 1.30 | 2.13 | 1.90 | 1.23 | 0.65 | 0.30 | 0.19 | 7.69 |
| | BGP | 1.35 | 2.18 | 1.92 | 1.19 | 0.56 | 0.22 | 0.12 | 7.53 |
| | BCMPtriv | 1.42 | 2.17 | 1.93 | 1.25 | 0.61 | 0.25 | 0.14 | 7.77 |
| | BCMPcomp | 1.34 | 2.21 | 1.97 | 1.26 | 0.64 | 0.28 | 0.16 | 7.85 |
| | BCMPsar1 | 2.54 | 2.37 | 1.59 | 0.83 | 0.36 | 0.14 | 0.05 | 7.87 |
| | BCMPsar2 | 1.94 | 2.65 | 1.84 | 0.94 | 0.39 | 0.14 | 0.05 | 7.95 |

*(continued)*

Table 4.3 *(cont.)*

| $x$ | | $y=0$ | $y=1$ | $y=2$ | $y=3$ | $y=4$ | $y=5$ | $y=6+$ | |
|---|---|---|---|---|---|---|---|---|---|
| 4 | OBS | 0 | 0 | 1 | 1 | – | – | – | 2 |
| | BP | 0.18 | 0.45 | 0.51 | 0.35 | 0.17 | 0.06 | 0.02 | 1.73 |
| | BNB | 0.33 | 0.64 | 0.66 | 0.49 | 0.29 | 0.15 | 0.11 | 2.66 |
| | BGP | 0.45 | 0.73 | 0.67 | 0.44 | 0.23 | 0.10 | 0.06 | 2.67 |
| | BCMPtriv | 0.40 | 0.66 | 0.64 | 0.46 | 0.26 | 0.12 | 0.08 | 2.60 |
| | BCMPcomp | 0.35 | 0.67 | 0.68 | 0.48 | 0.27 | 0.13 | 0.08 | 2.66 |
| | BCMPsar1 | 0.88 | 0.82 | 0.56 | 0.30 | 0.13 | 0.05 | 0.02 | 2.76 |
| | BCMPsar2 | 0.65 | 0.91 | 0.64 | 0.33 | 0.14 | 0.05 | 0.02 | 2.74 |

the perceived goodness of fit among the BCMP models appears best with the BCMPtriv and BCMPcomp models, followed by the BCMPsar1 and BCMPsar2 distributions. These assessments are validated through the approximate goodness-of-fit statistics provided in Table 4.2, where all of the reported goodness-of-fit statistics stem from an approximate chi-squared distribution with $41 - c$ degrees of freedom, where $c$ equals the number of estimated parameters; see Weems et al. (2021) for details. As assessed via Tables 4.3 and 4.4, the BCMPtriv (GOF $= 23.36$; $p = 0.96$) and BCMPcomp (GOF $= 22.16$; $p = 0.95$) were better models with regard to the goodness of fit than the BCMP models derived via the Sarmanov family (BCMPsar1: GOF $= 37.06$; $p = 0.37$; BCMPsar2: GOF $= 31.46$; $p = 0.68$).

### *4.5.2 Under-dispersed Example: Number of All-Star Basketball Players*

The `Kaggle` site (`https://www.kaggle.com/datasets/fmejia21/nba-all-star-game-20002016`) contains data regarding the NBA players who were selected as All-Stars from 2000 to 2016, along with their respective positions and other information. Summaries of these data allow for analysts to note the number of All-Star players represented in the respective positions on the basketball court (Sellers et al., 2021b; Weems et al., 2021). Focusing on the number of Forwards (F) or Centers (C) over 2000–2016, Weems et al. (2021) found the resulting data to be an under-dispersed bivariate dataset, while Sellers et al. (2021b) likewise determined that the resulting trivariate dataset stemming from the additional inclusion of the number of Forward-centers (FC) is also under-dispersed. Here,

Table 4.4 *Observed accident data (continued from Table 4.3) among 122 shunters.*

| $x$ | | $y = 0$ | $y = 1$ | $y = 2$ | $y = 3$ | $y = 4$ | $y = 5$ | $y = 6+$ | |
|---|---|---|---|---|---|---|---|---|---|
| 5 | OBS | – | – | – | – | – | – | – | 0 |
| | BP | 0.03 | 0.07 | 0.10 | 0.08 | 0.04 | 0.02 | 0.01 | 0.34 |
| | BNB | 0.08 | 0.17 | 0.20 | 0.17 | 0.11 | 0.06 | 0.05 | 0.84 |
| | BGP | 0.15 | 0.24 | 0.22 | 0.15 | 0.08 | 0.04 | 0.03 | 0.90 |
| | BCMPtriv | 0.10 | 0.18 | 0.18 | 0.14 | 0.09 | 0.05 | 0.03 | 0.78 |
| | BCMPcomp | 0.08 | 0.18 | 0.20 | 0.16 | 0.10 | 0.05 | 0.03 | 0.80 |
| | BCMPsar1 | 0.27 | 0.26 | 0.18 | 0.09 | 0.04 | 0.02 | 0.01 | 0.86 |
| | BCMPsar2 | 0.20 | 0.28 | 0.20 | 0.10 | 0.04 | 0.02 | 0.01 | 0.83 |
| 6 | OBS | – | – | – | – | – | – | – | 0 |
| | BP | 0.00 | 0.01 | 0.01 | 0.01 | 0.01 | 0.00 | 0.00 | 0.06 |
| | BNB | 0.02 | 0.04 | 0.06 | 0.05 | 0.04 | 0.02 | 0.02 | 0.25 |
| | BGP | 0.05 | 0.08 | 0.07 | 0.05 | 0.03 | 0.01 | 0.01 | 0.30 |
| | BCMPtriv | 0.02 | 0.04 | 0.05 | 0.04 | 0.03 | 0.02 | 0.01 | 0.21 |
| | BCMPcomp | 0.02 | 0.04 | 0.05 | 0.05 | 0.03 | 0.02 | 0.01 | 0.22 |
| | BCMPsar1 | 0.08 | 0.07 | 0.05 | 0.03 | 0.01 | 0.00 | 0.00 | 0.24 |
| | BCMPsar2 | 0.05 | 0.08 | 0.05 | 0.03 | 0.01 | 0.00 | 0.00 | 0.23 |
| 7+ | OBS | – | 1 | 0 | – | – | – | – | 1 |
| | BP | 0.00 | 0.00 | 0.00 | 0.00 | 0.00 | 0.00 | 0.00 | 0.01 |
| | BNB | 0.00 | 0.01 | 0.02 | 0.02 | 0.02 | 0.01 | 0.02 | 0.10 |
| | BGP | 0.02 | 0.04 | 0.03 | 0.02 | 0.01 | 0.01 | 0.00 | 0.14 |
| | BCMPtriv | 0.01 | 0.01 | 0.02 | 0.01 | 0.01 | 0.01 | 0.00 | 0.07 |
| | BCMPcomp | 0.00 | 0.01 | 0.02 | 0.01 | 0.01 | 0.01 | 0.00 | 0.07 |
| | BCMPsar1 | 0.02 | 0.02 | 0.01 | 0.01 | 0.00 | 0.00 | 0.27 | 0.34 |
| | BCMPsar2 | 0.01 | 0.02 | 0.01 | 0.01 | 0.00 | 0.00 | 0.25 | 0.31 |
| | OBS | 40 | 39 | 26 | 8 | 6 | 2 | 1 | 122 |
| | BP | 34.24 | 43.51 | 27.64 | 11.70 | 3.72 | 0.94 | 0.24 | 121.98 |
| | BNB | 40.65 | 38.90 | 23.41 | 11.32 | 4.80 | 1.87 | 1.02 | 122.00 |
| | BGP | 38.90 | 40.32 | 24.38 | 11.36 | 4.55 | 1.65 | 0.83 | 122.00 |
| | BCMPtriv | 40.54 | 38.67 | 23.66 | 11.54 | 4.85 | 1.82 | 0.91 | 121.96 |
| | BCMPcomp | 40.84 | 38.27 | 23.62 | 11.62 | 4.91 | 1.84 | 0.90 | 122.00 |
| | BCMPsar1 | 41.21 | 37.99 | 23.61 | 11.65 | 4.87 | 1.81 | 0.87 | 122.01 |
| | BCMPsar2 | 40.58 | 38.29 | 23.92 | 11.76 | 4.86 | 1.75 | 0.83 | 121.99 |

we illustrate model flexibility by comparing various associated bivariate discrete distributions and their ability to model the number of Forwards and Centers from 2000 to 2016 in the All-Star game: the BP, BNB, BGP, BCMPtriv, BCMPcomp, BCMPsar1, and BCMPsar2.

Table 4.5 *Respective maximum likelihood estimates (MLEs), log-likelihood (ln* L*) values, Akaike information criterion (AIC), and* $\Delta_i = AIC_i - AIC_{min}$ *values for various bivariate models, namely the bivariate Poisson distribution (BP), bivariate negative binomial (BNB), bivariate generalized Poisson (BGP), and four BCMP models attained via trivariate reduction (BCMPtriv), compounding (BCMPcomp), and two Sarmanov family approaches (BCMPsar1 and BCMPsar2), respectively, on the number of Forward and Center players dataset.*

| Model | Parameter MLEs | $\ln L$ | No. of param. | AIC | $\Delta_i$ |
|---|---|---|---|---|---|
| BP | $\hat{\lambda}_1 = 2.941$, $\hat{\lambda}_2 = 2.647$, $\hat{\lambda}_3 = 0$ | −54.395 | 3 | 114.790 | 14.269 |
| BNB | $\hat{m} = 2.938$, $\hat{r} = 100{,}000$, $\hat{\alpha}_1 = 0.899$, $\hat{\alpha}_2 = 0$ | −54.397 | 4 | 116.794 | 16.273 |
| BGP | $\hat{\theta}_1 = 0.560$, $\hat{\theta}_2 = 0.605$, $\hat{\theta}_3 = 4.048$, $\hat{\lambda}_1 = 0.324$, $\hat{\lambda}_2 = \text{-}0.133$, $\hat{\lambda}_3 = \text{-}1.000$ | −46.661 | 6 | 105.322 | 4.801 |
| BCMPtriv | $\hat{\lambda}_1 = 67.249$, $\hat{\lambda}_2 = 48.573$, $\hat{\lambda}_3 = 0$, $\hat{\nu} = 3.515$ | −46.262 | 4 | 100.521 | – |
| BCMPcomp | $\hat{\lambda} = 1,082,035$, $\hat{\nu} = 8.370$, $\hat{p}_{00} = 0$, $\hat{p}_{01} = 0.158$, $\hat{p}_{10} = 0.185$, $\hat{p}_{11} = 0.658$ | −47.986 | 5 | 105.972 | 5.451 |
| BCMPsar1 | $\hat{\lambda}_1 = 10.000$, $\hat{\nu}_1 = 2.193$, $\hat{\lambda}_2 = 10.000$, $\hat{\nu}_2 = 2.015$, $\hat{\alpha} = 0.208$, $\hat{\beta} = -1.000$ | −46.800 | 6 | 105.600 | 5.079 |
| BCMPsar2 | $\hat{\lambda}_1 = 10.000$, $\hat{\nu}_1 = 2.196$, $\hat{\lambda}_2 = 10.000$, $\hat{\nu}_2 = 2.018$, $\hat{\beta} = 1.000$ | −48.067 | 5 | 106.134 | 5.613 |

Table 4.5 provides the reported MLEs for the respective models, along with their associated log-likelihood, number of model parameters, Akaike information criterion (AIC), and associated AIC difference values $\Delta_i$ for model comparison as described in Chapter 1; again, see Table 1.2 for inferred model support levels based on $\Delta_i$ as described in Burnham and Anderson (2002). The BCMPtriv model performs optimally among the considered models, producing the largest log-likelihood (−46.262) and smallest AIC ($AIC_{min}$ = 100.521). The BGP and BCMPsar1 models

meanwhile, respectively, produce log-likelihoods that are close to that from the BCMPtriv model ($-46.661$ and $-46.8$, respectively); however, they each require two more model parameters; thus, their respective $\Delta_i$ values (4.801 and 5.079) imply that there is considerably less support for these models than the BCMPtriv model. Similarly, there is considerably less support for the BCMPcomp and BCMPsar2 models ($\Delta_i = 5.451$ and 5.613, respectively) than the BCMPtriv model.

All of the BCMP models recognize that these data are under-dispersed; in all cases, the relevant dispersion parameter(s) is (are) greater than 1. The BCMPtriv and BCMPcomp models, for example, detect statistically significant under-dispersion (Weems et al., 2021). Thus, all of the associated BCMP models report larger log-likelihood values than those of the BP and BNB models (both approximately $-54.4$). In fact, it makes sense that the BP and BNB models would report equal log-likelihoods because neither model can sufficiently account for data under-dispersion. The best that the BNB model can do is perform in-kind with the BP model by $\hat{r} \to \infty$ in that the BNB can only accommodate data equi- or over-dispersion. Neither the BP nor BNB models can address data under-dispersion; yet, the BNB model requires an extra parameter; thus, even though the BP and BNB models produce the same log-likelihood, the BNB reports a larger AIC than the BP. Regardless, there is essentially no empirical support associated with either the BP or BNB models ($\Delta_i > 10$ in both cases) when compared with the BCMPtriv model. In this example, the BCMPtriv proves itself to offer a simple, optimal form.

## 4.6 Summary

This chapter summarizes the various approaches that have been implemented to establish a BCMP or MultiCMP distribution. While each of these methods is motivated by the univariate CMP distribution, they each contain different attributes and qualities associated with its multivariate form; Table 4.6 summarizes the BCMP distributions described in this chapter, along with their respective characteristics.

Among the qualities featured, the correlation is arguably the most pertinent measure. The correlation receives prominent attention because the ideal bivariate or multivariate model would have a form that attains the $[-1,1]$ correlation structure that is achieved for the multivariate continuous (e.g. the multivariate Gaussian) distributions. To that end, the Sarmanov and copula approaches both produce such desired correlations, while the

Table 4.6 *Bivariate CMP development approaches (trivariate reduction; compounding; the Sarmanov families considering the CMP distribution as a weighted Poisson (Sarmanov 1) or based on the CMP probability generating function (Sarmanov 2), respectively; and copulas) and associated qualities. For each of the considered approaches, the correlation range and reported special-case distributions attainable for the bivariate (Biv.) and marginal (Marg.) distributions are supplied.*

| | Method | | | | |
|---|---|---|---|---|---|
| | Triv Red. | Compounding | Sarm. 1 | Sarm. 2 | Copulas |
| Correlation | [0,1] | [0,1] | [−1,1] | [−1,1] | [−1,1] |
| Marg. special cases | sCMP | Poisson | CMP | CMP | CMP |
| Biv. special cases | Holgate (1964) Poisson | Holgate (1964) Poisson, Bern, Geom | – | Lee (1996) Poisson | See Discussion |

trivariate reduction and compounding approaches are limited to nonnegative correlations. Another property of interest is to establish a bivariate or multivariate distribution whose marginal structure is a familiar form. To some, the ideal distribution has marginals that are themselves the univariate form of the multivariate distribution. Among the featured bivariate distributions in this chapter, the copula approach is the only means by which to establish a BCMP distribution that contains univariate CMP marginal distributions by definition (and, hence, its special-case distributions, namely the Poisson, Bernoulli, and geometric models). The BCMPtriv model, however, produces sCMP (discussed in Section 3.2) marginals; hence, its marginal distributions can be Poisson, negative binomial, or binomial distributions as special cases. Finally, the BCMPsar1 and BCMPsar2 models do not have known marginal distributional structures.

For all of the considered BCMP distributions, an appealing quality is their flexibility to accommodate count data containing dispersion. Accordingly, various BCMP distributions contain well-studied discrete bivariate models as special cases. Both the BCMPtriv and BCMPcomp models contain the Holgate (1964) BP distribution as a special case; however, the BCMPcomp further contains the bivariate Bernoulli and bivariate geometric distributions (Johnson et al., 1997; Marshall and Olkin, 1985). While Ong et al. (2021) do not report the existence of any special distributions ascertained from the BCMPsar1 model; the BCMPsar2 contains the Lee (1996) BP distribution as a special case. The copula approach does allow for BP distributions as special cases, although they are not contained in Table 4.6. These cases are not included here because identifying the special-case distribution is contingent on the choice of copula function used for BP and BCMP development. Given the numerous choices of copula function, we refer the interested reader to Joe (2014) and Nelsen (2010) for discussion.

Each of these approaches appears to be naturally generalizable, thus allowing for the development of analogous MultiCMP distributions; however, the literature does not yet exist regarding their explicit developments and associated statistical properties; to date, only the MultiCMPcomp distribution has been developed and studied (Sellers et al., 2021b). Further generalizations can likewise be considered, for example, the development of a bivariate and multivariate zero-inflated CMP distribution (Santana et al., 2021). All cases generalizing from the bivariate to multivariate CMP, however, introduce computational complexity as the dimensionality increases.

# 5

# COM–Poisson Regression

Regression modeling is a fundamental statistical method for describing the association between a (set of) predictor variable(s) and a response variable. For scenarios where the response variable describes count data, one must further account for the conditional data dispersion of the response variable given the explanatory variable(s). This chapter introduces the Conway–Maxwell–Poisson[1] (COM–Poisson) regression model, along with adaptations of the model to account for zero-inflation, censoring, and data clustering. Section 5.1 motivates the consideration and development of the various COM–Poisson regressions considered in this chapter by first providing a broader context of generalized linear models, and then the constraints associated with several such models that are popular for analysis. Section 5.2 introduces the COM–Poisson regression model and discusses related issues, including parameter estimation, hypothesis testing, and statistical computing in R (R Core Team, 2014). Section 5.3 advances that work to address excess zeroes, while Section 5.4 describes COM–Poisson models that incorporate repeated measures and longitudinal studies. Section 5.5 focuses attention on the R statistical packages and functionality associated with regression analysis that accommodates excess zeroes and/or clustered data as described in the two previous sections. Section 5.6 considers a general additive model based on COM–Poisson. Finally, Section 5.7 informs readers of other statistical computing software

[1] "Conway–Maxwell–Poisson" is abbreviated in various literatures as COM–Poisson or CMP. To avoid confusion for the reader, this reference will maintain the use of "COM–Poisson" for general referencing of Conway–Maxwell–Poisson models, and "CMP" will be reserved for discussion that uses the CMP($\lambda, \nu$) parametrization as described in Section 2.2. Other COM–Poisson parametrizations are similarly defined, and their respective abbreviations are noted in Section 2.6.

applications that are also available to conduct the COM–Poisson regression, discussing their associated functionality. This section acknowledges the broader impact attained by the COM–Poisson regression as it relates to statistical computing and its recognized flexibility and significance to statistical analysis. Section 5.8 summarizes and concludes the chapter with discussion.

## 5.1 Introduction: Generalized Linear Models

Generalized linear modeling describes a class of models that associate the mean of a collection of independent response values with predictor variables via a specified link function. Consider a collection of $n$ independent responses $\boldsymbol{Y} = (Y_1, \ldots, Y_n)'$ that stem from an exponential family distribution with mean $\mu_i$, $i = 1, \ldots, n$. Further, consider a linear predictor based on explanatory variables $\boldsymbol{X}_i = (X_{i1}, \ldots, X_{i,p_1-1})$, $p_1$ parameters $\boldsymbol{\beta} = (\beta_0, \ldots, \beta_{p_1-1})'$, and some link function $g(\cdot)$ such that $g(\mu_i) = \boldsymbol{X}_i\boldsymbol{\beta} = \beta_0 + \beta_1 X_{i1} + \cdots + \beta_{p_1-1}X_{i,p_1-1}$. Table 5.1 provides the common link functions associated with the usual Gaussian/normal regression, as well as the logistic and Poisson regression. Sections 5.1.1 through 5.1.3 highlight generalized linear models that motivate and develop consideration for a COM–Poisson regression – logistic (Section 5.1.1), Poisson (Section 5.1.2), and negative binomial (NB) (Section 5.1.3) regressions. Each of these regressions serve as fundamental tools for analyzing associations between a discrete response variable of a type $Y$ and a collection of explanatory variables. The popularity of these models is demonstrated in part through their available functionality via many statistical software packages (e.g. R and SAS).

### *5.1.1 Logistic Regression*

Logistic regression is an example of a generalized linear model where the response variable is binary (typically $\{0, 1\}$). This representation naturally

Table 5.1 *Structure of various generalized linear models.*

| Regression | Link function | $g(\mu)$ |
|---|---|---|
| Gaussian | Identity | $g(\mu) = \mu$ |
| Logistic | Logit | $g(\mu) = \text{logit}(\mu) = \ln\left(\frac{\mu}{1-\mu}\right)$ |
| Poisson | Log | $g(\mu) = \ln(\mu)$ |

allows for the Bernoulli distribution with associated success probability $E(\boldsymbol{Y}) = \boldsymbol{p}$ to describe the distribution. A popular link function for modeling the relationship between the success probability of interest and the covariates is the logit link function as defined in Table 5.1. This link's popularity in association with modeling logistic regression functions stems from its ability to transform the parameter space from the unit interval to the unconstrained real line for easy association with the linear model. The mean of the Bernoulli distribution equals $\boldsymbol{p} \in (0, 1)$; the logit function transforms the parameter space to $\ln\left(\frac{p}{1\text{-}p}\right) \in (-\infty, \infty)$, ensuring that the model relationship describing $\boldsymbol{p}$ remains confined within the unit interval. Another benefit to this construct lies in its ability for coefficient interpretability based on the odds of an outcome.

Given the $n$ independent Bernoulli random variables $Y_i, i = 1, \ldots, n$, with respective success probability $p_i$, the associated likelihood function is

$$L(\beta_0, \beta_1; \boldsymbol{y}) = \prod_{i=1}^{n} p_i^{y_i}(1 - p_i)^{1-y_i},$$

where $\ln\left(\frac{p}{1\text{-}p}\right) = \boldsymbol{X\beta}$; hence, the log-likelihood is

$$\ln L(\beta_0, \beta_1; \boldsymbol{y}) = \sum_{i=1}^{n} [y_i(\beta_0 + \beta_1 X_i)] + \sum_{i=1}^{n} \ln[1 + \exp(\beta_0 + \beta_1 X_i)]. \quad (5.1)$$

Equation (5.1) does not have a closed-form solution for the corresponding normal equations; thus, the maximum likelihood estimators can be determined via numerical optimization procedures. Statistical computing in `R` is achieved via the `glm` function (contained in the `stats` package) with the specified input `family = logit`.

### 5.1.2 Poisson Regression

Poisson regression is another generalized linear model – the response variable is now represented by discrete count values $\mathbb{N} = \{0, 1, 2, \ldots\}$. Count data are naturally represented by the Poisson distribution with associated intensity or rate parameter $E(Y) = \lambda$ as described in Chapter 1. A popular choice for modeling the relationship between this parameter and the covariates is the log link function $\ln(\boldsymbol{\lambda}) = \boldsymbol{X\beta}$ so that the parameter space for $\boldsymbol{\lambda} \in (0, \infty]$ transforms to the real line $\mathbb{R}$. The loglinear model likewise allows for understandable coefficient interpretation as a multiplicative effect associating with the mean outcome. Several references provide detailed insights regarding Poisson regression, e.g. McCullagh and Nelder (1997), Hilbe (2014), Dobson and Barnett (2018); interested readers are encouraged to refer to these references. The `glm` function in `R` allows analysts to conduct Poisson regression by specifying the input `family = poisson`.

### *5.1.3 Addressing Data Over-dispersion: Negative Binomial Regression*

A significant criticism of the Poisson distribution is that its mean and variance equal, i.e. data equi-dispersion holds; see Chapter 1 for details. In the regression context, this implies a constraint that the conditional mean and variance equal; however, real data do not generally conform to this assumption. Over-dispersion occurs more prevalently in count data and is caused by any number of factors, including the positive correlation between responses, excess variation between response probabilities or counts, or other violations in distributional assumptions associated with the data (Hilbe, 2007). This is a serious problem in data analysis because it can cause statistical significance to be overestimated (i.e. explanatory variable associations identified as statistically significant may not actually be so, or at least not to the same level as initially noted from the perspective of a Poisson regression).

A popular approach to address data over-dispersion is via NB regression. Under this construct, the response variables $Y_i$ are NB($r$, $\frac{r}{r+\mu_i}$) distributed as described in Equation (1.9) with $r > 0$ and success probabilities $\frac{r}{r+\mu_i}$, where $\mu_i > 0$ denotes the mean of the response random variable $Y_i$, and we assume the log link relation, $\ln(\mu_i) = \boldsymbol{X}_i\boldsymbol{\beta}$, $i = 1, \ldots, n$; see Section 1.2 for details. With the mean $\mu_i$ and variance $\mu_i + \frac{\mu_i^2}{r}$, respectively, the NB distribution allows for data over-dispersion because the variance clearly exceeds the mean. The NB regression is well studied and, because of its nice properties, widely used for analyzing over-dispersed data. Statistical computing in `R` is achieved via the `glm.nb` function (contained in the `MASS` package).

There are significant drawbacks, however, associated with the NB model. As described, the variance is always greater than or equal to the mean; thus, the NB distribution is unable to address data under-dispersion. Further, the NB regression requires fixing $r$ in order to express the log-likelihood in the form of a generalized linear model (McCullagh and Nelder, 1997). Focusing their attention on crash data, Lord et al. (2008) likewise note several documented limitations associated with the Poisson and NB models. For data containing a small sample mean, goodness-of-fit methods associated with either Poisson or NB generalized linear models where the method of maximum likelihood is used for parameter estimation have been found to be biased and therefore unreliable. In particular, the dispersion parameter associated with NB models is significantly biased for datasets with small sample size and/or small sample mean, whether determined via maximum likelihood (Clark and Perry, 1989; Lord, 2006; Piegorsch, 1990) or Bayesian estimation (Airoldi et al., 2006; Lord and Miranda-Moreno, 2008). Such errors impact empirical Bayesian estimation and constructions of confidence intervals.

### *5.1.4 Addressing Data Over- or Under-dispersion: Restricted Generalized Poisson Regression*

Data under-dispersion, while often believed to be the result of some anomaly, is surfacing more frequently in some applications. For example, some crash data have expressed under-dispersion, particularly when the sample mean is small (Oh et al., 2006). An alternative to the Poisson model that accounts for data over- or under-dispersion is the restricted generalized Poisson (RGP) regression (Famoye, 1993). This model has the form

$$P(Y_i = y_i \mid \mu_i, \alpha) = \left(\frac{\mu_i}{1+\alpha\mu_i}\right)^{y_i} \frac{(1+\alpha y_i)^{y_i-1}}{y_i!} \exp\left(\frac{-\mu_i(1+\alpha y_i)}{1+\alpha\mu_i}\right),$$
$$y_i = 0, 1, 2, \ldots, \tag{5.2}$$

where the link function is $\ln(\mu_i) = \boldsymbol{X}_i\boldsymbol{\beta}$ and $\alpha$ is restricted such that $1 + \alpha\mu_i > 0$ and $1 + \alpha y_i > 0$. Poisson regression is noted as the special case of RGP regression where $\alpha = 0$, while $\alpha > 0$ and $\frac{-2}{\mu_i} < \alpha < 0$, respectively, address data over- and under-dispersion. The obvious benefit of this regression model lies in its ability to address data over- or under-dispersion; however, the RGP regression can only handle under-dispersion in a limited sense because of the constrained space for $\alpha$. For such data, the RGP can violate probability axioms in that the probability mass function "gets truncated and does not necessarily sum to one" (Famoye et al., 2004). Further, the regression model belongs to an exponential family only for a constant $\alpha$. Accordingly, a model with observation-specific dispersion no longer belongs to the exponential family (Sellers and Shmueli, 2010).

## 5.2 Conway–Maxwell–Poisson (COM–Poisson) Regression

Conway–Maxwell–Poisson (COM–Poisson) regression is an alternative model that addresses data over- or under-dispersion that avoids the performance constraints and associated implications that exist with the aforementioned count regression models. This section utilizes the statistical qualities of the COM–Poisson distribution as outlined in Chapter 2 to describe an associated regression framework. The COM–Poisson generalized linear model is a flexible model that can better accommodate under-dispersed data and model over-dispersed data in a manner comparable in performance with the NB regression (Guikema and Coffelt, 2008; Sellers and Shmueli, 2010).

### 5.2.1 Model Formulations

Several model formulations have been proposed for COM–Poisson regression, motivated by the respective parametrizations described in Chapter 2 (particularly Section 2.6). Early work in this area assumes a constant dispersion parameter, focusing the model association on the relevant intensity parameter for the COM–Poisson parametrization under consideration. COM–Poisson regression analysis, however, can generalize to allow for both observation-level intensity and dispersion modeling.

Sellers and Shmueli (2010) work with the original CMP($\boldsymbol{\lambda}, \nu$) parametrization and associate the respective parameters directly to the covariates, utilizing the log link function $\ln(\lambda_i) = \boldsymbol{X}_i\boldsymbol{\beta}$ to describe the relationship between the response and explanatory variables while assuming constant dispersion. This choice of formulation does not necessarily consistently allow for easy coefficient interpretation; however, it remains quite useful because it generalizes the formulations described for logistic and Poisson regression (see Sections 5.1.1 and 5.1.2). In Poisson regression, this link precisely associates the mean and explanatory variables, as described earlier. Meanwhile, in logistic regression, the success probability is $\boldsymbol{p} = \frac{\lambda}{1+\lambda}$, and the proposed CMP link function reduces to $\text{logit}(\boldsymbol{p}) = \ln(\boldsymbol{\lambda})$. The CMP regression link $\ln(\boldsymbol{\lambda})$ further leads to elegant estimation, inference, and diagnostics, given the CMP properties described in Chapter 2.

Some researchers argue against CMP regression because any specified link function for generalized linear modeling does not offer a straightforward association between the mean response and explanatory variables, making coefficient interpretation difficult. Neither $\lambda$ nor $\nu$ offer a clear centering parameter; while $\lambda$ is close to the mean when $\nu$ is near 1, this is not true for cases of significant dispersion (Guikema and Coffelt, 2008). Particularly, in cases of data over-dispersion, Lord et al. (2008) argue that the small $\nu$ makes a CMP model difficult with regard to parameter estimation. They instead use the ACMP($\boldsymbol{\mu}_*, \nu$) parametrization with $\ln(\boldsymbol{\mu}_*) = \ln\left(\boldsymbol{\lambda}^{1/\nu}\right) = \boldsymbol{X\beta}$, arguing that the approximate COM–Poisson (ACMP) regression better allows analysts to interpret coefficients by describing their association with the centering parameter $\boldsymbol{\mu}_*$; the dispersion $\boldsymbol{\nu}$ can meanwhile be assumed constant or varying with the link function $\ln(\boldsymbol{\nu}) = \boldsymbol{G\gamma}$ (Guikema and Coffelt, 2008; Lord and Guikema, 2012; Lord et al., 2008). Other works consider a broader loglinear model for $\boldsymbol{\mu}_*$, namely

$$\ln\left(\frac{\mu_{*i}}{h_i}\right) = \boldsymbol{x}_i\boldsymbol{\beta} + \phi_i \tag{5.3}$$

while defining the loglinear model allowing for varying dispersion as $\ln(\nu_i) = -\boldsymbol{g}_i\boldsymbol{\gamma}$ (Chanialidis et al., 2014, 2017). As represented in Equation (5.3), $\boldsymbol{\mu}_*$ can further allow for varying offset $h_i$ and spatial autocorrelation through random effects $\phi_i$; see Section 5.2.2 for further discussion. Meanwhile, considering the latter approach where the model for $\boldsymbol{\nu}$ is negated allows for positive (negative) values for $\boldsymbol{\gamma}$ to associate now with potential data over- (under-)dispersion. While these ACMP formulations offer some promising considerations, analysts should use them with caution. Lord et al. (2008) report that this configuration links to the mean or variance; thus, the regression can model data that are over- or under-dispersed or a mix of the two (when allowing for a varying dispersion model). Chanialidis et al. (2017) likewise stress that the approximations to the mean and variance achieved via ACMP regression are reasonable when $\boldsymbol{\mu}_*$ and $\boldsymbol{\nu}$ are sufficiently large; however, this implies that ACMP regression is not necessarily appropriate on under-dispersed response data. The underlying reparametrization as $\boldsymbol{\mu}_* = \boldsymbol{\lambda}^{1/\nu}$ further approximates the already approximate closed-form representation of the expected value provided in Equation (2.14) that holds only if the dataset is conditionally over-dispersed ($\nu \leq 1$) or for $\boldsymbol{\lambda}$ sufficiently large ($\boldsymbol{\lambda} > 10^{\nu}$); thus, this refined approximation is satisfied only in an even more constrained space. Francis et al. (2012), in fact, note that the ACMP-generalized linear model inferences display bias when the data have a low mean and are under-dispersed. Further, while the ACMP parametrization views $\mu_*$ as a centering parameter, it is not the distribution mean. Thus, analysts must bear in mind these issues when conducting ACMP regression, particularly with regard to accuracy and coefficient interpretation.

Alternative parametrizations based on better approximations of the mean have likewise resulted in the development of associated regression models. The MCMP1 regression considers the model where $\boldsymbol{Y}$ (conditioned on the design matrix $\boldsymbol{X}$) has an MCMP1($\boldsymbol{\mu}, \nu$) distribution (Equation (2.51)) with link function $\ln(\boldsymbol{\mu}) = \boldsymbol{X\beta}$ and constant dispersion $\nu$. The underlying motivation for this construct stems from the fact that the MCMP1 parametrization produces mean and dispersion parameters ($\mu$ and $\nu$) that are orthogonal, a beneficial quality in a regression setting. The maximum likelihood estimators for the regression coefficients $\hat{\boldsymbol{\beta}} = (\hat{\beta}_0, \hat{\beta}_1, \ldots, \hat{\beta}_{p_1})$ will be asymptotically efficient and asymptotically independent of $\hat{\nu}$. This reparametrization further allows one to easily incorporate offsets into the model where the mean is updated accordingly to $\exp(\textbf{offset}) \times \mu(\boldsymbol{X\beta})$, i.e. the MCMP1 distribution is now updated to MCMP1($\exp(\textbf{offset}) \times \mu(\boldsymbol{X\beta}), \nu$) and (say) assuming a log linear link function; we now have $\ln(\boldsymbol{\mu}) = \boldsymbol{X\beta} + \textbf{offset}$. MCMP2 regression is analogously developed assuming that the response vector $\boldsymbol{Y}$ has an MCMP2($\boldsymbol{\mu}, \phi$) distribution (Equation (2.51)), where $\ln(\boldsymbol{\mu}) = \boldsymbol{X\beta}$ for $\boldsymbol{\mu} \doteq \boldsymbol{\lambda}^{1/\nu} - \frac{\nu-1}{\nu}$ and $\phi = \ln(\nu)$ is constant. Like MCMP1, MCMP2 regression converges quickly because of the orthogonality of the resulting parameters and provides natural coefficient interpretation under the usual generalized linear model construct. Convergence via MCMP2 regression is arguably faster than that for MCMP1 because $\mu$ under the MCMP2 model is obtained algebraically (thus providing a simpler form) rather than computationally as is done with MCMP1 regression (Ribeiro Jr. et al., 2019). Analysts, however, should utilize MCMP2 regression with caution. While its approximation for the mean is more precise than that provided for the ACMP regression, it still relies on the constraints for the mean approximation described in Equation (2.23) to hold.

While all of the model formulations are introduced assuming a constant dispersion parameter, data dispersion can likewise vary and thus be modeled via an appropriate link function (e.g. $\ln(\nu_i) = \boldsymbol{G}_i\boldsymbol{\gamma}$, where $\boldsymbol{\gamma} = (\gamma_0, \gamma_1, \ldots, \gamma_{p_2-1})$ ) to associate the amount of dispersion with explanatory variables of interest across measurements (Sellers and Shmueli, 2009, 2013). This approach of considering varying dispersion is consistent with that for the NB regression described in Miaou and Lord (2003) and offers an insight into the relative effects of different covariates for both the mean and variance of the counts.

### *5.2.2 Parameter Estimation*

Coefficient and parameter estimations in the COM–Poisson regression format have been achieved via the method of maximum likelihood, moment-based estimation methods, including the marginal and joint generalized quasi-likelihood, and Bayesian estimation via the Markov Chain Monte Carlo and Metropolis–Hastings. The following subsections detail these three ideas.

*Maximum Likelihood Estimation*

The most popular approach for COM–Poisson regression coefficient estimation is the method of maximum likelihood; it is the only method pursued for each of the respective COM–Poisson parametrizations. Consider first a CMP($\lambda_i, \nu$) regression model that allows for constant dispersion. The log-likelihood provided in Equation (2.10) is thus modified to allow for varying $\lambda_i$, updating the log-likelihood to

$$\ln L(\boldsymbol{\beta}, \nu \mid \boldsymbol{x}, \boldsymbol{y}) = \sum_{i=1}^{n} y_i \ln(\lambda_i) - \nu \sum_{i=1}^{n} \ln(y_i!) - \sum_{i=1}^{n} \ln(Z(\lambda_i, \nu)), \quad (5.4)$$

where we assume $\ln(\boldsymbol{\lambda}) = \boldsymbol{X\beta}$ with an $n \times p_1$ design matrix $\boldsymbol{X}$ and $p_1$-parameter vector $\boldsymbol{\beta}$. Differentiating Equation (5.4) with respect to $\beta_j$, $j = 0, \ldots, p_1 - 1$, and $\nu$, the corresponding score equations satisfy

$$\sum_{i=1}^{n} y_i x_{ij} = \sum_{i=1}^{n} \left\{ x_{ij} \frac{\sum_{s=0}^{\infty} s e^{sX_i\beta} / (s!)^{\nu}}{\sum_{s=0}^{\infty} e^{sX_i\beta} / (s!)^{\nu}} \right\}, \qquad j = 0, \ldots, p_1 - 1 \quad (5.5)$$

$$\sum_{i=1}^{n} \ln y_i! = \sum_{i=1}^{n} \left\{ \frac{\sum_{s=0}^{\infty} \ln(s!) e^{sX_i\beta} / (s!)^{\nu}}{\sum_{s=0}^{\infty} e^{sX_i\beta} / (s!)^{\nu}} \right\}. \quad (5.6)$$

Similarly, the ACMP($\mu_{*i}, \nu$) reparameterized form (Equation (2.48)) produces the log-likelihood

$$\ln L(\boldsymbol{\beta}, \nu \mid \boldsymbol{x}, \boldsymbol{y}) = \nu \sum_{i=1}^{n} y_i \eta_i - \nu \sum_{i=1}^{n} \ln(y_i!) - \sum_{i=1}^{n} \ln(Z_1(\eta_i, \nu)), \quad (5.7)$$

where $\eta_i = \ln(\mu_{*i}) = \boldsymbol{x}_i \boldsymbol{\beta}$, while the MCMP1($\mu_i, \nu$) log-likelihood resembles Equation (5.4) with $\lambda_i$ updated to $\lambda(\mu_i, \nu)$, and the MCMP2($\mu_i, \phi$) log-likelihood is

$$\ln L(\boldsymbol{\beta}, \phi \mid \boldsymbol{x}, \boldsymbol{y}) = e^{\phi} \left[ \sum_{i=1}^{n} y_i \ln\left( \mu_i + \frac{e^{\phi} - 1}{2e^{\phi}} \right) - \sum_{i=1}^{n} \ln(y_i!) \right] - \sum_{i=1}^{n} \ln(Z_2(\mu_i, \phi)), \quad (5.8)$$

where $\ln(\mu_i) = \boldsymbol{x}_i \boldsymbol{\beta}$ and $\phi = \ln(\nu)$. Each of the respective log-likelihoods can be used to produce their corresponding score equations.

As is the case for all of these COM–Poisson parametrizations, the resulting score equations are nonlinear with respect to the coefficients and do not have a closed form; thus, numerical approaches must be considered to obtain the maximum likelihood estimates (MLEs). Sellers and Shmueli (2010) suggest using the generalized linear model framework to formulate the CMP likelihood maximization as a weighted least-squares procedure and solving it iteratively. Consider a reweighted least squares of the form

$$\boldsymbol{B}'\boldsymbol{W}\boldsymbol{B}\boldsymbol{\theta}^{(m)} = \boldsymbol{B}'\boldsymbol{W}\boldsymbol{T}, \tag{5.9}$$

where $\boldsymbol{B} = [\boldsymbol{X} \mid h(\boldsymbol{Y})]$ is an $n \times (p_1 + 1)$ matrix that binds the design matrix $\boldsymbol{X}$ with $h(\boldsymbol{Y}) = \left(\frac{-\ln(Y_1!)+E(\ln(Y_1!))}{Y_1-E(Y_1)}, \ldots, \frac{-\ln(Y_n!)+E(\ln(Y_n!))}{Y_n-E(Y_n)}\right)'$, $\boldsymbol{W} = \text{diag}(V(Y_1), \ldots, V(Y_n))$ is an $n \times n$ diagonal matrix, $\boldsymbol{\theta}^{(m)} = (\hat{\beta}_0^{(m)}, \ldots, \hat{\beta}_{p_1-1}^{(m)}, \hat{\nu}^{(m)})'$ is the $m$th iteration of the estimated coefficient vector of length $(p_1 + 1)$, and $\boldsymbol{T}$ is a vector of length $n$, with elements involving coefficients estimated in the previous step,

$$t_i = \boldsymbol{x}_i'\boldsymbol{\beta}^{(m-1)} + h(y_i)\nu^{(m-1)} + \frac{Y_i - E(Y_i)}{V(Y_i)}. \tag{5.10}$$

Standard errors associated with the estimated coefficients are meanwhile derived via the $(p_1 + 1) \times (p_1 + 1)$ Fisher information matrix with the block form

$$\boldsymbol{I} = \begin{pmatrix} \boldsymbol{I}^{\boldsymbol{\beta}} & \boldsymbol{I}^{\boldsymbol{\beta},\nu} \\ \boldsymbol{I}^{\boldsymbol{\beta},\nu} & I^{\nu} \end{pmatrix}, \tag{5.11}$$

where $\boldsymbol{I}^{\boldsymbol{\beta}}$ is a $p_1 \times p_1$ matrix denoting the estimated variance–covariance matrix associated with $\hat{\boldsymbol{\beta}}$, $I^{\nu}$ equals the estimated variance for $\hat{\nu}$, and $\boldsymbol{I}^{\boldsymbol{\beta},\nu}$ is a $p_1$-length vector of covariance estimates between $\hat{\boldsymbol{\beta}}$ and $\hat{\nu}$. These components have the form

$$I^{\boldsymbol{\beta}}_{j,k} = \sum_{i=1}^{n} x_{ij}x_{ik} \left\{ \frac{\sum_{s=0}^{\infty} s^2 e^{sX_i\boldsymbol{\beta}}/(s!)^{\nu}}{\sum_{s=0}^{\infty} e^{sX_i\boldsymbol{\beta}}/(s!)^{\nu}} - \left[\frac{\sum_{s=0}^{\infty} s e^{sX_i\boldsymbol{\beta}}/(s!)^{\nu}}{\sum_{s=0}^{\infty} e^{sX_i\boldsymbol{\beta}}/(s!)^{\nu}}\right]^2 \right\},$$
$$j,k = 0, \ldots, p_1 - 1$$

$$I^{\nu} = \sum_{i=1}^{n} x_{ij}x_{ik} \left\{ \frac{\sum_{s=0}^{\infty} (\ln(s!))^2 e^{sX_i\boldsymbol{\beta}}/(s!)^{\nu}}{\sum_{s=0}^{\infty} e^{sX_i\boldsymbol{\beta}}/(s!)^{\nu}} - \left[\frac{\sum_{s=0}^{\infty} (\ln(s!)) e^{sX_i\boldsymbol{\beta}}/(s!)^{\nu}}{\sum_{s=0}^{\infty} e^{sX_i\boldsymbol{\beta}}/(s!)^{\nu}}\right]^2 \right\}$$

$$I^{\beta_j,\nu} = \sum_{i=1}^{n} x_{ij}x_{ik} \left\{ \frac{\sum_{s=0}^{\infty} s(\ln(s!)) e^{sX_i\boldsymbol{\beta}}/(s!)^{\nu}}{\sum_{s=0}^{\infty} e^{sX_i\boldsymbol{\beta}}/(s!)^{\nu}} - \left(\frac{\sum_{s=0}^{\infty} s e^{sX_i\boldsymbol{\beta}}/(s!)^{\nu}}{\sum_{s=0}^{\infty} e^{sX_i\boldsymbol{\beta}}/(s!)^{\nu}}\right) \left(\frac{\sum_{s=0}^{\infty} (\ln(s!)) e^{sX_i\boldsymbol{\beta}}/(s!)^{\nu}}{\sum_{s=0}^{\infty} e^{sX_i\boldsymbol{\beta}}/(s!)^{\nu}}\right) \right\}.$$

The iterative reweighted least-squares form provided in Equation (5.9) stems from a Newton–Raphson iterative formulation, $\boldsymbol{\theta}^{(m)} = \boldsymbol{\theta}^{(m-1)} + \boldsymbol{I}^{-1}\boldsymbol{U}$, where $\boldsymbol{I}$ is the Fisher information matrix defined in Equation (5.11) and $\boldsymbol{U} = (U_0, \ldots, U_{p_1-1}, U_\nu)'$ is a vector of length $(p_1 + 1)$ with components $U_j = \sum_{i=1}^n \left\{ x_{ij} \frac{\sum_{s=0}^\infty s e^{s\boldsymbol{X}_i\boldsymbol{\beta}}/(s!)^\nu}{\sum_{s=0}^\infty e^{s\boldsymbol{X}_i\boldsymbol{\beta}}/(s!)^\nu} \right\}$, for $j = 0, \ldots, p_1 - 1$, and $U_\nu = \sum_{i=1}^n \left\{ \frac{\sum_{s=0}^\infty \ln(s!) e^{s\boldsymbol{X}_i\boldsymbol{\beta}}/(s!)^\nu}{\sum_{s=0}^\infty e^{s\boldsymbol{X}_i\boldsymbol{\beta}}/(s!)^\nu} \right\}$. Left multiplying both sides by $\boldsymbol{I} = \boldsymbol{B}'\boldsymbol{W}\boldsymbol{B}$ produces Equation (5.9). This iterative reweighted least-squares approach is one of several Newton-type iteration procedures that can be considered for optimization and can be directly programmed for statistical computing. Other Newton-type numerical optimization tools are readily available in R, e.g. `optim`, `nlminb`, or `nlm`. Ribeiro Jr. et al. (2019), for example, adopt the `optim` function's `BFGS` algorithm to obtain the MCMP2 regression coefficient MLEs. The added benefit to `optim` is that it can supply the corresponding Hessian matrix that can be used to produce the Fisher information matrix in order to approximate the corresponding estimate standard errors. Analysts can represent the dispersion parameter on a log-scale in order to allow for unconstrained optimization procedures to be considered for parameter estimation; this approach circumvents any potential convergence issues that would arise if optimization procedures are conducted on the original dispersion scale (Francis et al., 2012). Constrained optimizers such as `nlminb` (in the `stats` package), however, also exist to address such situations.

While the above framework assumes a constant dispersion $\nu$, one can likewise allow for varying dispersion that assumes (say) a loglinear relationship, i.e. $\ln(\boldsymbol{\nu}) = \boldsymbol{G}\boldsymbol{\gamma}$, where $\boldsymbol{G} = (g_1, \ldots, g_n)'$ is an $n \times p_2$ design matrix whose predictors can be either shared with or distinct from $\boldsymbol{X}$, and $\boldsymbol{\gamma} = (\gamma_1, \ldots, \gamma_{p_2})$. Focusing our attention first on the CMP regression, the weighted least-squares equation (Equation (5.9)) now has $\boldsymbol{B} = (\boldsymbol{X} \mid (h(\boldsymbol{Y}) \cdot \boldsymbol{G}))'$, $\boldsymbol{\theta}^{(m)} = (\boldsymbol{\beta}^{(m)} \mid \boldsymbol{\gamma}^{(m)})'$, and $\boldsymbol{T} = (t_1, \ldots, t_n)'$ with components

$$t_i = \boldsymbol{x}_i'\boldsymbol{\beta}^{(m-1)} + h(Y_i)\boldsymbol{g}_i'\boldsymbol{\gamma}^{(m-1)} + \frac{Y_i - E(Y_i)}{V(Y_i)}.$$

Motivated by the above constant dispersion iterative reweighted least squares, Chatla and Shmueli (2018) suggest an analogous two-step algorithm for parameter estimations of both $\boldsymbol{\beta}$ and $\boldsymbol{\gamma}$ that uses the expected information matrix to update estimates efficiently and allows for the CMP regression to be extended to include additive components or least absolute shrinkage and selection operator, i.e. LASSO (see Section 5.6 for discussion); Algorithm 4 provides the iterative reweighted least-squares algorithm. The authors offer three tips for performing this algorithm. First, set the suggested starting values for the algorithm to an approximate method

1. Initialize values $\nu_i^{(0)}$ and $\lambda_i^{(0)} = (y_i + 0.1)^{\nu_i^{(0)}}$, $i = 1, \ldots, n$.
2. Compute $\ln(\lambda_i^{(0)})$ and $\ln(\nu_i^{(0)})$, $i = 1, \ldots, n$.
3. Compute initial deviance,
   $D^{(0)}(\boldsymbol{\lambda}^{(0)}, \boldsymbol{\nu}^{(0)}) = -2\sum_{i=1}^{n} \ln L(y_i; \lambda_i^{(0)}, \nu_i^{(0)})$.
4. Compute $[E(y_i)]^{(0)}$ and $[V(y_i)]^{(0)}$, $i = 1, \ldots, n$.
5. For $k$ in 1 : itermax,
   (a) Compute $t_{i1}^{(k)} = \ln(\lambda_i^{(k-1)}) + \frac{y_i - [E(y_i)]^{(k-1)}}{[V(y_i)]^{(k-1)}}$, $i = 1, \ldots, n$.
   (b) Regress $\boldsymbol{T}_1^{(k)} = (t_{11}, \ldots, t_{n1})'$ on $\boldsymbol{X}$ via weighted least-squares regression with weights
   $\boldsymbol{W}_1^{(k-1)} = \text{diag}([V(y_1)]^{(k-1)}, \ldots, [V(y_n)]^{(k-1)})$ to determine $\boldsymbol{\beta}^{(k)}$.
   (c) Compute $\ln(\lambda_i^{(k)}) = \boldsymbol{x}_i'\boldsymbol{\beta}^{(k)}$ and $\lambda_i^{(k)}$, $i = 1, \ldots, n$.
   (d) Compute $[E(\ln(y_i!))]^{(k-1)}$ and $[V(\ln(y_i!))]^{(k-1)}$, $i = 1, \ldots, n$.
   (e) Compute $t_{i2}^{(k)} = \nu_i^{(k-1)} \ln(\nu_i^{(k-1)}) + \frac{-\ln(y_i!) + [E(\ln(y_i!))]^{(k-1)}}{[V(\ln(y_i!))]^{(k-1)}} \nu_i^{(k-1)}$.
   (f) Regress $\boldsymbol{T}_2^{(k)} = (t_{12}, \ldots, t_{n2})'$ on $\boldsymbol{\nu}^{(k-1)} \cdot \boldsymbol{G} = (\nu_1^{(k-1)} g_1, \ldots,$ $\nu_n^{(k-1)} g_n)$ via weighted least-squares regression with weights $\boldsymbol{W}_2^{(k-1)} = \text{diag}([V(\ln(y_1!))]^{(k-1)}, \ldots, [V(\ln(y_n!))]^{(k-1)})$ to determine $\boldsymbol{\gamma}^{(k)}$
   (g) Compute $\ln(\nu_i^{(k)}) = \boldsymbol{g}_i'\boldsymbol{\gamma}^{(k)}$ and $\nu_i^{(k)}$, $i = 1, \ldots, n$.
   (h) Compute deviance, $D^{(k)}(\boldsymbol{\lambda}^{(k)}, \boldsymbol{\nu}^{(k)}) = -2\sum_{i=1}^{n} \ln L(y_i; \lambda_i^{(k)}, \nu_i^{(k)})$.
      i. if $\frac{D^{(k)} - D^{(k-1)}}{D^{(k)}} > 10^{-6}$, then initiate step size optimization; end if.
      ii. if $\left|\frac{D^{(k)} - D^{(k-1)}}{D^{(k)}}\right| < 10^{-6}$, then convergence achieved; end loop. Else compute $[E(y_i)]^{(k)}$ and $[V(y_i)]^{(k)}$, $i = 1, \ldots, n$; end if.

   end for

**Algorithm 4** Iterative reweighted least-squares algorithm proposed by Chatla and Shmueli (2018) to conduct parameter estimation to determine the CMP regression coefficients, $\hat{\beta}$ and $\hat{\gamma}$.

of moments estimator for $\lambda_i$ (i.e., $\lambda_i^{(0)} = (y_i + 0.1)^{\nu_i^{(0)}}$) and $\nu_i^{(0)}$ close to zero (e.g. $\nu_i^{(0)} = 0.25$). Second, a modification of the deviance criterion (namely $-2\sum_{i=1}^{n} \ln L(y_i; \hat{\lambda}_i, \hat{\nu}_i)$) is proposed as the stopping rule in order to circumvent computational complexities associated with the estimates for $\lambda$ and $\nu$ under the saturated model. Finally, in order to address common convergence issues associated with iterative reweighted least squares, the step-halving approach of Marschner (2011) is encouraged; this approach uses step-halving at the boundary or when experiencing an increasing deviance in order to ensure staying within the interior support space for convergence (Chatla and Shmueli, 2018). The optimization procedure described above as well as those in R are analogously applicable for the varying dispersion framework under any parametrization.

Simulation studies show that each of the COM–Poisson generalized linear models (i.e. assuming any parametrization) can effectively model data under-, equi-, or over-dispersion through the flexible model and produce reasonable (properly scaled) parameter estimates and fitted values. The COM–Poisson models all likewise produce comparable confidence intervals that are larger (smaller) than Poisson confidence intervals when the data are over- (under-) dispersed. The MCMP1 and MCMP2 regressions are particularly appealing; however, because they can produce coefficient estimates that are similar across various generalized linear models (e.g. Poisson, NB, and GP). This is to be expected because these parametrizations better allow for generalized linear models where the function associating the mean to the explanatory variables is roughly identical across regression constructs. Accordingly, MCMP1 and MCMP2 most easily allow for an apples-to-apples comparison of coefficient estimates (followed by the ACMP regression), whereas any reported CMP coefficient estimates are not easily comparable to these reparametrized forms. Nonetheless, analysts should remain cautiously optimistic when utilizing the ACMP or MCMP2 parametrizations as they rely on the moment approximations and associated constraints for accuracy. As such, while asymptotic efficiency and normality hold for the coefficients under any parametrization, the orthogonality property attained via MCMP1 regression further implies that the coefficient estimates $\hat{\boldsymbol{\beta}}$ are asymptotically independent of $\hat{\nu}$ (Huang, 2017); analogous results hold for the case with varying dispersion. Simulation studies further show all regression MLEs (under any parametrization) to be at least asymptotically unbiased and consistent. The empirical coverage rates of the confidence intervals stemming from the asymptotic distribution of the MLEs are likewise close to their nominal levels. Some anamolies, however, can surface for various parametrizations. For small sample sizes, for example, the MCMP2 dispersion estimator is over estimated with an empirical distribution that is right-skewed and lower than nominal empirical coverage rates; the lowest empirical coverage rates occur in scenarios with small sample size and strong over-dispersion (Ribeiro Jr. et al., 2019).

### *Moment-based Estimation*

The method of maximum likelihood is arguably a computationally expensive process for parameter estimation because evaluating the likelihood function depends on the infinite sum that can be slow to converge. Some of the alternative approaches that have garnered attention for COM–Poisson regression coefficient estimation are based on the method of generalized quasi-likelihood. Jowaheer and Mamode Khan (2009) conduct joint generalized quasi-likelihood estimation where the respective COM–Poisson mean and variance approximations (Equations (2.23) and (2.24)) along with the moment recursion formula (Equation (2.16)) aid in deriving the joint quasi-likelihood equations. Assuming the approximations to be exact, the joint quasi-likelihood equations are

$$\sum_{i=1}^{n} \boldsymbol{D}_i' \boldsymbol{V}_i^{-1} (\boldsymbol{f}_i - \boldsymbol{\tau}_i) = \boldsymbol{0}, \tag{5.12}$$

where $\boldsymbol{V}_i$ is a $2 \times 2$ variance–covariance matrix of $\boldsymbol{f}_i' = (Y_i, Y_i^2)$, $\boldsymbol{\tau}_i = (\tau_{1i}, \tau_{2i})'$ such that $\tau_{ki}$ approximates $E(Y_i^k)$ as determined from Equations (2.23) and (2.24), and

$$\boldsymbol{D}_i = \begin{pmatrix} \frac{\partial \tau_{1i}}{\partial \beta_0} & \cdots & \frac{\partial \tau_{1i}}{\partial \beta_{p-1}} & \frac{\partial \tau_{1i}}{\partial \nu} \\ \frac{\partial \tau_{2i}}{\partial \beta_0} & \cdots & \frac{\partial \tau_{2i}}{\partial \beta_{p-1}} & \frac{\partial \tau_{2i}}{\partial \nu} \end{pmatrix}.$$

The Newton–Raphson method

$$\begin{pmatrix} \boldsymbol{\beta}_{(m+1)} \\ \nu_{(m+1)} \end{pmatrix} = \begin{pmatrix} \boldsymbol{\beta}_{(m)} \\ \nu_{(m)} \end{pmatrix} + \left[ \sum_{i=1}^{n} \boldsymbol{D}_{i,(m)}' \boldsymbol{V}_{i,(m)}^{-1} \boldsymbol{D}_{i,(m)} \right]^{-1} \left[ \sum_{i=1}^{n} \boldsymbol{D}_{i,(m)}' \boldsymbol{V}_{i,(m)}^{-1} (\boldsymbol{f}_{i,(m)} - \boldsymbol{\tau}_{i,(m)}) \right] \tag{5.13}$$

is used until convergence is reached to solve Equation (5.12). The resulting joint quasi-likelihood estimators $(\boldsymbol{\beta}_{\text{JQ}}, \nu_{\text{JQ}})$ are consistent, and (under mild regularity conditions) $\sqrt{n}((\boldsymbol{\beta}_{\text{JQ}}, \nu_{\text{JQ}}) - (\boldsymbol{\beta}, \nu))'$ is asymptotically normal as $n \to \infty$.

Mamode Khan and Jowaheer (2010) instead consider the marginal generalized quasi-likelihood equations

$$\begin{cases} \sum_{i=1}^{n} \boldsymbol{D}_{i,\boldsymbol{\beta}}' \boldsymbol{V}_{i,\boldsymbol{\beta}}^{-1} (y_i - \boldsymbol{\tau}_{1i}) = \boldsymbol{0} \\ \sum_{i=1}^{n} \boldsymbol{D}_{i,\nu}' \boldsymbol{V}_{i,\nu}^{-1} (y_i^2 - \boldsymbol{\tau}_{2i}) = \boldsymbol{0} \end{cases} \tag{5.14}$$

to estimate the coefficient parameters, $\boldsymbol{\beta}_{\text{MQ}}$ and $\nu_{\text{MQ}}$, where $\boldsymbol{\tau}_{1i}$ and $\boldsymbol{\tau}_{2i}$ again denote the approximated first and second CMP moments, $\boldsymbol{D}_{i,\beta} = \frac{\partial \tau_{1i}}{\partial \boldsymbol{\beta}'}$ has dimension $p \times 1$, $V_{i,\beta} = \frac{\lambda_i^{1/\nu}}{\nu}$,

$$D_{i,\nu} = \frac{1}{2\nu^3}\left(2\lambda^{1/\nu}\nu\ln(\lambda_i) + \nu - 1 - 4\lambda_i^{2/\nu}\ln(\lambda_i)\nu - 4\lambda_i^{1/\nu}\nu - r\lambda_i^{1/\nu}\ln(\lambda_i)\right), \tag{5.15}$$

and $V_{i,\nu} = \tau_{4i} - (\tau_{2i})^2$, where

$$\tau_{4i} = \frac{1}{\nu^3}\left(\lambda_i^{1/\nu}\nu^2 + 4\lambda_i^{3/\nu}\nu^2 + 10\lambda_i^{2/\nu}\nu - 4\lambda_i^{1/\nu}\nu + 4\lambda_i^{1/\nu} - 4\lambda_i^{2/\nu}\nu^2\right) + \left[\frac{\lambda_i^{1/\nu}}{\nu}\left(\lambda_i^{1/\nu} - \frac{\nu-1}{2\nu}\right)^2\right]^2. \tag{5.16}$$

Again using the Newton–Raphson method produces the scheme

$$\boldsymbol{\beta}_{(m+1)} = \boldsymbol{\beta}_{(m)} + \left[\sum_{i=1}^{n} \boldsymbol{D}'_{i,\boldsymbol{\beta},(m)} \boldsymbol{V}^{-1}_{i,\boldsymbol{\beta},(m)} \boldsymbol{D}_{i,\boldsymbol{\beta},(m)}\right]^{-1}_{(m)} \left[\sum_{i=1}^{n} \boldsymbol{D}'_{i,\boldsymbol{\beta},(m)} \boldsymbol{V}^{-1}_{i,\boldsymbol{\beta},(m)} (\boldsymbol{y}_{i,(m)} - \boldsymbol{\tau}_{1i,(m)})\right]$$

$$\nu_{(m+1)} = \nu_{(m)} + \left[\sum_{i=1}^{n} \boldsymbol{D}'_{i,\nu,(m)} \boldsymbol{V}^{-1}_{i,\nu,(m)} \boldsymbol{D}_{i,\nu,(m)}\right]^{-1}_{(m)} \left[\sum_{i=1}^{n} \boldsymbol{D}'_{i,\nu,(m)} \boldsymbol{V}^{-1}_{i,\nu,(m)} (\boldsymbol{y}^2_{i,(m)} - \boldsymbol{\tau}_{2i,(m)})\right],$$

where the convergence scheme oscillates for a fixed $\boldsymbol{\beta}$ or $\nu$ to optimize either of the respective equations. This pattern repeats until convergence is reached for both equations and the marginal generalized quasi-likelihood estimators are obtained. Again, these estimators are consistent where, as $n \rightarrow \infty$ and under mild regularity conditions, $\sqrt{n}(\boldsymbol{\beta}_{\text{MQ}} - \boldsymbol{\beta})$ and $\sqrt{n}(\nu_{\text{MQ}} - \nu)$ are asymptotically normally distributed.

Jowaheer and Mamode Khan (2009) perform simulation studies from which they conclude that the joint generalized quasi-likelihood approach outperforms maximum likelihood estimation. They find that the joint quasi-likelihood approach produces estimates that are nearly as efficient as those attained via maximum likelihood estimation (showing a loss of no more than 1% even for small samples) while experiencing fewer convergence issues, particularly for $\nu > 1$ and $n \leq 100$. However, their study only considers values of $\nu$ within the dual space (i.e. $0.5 \leq \nu \leq 2$); thus, given the required assumptions underlying their approach ($\nu \leq 1$ or $\lambda > 10^{\nu}$), it is unclear whether the same level of success would hold for $\nu > 2$. Meanwhile, Mamode Khan and Jowaheer (2010) report that the marginal generalized quasi-likelihood estimates are comparable to the joint generalized quasi-likelihood estimates. The larger concern for analysts, however, remains that the method of quasi-likelihood estimation presumes the existence of data over-dispersion; under that scenario, it is reasonable to use the moment approximations considered here. For cases of data under-dispersion, however, the approximations rely on $\lambda_i > 10^{\nu}$; thus, $\lambda_i$ must be considerably large for $\nu > 2$. Accordingly, unless analysts are already informed of the underlying type of data dispersion present in their study, caution should be used when performing either of the proposed quasi-likelihood methods.

### *Bayesian Estimation*

There do not exist closed-form solutions for the MLEs because of the complex nature of the score equations stemming from the resulting log-likelihood function in Equation (5.4). Some researchers argue that numerical approaches toward addressing maximum likelihood estimation can be challenging, and thus instead they propose conducting Bayesian COM–Poisson regression. This not only lets analysts update probabilities based on prior knowledge but further allows for posterior predictive distributions to be determined for a new observation associated with the covariates of interest, given the observed response data. As noted in Section 5.2.1, the first works in Bayesian COM–Poisson regression considered an ACMP($\mu_{*i}, \nu_i$) regression where $\ln(\mu_*) = \boldsymbol{X\beta}$ and either constant or varying dispersion via $\ln(\boldsymbol{\nu}) = \boldsymbol{G\gamma}$ (Guikema and Coffelt, 2008; Lord and Guikema, 2012; Lord et al., 2008), while other works further allow for an offset and spatial autocorrelation through random effects $\phi_i$ (Equation (5.3)) and instead consider $\ln(\boldsymbol{\nu}) = -\boldsymbol{G\gamma}$. Under the broader construct, for example, $\phi_i$ can be assumed to have a normally distributed conditional autoregressive prior (Chanialidis et al., 2014, 2017). Huang and Kim (2019) meanwhile developed a Bayesian MCMP1($\boldsymbol{\mu}$, $\nu$) regression where $\ln(\boldsymbol{\mu}) = \boldsymbol{X\beta}$, while $\nu$ is assumed an additional (constant) dispersion parameter.

One would think the COM–Poisson conjugate prior (Equation (2.34)) to be a natural choice for COM–Poisson Bayesian analysis; however, arguments against its use include that (1) computing the normalizing constant associated with the resulting posterior distribution requires numerical integration, Markov Chain Monte Carlo (MCMC) sampling, or another estimation method; (2) the process for determining proper hyperparameters is not transparent; and (3) the conjugacy property only holds true for regression models involving only the intercept (Huang and Kim, 2019). Instead, early works first assumed the regression coefficients to have noninformative normally distributed priors[2] and are subsequently estimated via MCMC sampling. The resulting procedure is implemented via WinBUGS[3] for ACMP regression with $Z_1(\mu_*, \nu)$ approximated by bounding it via a geometric series comparable to that described in Minka et al. (2003), where the level of precision is inputted by the analyst (Guikema and Coffelt, 2008; Lord et al., 2008). While the Bayesian approach provides added flexibility allowing analysts to incorporate (non)informative priors that represent expert knowledge and/or opinion, this approach significantly impacts computation time. In particular, because the normalizing constant does not have a closed form, there exists a computational expense associated with determining the likelihood, and sampling from a posterior distribution of the COM–Poisson regression model parameters can prove difficult.

Various alternatives exist for approximating the ACMP normalizing constant $Z_1(\mu_*, \nu)$ for use in an MCMC algorithm; see Section 2.8 for details. Alternatively, approximate Bayesian computation methods can circumvent evaluating $Z_1(\mu_*, \nu)$ precisely. Chanialidis et al. (2014), for example, propose a modified version of retrospective sampling where probability bounds are based on approximated bounds for $Z_1(\mu_*, \nu)$. Thus, the bounds on the acceptance probability stem from upper and lower bounds on the likelihood function. Such algorithms, however, may not sample from the distribution of interest and are considerably less efficient than more standard MCMC approaches. Because each of these approaches suffers from significant drawbacks, Chanialidis et al. (2017) instead consider an MCMC algorithm that utilizes the exchange algorithm in order to conduct the Bayesian ACMP regression; thus, the regression analysis can be performed without computing the normalizing constant (thereby improving the computational speed and accuracy of the analysis); see Algorithm 5. This algorithm likewise relies on constructing a piecewise truncated geometric envelope distribution to enclose the ACMP distribution, while the sampling

[2] Guikema and Coffelt (2008) initializes the MCMC procedure with a normal prior distribution with mean zero and variance $1 \times 10^6$. Lord et al. (2008) assume normal(0,100) distributed $\beta$ coefficients associated with $\mu_*$, and a gamma(0.03, 0.1) prior for $\nu$, but note that the choice of prior did not significantly impact posterior parameter estimates.

[3] The interested reader can locate the associated codes at the WinBUGS developer web page, `www.winbugsdevelopment.org.uk/`.

1. Denote $M = \lfloor \mu_* \rfloor$ the ACMP mode and $s = \lceil \sqrt{\mu_*}/\sqrt{\nu} \rceil$ the approximate standard deviation.
2. Construct an ACMP upper bound $r_{\mu_*,\nu}(y)/Z_g(\mu_*, \nu)$ based on a piecewise geometric distribution with three cut-offs ($M - s$, $M$, $M + s$, where without loss of generality assume $M - s \geq 0$), where

$$r_{\mu_*,\nu}(y) = \begin{cases} q_{\mu_*,\nu}(M-s) \cdot \left(\frac{M-s}{\mu_*}\right)^{\nu \cdot (M-s-y)} & \text{for } y = 0, \ldots, M - s \\ q_{\mu_*,\nu}(M-1) \cdot \left(\frac{M-1}{\mu_*}\right)^{\nu \cdot (M-1-y)} & \text{for } y = M - s + 1, \ldots, M - 1 \\ q_{\mu_*,\nu}(M) \cdot \left(\frac{\mu_*}{M+1}\right)^{\nu \cdot (y-M)} & \text{for } y = M, \ldots, M + s - 1 \\ q_{\mu_*,\nu}(M+s) \cdot \left(\frac{\mu_*}{M+s+1}\right)^{\nu \cdot (y-M-s)} & \text{for } y = M + s, M + s + 1, \ldots \end{cases} \tag{5.17}$$

and $Z_g(\mu_*, \nu) = \sum_{y=0}^{\infty} r_{\mu_*,\nu}(y)$, where $q_{\mu_*,\nu}(y) = \left(\frac{\mu_*^y}{y!}\right)^{\nu}$ and $r_{\mu_*,\nu}(y) \geq q_{\mu_*,\nu}(y)$.
3. Sample from $p(y \mid \mu_*, \nu)$ via the rejection method with rejection envelope $\frac{Z_g(\mu_*,\nu)}{Z_1(\mu_*,\nu)} g_{\mu_*,\nu}(y)$, where $g_{\mu_*,\nu}(y) = \frac{r_{\mu_*,\nu}(y)}{Z_g(\mu_*,\nu)}$. In other words, draw an outcome $y$ from $g_{\mu_*,\nu}(y)$ and accept that outcome with probability $\frac{q_{\mu_*,\nu}(y)}{r_{\mu_*,\nu}(y)}$; otherwise, reject it.

**Algorithm 5** Chanialidis et al. (2017) algorithm to generate random data from ACMP($\mu_*$, $\nu$).

method only requires the unnormalized densities, circumventing the need for $Z_1(\mu_*, \nu)$.

Benson and Friel (2017) advance the Chanialidis et al. (2017) work, introducing a new/different rejection sampling algorithm that significantly increases computational speed. This algorithm suggests a "less intensive" sampler that stems from a single-envelope distribution that depends on the ACMP dispersion parameter. While this distribution infers a higher rejection rate than the Chanialidis et al. (2017) envelope, it circumvents the setup and sampling costs associated with producing the truncated geometric envelope distribution. Under the ACMP regression allowing for observation-level dispersion, the associated likelihood function involves multiple normalizing constants that each have a complex form, and the posterior distribution cannot be normalized. While rejection sampling is not applicable to all complicated likelihoods, Benson and Friel (2017) show that an unbiased estimate of the likelihood is guaranteed to be positive and can be obtained via an unbiased estimate of the normalizing constant when rejection sampling is available.

Bayesian MCMP1 regression can be conducted via a Metropolis–Hastings algorithm, assuming (for example) a multivariate normal (MVN) $N(\boldsymbol{\mu}_{\boldsymbol{\beta}}, \boldsymbol{\Sigma}_{\boldsymbol{\beta}})$ distribution for $\boldsymbol{\beta}$ and a log-normal$(\mu_\nu, \sigma_\nu^2)$ for $\nu$, where large $\boldsymbol{\Sigma}_{\boldsymbol{\beta}}$ and $\sigma_\nu^2$ produce improper flat priors (Huang and Kim, 2019). The MCMP1 structure and assumed MVN and lognormal priors infer that the joint posterior distribution for $\boldsymbol{\beta}$ and $\nu$ given a sample $\boldsymbol{y} = (y_1, \ldots, y_n)$ is

$$p(\boldsymbol{\beta}, \nu \mid \boldsymbol{y}, \boldsymbol{X}) \propto \prod_{i=1}^{n} \frac{\lambda(\exp(\boldsymbol{x}_i\boldsymbol{\beta}), \nu)^{y_i}}{(y_i!)^\nu Z(\lambda(\exp(\boldsymbol{x}_i\boldsymbol{\beta}), \nu), \nu)} \cdot \exp\left[-\frac{1}{2}(\boldsymbol{\beta} - \boldsymbol{\mu}_{\boldsymbol{\beta}})' \boldsymbol{\Sigma}_{\boldsymbol{\beta}}{}^{-1} (\boldsymbol{\beta} - \boldsymbol{\mu}_{\boldsymbol{\beta}})\right] \cdot \frac{1}{\nu} \exp\left[-\frac{(\ln(\nu) - \mu_\nu)^2}{2\sigma_\nu^2}\right]. \tag{5.18}$$

The posterior predictive distribution of a new observation $y^*$ given a predictor matrix of interest $\boldsymbol{X}^*$ and response outcomes $\boldsymbol{y}$, namely

$$p(y^* \mid \boldsymbol{X}^*, \boldsymbol{y}) = \int p(y^* \mid \boldsymbol{\beta}, \nu, \boldsymbol{X}^*) p(\boldsymbol{\beta}, \nu \mid \boldsymbol{y}, \boldsymbol{X}) \mathrm{d}(\boldsymbol{\beta}, \nu), \tag{5.19}$$

is approximated by first obtaining the posterior distribution (Equation (5.18)) via the Metropolis–Hastings algorithm, and then by conducting Monte Carlo averaging of $p(y^* \mid \boldsymbol{\beta}, \nu, \boldsymbol{X}^*)$ evaluated at draws $(\boldsymbol{\beta}, \nu)$ from the said posterior distribution.

The Metropolis–Hastings algorithm (Algorithm 6) uses the MLEs as starting values ($\boldsymbol{\beta}_{(0)} = \hat{\boldsymbol{\beta}}$ and $\nu_{(0)} = \hat{\nu}$) in lieu of a burn-in period for the algorithm and alternates updates for each parameter in order to speed up convergence in the Markov chain. The estimated variance–covariance associated with $\hat{\boldsymbol{\beta}}$ serves as the variance matrix, $S_{\boldsymbol{\beta}}$, for the MVN distribution from which proposed draws are obtained. While any proposal densities can be considered to determine the acceptance probabilities, Huang and Kim (2019) illustrate the procedure using an MVN, $N(\boldsymbol{\beta}_{(0)}, S_{\boldsymbol{\beta}})$, for the proposed coefficients $\boldsymbol{\beta}_{(1)}$ associated with the MCMP1 means, and the exponential distribution Exp$(1/\nu_{(0)})$ with pdf $p(\nu_{(1)}) = \frac{1}{\nu_{(0)}} \exp(-\nu_{(1)}/\nu_{(0)})$ for the proposed dispersion parameters $\nu_{(1)}$; analysts are not restricted, however, to these proposal densities. The respective acceptance probabilities are

Given data $\boldsymbol{X}$ and $\boldsymbol{y}$, and starting points $\boldsymbol{\beta}_{(0)}$ and $\nu_{(0)}$,

1. Draw a sample from $\boldsymbol{\beta}_{(1)} \sim N(\boldsymbol{\beta}_{(0)}, \boldsymbol{S}_{\beta})$.
2. Accept $\boldsymbol{\beta}_{(1)}$ with probability $\min\left(1, \alpha_{\beta} = \frac{p(\boldsymbol{\beta}_{(1)},\nu_{(0)}|\boldsymbol{y},\boldsymbol{X})}{p(\boldsymbol{\beta}_{(0)},\nu_{(0)}|\boldsymbol{y},\boldsymbol{X})}\right)$; otherwise, maintain $\boldsymbol{\beta}_{(0)}$.
3. Draw sample from $\nu_{(1)} \sim \text{Exp}(1/\nu_{(0)})$, i.e. an exponential distribution with pmf $p(\nu_{(1)}) = \frac{1}{\nu_{(0)}} \exp(-\nu_{(1)}/\nu_{(0)})$.
4. Accept $\nu_{(1)}$ with probability $\min\left(1, \alpha_{\nu} = \frac{p(\boldsymbol{\beta}_{(0)},\nu_{(1)}|\boldsymbol{y},\boldsymbol{X})}{p(\boldsymbol{\beta}_{(0)},\nu_{(0)}|\boldsymbol{y},\boldsymbol{X})} \frac{\nu_{(1)}}{\nu_{(0)}} \exp\left(\frac{\nu_{(1)}}{\nu_{(0)}} - \frac{\nu_{(0)}}{\nu_{(1)}}\right)\right)$; otherwise, maintain $\nu_{(0)}$.
5. Repeat Steps 1–4 until $N$ MCMC samples are produced.

**Algorithm 6** Huang and Kim (2019) Metropolis–Hastings algorithm for MCMP1 regression illustrated assuming MVN and exponential proposal distributions, respectively, for $\boldsymbol{\beta}$ and $\nu$.

$$\alpha_{\beta} = \frac{p(\boldsymbol{\beta}_{(1)}, \nu_{(0)} \mid \boldsymbol{y}, \boldsymbol{X})}{p(\boldsymbol{\beta}_{(0)}, \nu_{(0)} \mid \boldsymbol{y}, \boldsymbol{X})} \tag{5.20}$$

$$\alpha_{\nu} = \frac{p(\boldsymbol{\beta}_{(0)}, \nu_{(1)} \mid \boldsymbol{y}, \boldsymbol{X})}{p(\boldsymbol{\beta}_{(0)}, \nu_{(0)} \mid \boldsymbol{y}, \boldsymbol{X})} \frac{\nu_{(1)}}{\nu_{(0)}} \exp\left(\frac{\nu_{(1)}}{\nu_{(0)}} - \frac{\nu_{(0)}}{\nu_{(1)}}\right). \tag{5.21}$$

While this approach does not produce the most efficient algorithm, it serves as a starting point for Bayesian MCMP1 regression modeling; code reflecting this algorithm can be obtained as a plug-in with the `mpcmp` package (Huang and Kim, 2019).

The COM–Poisson regressions (under any parametrization and framework) produce estimates, fitted values, and intervals that remain impressive. In fact, whether considering confidence or credible intervals, the COM–Poisson models effectively reflect and account for data dispersion. For over-dispersed data, the COM–Poisson models perform in a manner close to the NB, both of which outperform the Poisson model. Meanwhile for under-dispersed data, the COM–Poisson confidence and credible intervals are more appropriate; Poisson and NB models produce confidence and credible intervals that are too wide because they do not account for the smaller amount of data variation. Analysts are warned against using the ACMP regressions, however, for analyzing considerably under-dispersed data. Under any parametrization, the COM–Poisson model has proven its ability to handle data dispersion both under correctly specified and mis-specified data constructs.

### 5.2.3 Hypothesis Testing

The hypothesis test for data dispersion described in Section 2.4.5 easily extends to a regression framework. Assuming a COM–Poisson model with constant dispersion (under any parametrization) implies the null and alternative hypotheses, $H_0 : \nu = 1$ versus $H_1 : \nu \neq 1$. The corresponding likelihood ratio test statistic is $C_\nu = -2\ln\Lambda_\nu = -2\left[\ln L\left(\hat{\boldsymbol{\beta}}^{(0)}, \hat{\nu} = 1\right) - \ln L\left(\hat{\boldsymbol{\beta}}, \hat{\nu}\right)\right]$, where $\hat{\boldsymbol{\beta}}^{(0)}$ are the Poisson regression coefficient estimates and $(\hat{\boldsymbol{\beta}}, \hat{\nu})$ are the COM–Poisson estimates under the appropriate parametrization. Under the null hypothesis, the test statistic $C_\nu$ has an asymptotic $\chi^2$ distribution with 1 degree of freedom; asymptotic confidence intervals can be attained accordingly. When allowing for the variable dispersion model, the aforementioned hypothesis test generalizes to $H_0\colon \gamma_0 = \cdots = \gamma_{p_2} = 0$ versus $H_1\colon$ at least one $\gamma_k \neq 0$ for some $k = 1, \ldots, p_2$. The likelihood ratio test statistic then becomes $C_\gamma = -2\ln\Lambda_\gamma = -2\left[\ln L\left(\hat{\boldsymbol{\beta}}^{(0)}, \hat{\boldsymbol{\gamma}} = \mathbf{0}\right) - \ln L\left(\hat{\boldsymbol{\beta}}, \hat{\boldsymbol{\gamma}}\right)\right]$, where $\hat{\boldsymbol{\beta}}^{(0)}$ still denotes the Poisson estimates, while $\left(\hat{\boldsymbol{\beta}}, \hat{\boldsymbol{\gamma}}\right)$ denote the COM–Poisson coefficient estimates under the variable dispersion model. Under the null hypothesis with the variable dispersion model, $C_\gamma$ now has an asymptotic chi-squared distribution with $p_2$ degrees of freedom, and asymptotic confidence intervals or $p$-values can be determined for the coefficients. When analyzing small samples under either construct, the test statistic distribution and confidence intervals can be obtained via bootstrapping (Ribeiro Jr. et al., 2019; Sellers and Shmueli, 2010).

### 5.2.4 `R` Computing

Five `R` packages conduct COM–Poisson regression. Four of the packages are available on the Comprehensive `R` Archive Network (R Core Team, 2014): `CompGLM` (Pollock, 2014a), `COMPoissonReg` (Sellers et al., 2019), `mpcmp` (Fung et al., 2020), and `DGLMExtPois` (Saez-Castillo et al., 2020). Functions in these packages perform maximum likelihood estimation via an optimization procedure where starting values for the optimization are assumed to be the Poisson model estimates unless otherwise specified by the analyst. The `mpcmp` package is also equipped to conduct Bayesian MCMP1 regression as described in Huang and Kim (2019), but it requires a plug-in that is only available through the first author. The fifth package, `combayes`, likewise conducts Bayesian COM–Poisson regression and is available via `GitHub` (Chanialidis, 2020).

All of the aforementioned packages contain functions that allow users to obtain a coefficient table and other summary results, including the resulting log-likelihood value and Akaike information criterion (AIC). `COMPoissonReg` can further supply the Bayesian information criterion (BIC), leverage, and parametric bootstrapping results, and can conduct a likelihood ratio test to account for statistically significant (constant or variable) data dispersion. The `CompGLM` package can meanwhile provide predicted outcomes, and the `mpcmp` package can supply various diagnostic plots. While all of these functions share similarities among them, the `COMPoissonReg::glm.cmp` function distinguishes itself by further allowing analysts to conduct zero-inflated CMP regression with varying $\boldsymbol{p}$ via $\ln(\boldsymbol{p}/(\mathbf{1}-\boldsymbol{p})) = \boldsymbol{D\delta}$ (Sellers and Raim, 2016; Sellers et al., 2019). This allows analysts to consider model relationships for both CMP parameters and the success probability associated with excess zeroes; see Sections 5.3 and 5.5 for further details.

*Maximum likelihood Estimation for CMP Regression*

The `glm.comp` function (contained in the `CompGLM` package) allows the analyst to conduct CMP-generalized linear modeling of the form

$$\ln(\boldsymbol{\lambda}) = \boldsymbol{X\beta} \text{ and} \tag{5.22}$$

$$\ln(\boldsymbol{\nu}) = \boldsymbol{G\gamma}, \tag{5.23}$$

where $\boldsymbol{X}$ and $\boldsymbol{G}$ are design matrices associated with $\boldsymbol{\lambda}$ and $\boldsymbol{\nu}$, respectively, through the parameters $\boldsymbol{\beta}$ and $\boldsymbol{\gamma}$. The model formulae for Equations (5.22) and (5.23) are inputted as usual in `lamFormula` and `nuFormula`, respectively, where the default setting for `nuFormula` assumes constant dispersion. As noted in Chapter 2, the loglinear relationship for $\boldsymbol{\nu}$ allows for unconstrained optimization to be performed via `optim` in order to determine the MLEs. The `optim` function assumes `method="BFGS"`, thus performing the Broyden–Fletcher–Goldfarb–Shanno algorithm (also referred to as a variable metric algorithm); however, analysts are able to modify the choice of Newton-type algorithm. Other method options are `"Nelder-Mead"`, `"CG"`, `"L-BFGS-B"`, `"SANN"`, or `"Brent"`; see the `optim` help page for a detailed discussion. In all cases, unless starting values are otherwise supplied by the user in `lamStart` and `nuStart`, the `glm.comp` function assumes Poisson estimates for $\boldsymbol{\beta}$ in Equation (5.22) and $\boldsymbol{\gamma} = \mathbf{0}$ in Equation (5.23). Another input for the `glm.comp` function is `sumTo` that determines the number of terms used to approximate the infinite summation for $Z(\lambda, \nu)$. As noted in Section 2.8, `sumTo = 100` is the default setting for this input, but it can be supplied by the analyst. As with the standard `glm` function, `glm.comp` can accommodate other `R` functions providing model analysis, including `summary`, `logLik`, `extractAIC`, and `predict`.

The `glm.cmp` function (contained in the `COMPoissonReg` package) operates in a similar manner to that described in `glm.comp` considering the loglinear models described in Equations (5.22) and (5.23); these model formulae are inputted in `formula.lambda` and `formula.nu`, respectively. This function likewise assumes a constant dispersion model but allows for flexible dispersion and uses `optim` to determine the MLEs. The `glm.cmp` function, however, uses the defaulted `"Nelder-Mead"` quasi-Newton algorithm in `optim` to perform the optimization, where starting values can be supplied by the user in `beta.init` (usually Poisson estimates) and `gamma.init` (presumably $\boldsymbol{\gamma} = \mathbf{0}$ to agree with the Poisson model estimates as starting values). Consistent with the discussion in Section 2.8, the $Z(\lambda, \nu)$ function is computed in a hybrid fashion to optimize computation speed and efficiency. For $\lambda$ sufficiently large and $\nu$ sufficiently small, the closed-form approximation (Equation (2.12)) is used to estimate $Z(\lambda, \nu)$. If those conditions are not satisfied, then the infinite summation is approximated to meet a desired accuracy level. The `glm.cmp` can accommodate other `R` functions providing model analysis, including `summary`, `logLik`, `AIC` and `BIC`, and `predict`. The `glm.cmp` model output can further be supplied as input in other `COMPoissonReg` functions such as `equi.test`, `leverage`, `deviance`, and `parametric_bootstrap`.

The `glm.cmp` (`COMPoissonReg`) and `glm.comp` (`CompGLM`) share several similarities. Both functions allow for a variable dispersion model via the log link, can accommodate an offset in the model formulation, and use `optim` to maximize the likelihood function and output gradient information in order to obtain corresponding standard errors. However, `glm.comp` computes these results at a faster rate than `glm.cmp` due to the way in which the support functions were implemented; support functions for `glm.cmp` were implemented in `R`, while those for `glm.comp` were implemented in `C++`. An unfortunate feature shared by both functions is their inability to handle larger counts. Even one large count value in a dataset can cause both methods to produce errors (Chatla and Shmueli, 2018). While both functions can conduct CMP regression, `glm.cmp` allows the analyst to also consider zero-inflated CMP models.

### *Maximum Likelihood Estimation for MCMP1 Regression*

The `glm.cmp` (`mpcmp`) and `glm.CMP` (`DGLMExtPois`) functions likewise conduct regression analysis and share several similarities; however these functions focus on MCMP1 regression (Fung et al., 2020; Saez-Castillo et al., 2020). Both functions consider a loglinear link $\ln(\boldsymbol{\mu}) = \boldsymbol{X\beta}$ for the mean with an additional ability to include an offset, and they likewise allow for variable dispersion via a loglinear model. Both functions use the form $\ln(\boldsymbol{\nu}) = \boldsymbol{G\gamma}$; however, `mpcmp::glm.cmp` follows this notation explicitly, while `DGLMExtPois::glm.CMP` replaces `gamma` with `delta` as their coefficient definition relating to the dispersion. Poisson estimates serve as reasonable starting values for these functions. They are provided for the `mpcmp::glm.cmp` and `DGLMExtPois::glm.CMP` inputs `betastart` and `init.beta`, respectively, while the associated corresponding starting values to account for dispersion (`gammastart` or `init.delta`, respectively) are set equal to zero. Both functions also allow analysts to address how to handle missing data via the `na.action` input and to specify a subset of observations via the `subset` input. Along with these similarities, however, differences exist between the two functions. For example, both functions include a `tol` input; however, they have different meanings in the respective functions; `mpcmp` uses this as an overall convergence threshold for $\boldsymbol{\lambda}$, while `tol` in `DGLMExtPois` is used to approximate the infinite sum calculation for the normalizing constant, $Z(\lambda(\mu, \nu), \nu)$. Broader differences between the two functions stem from additional respective inputs available in their respective codes. The `mpcmp::glm.cmp` function allows contrasts for $\boldsymbol{\mu}$ and $\boldsymbol{\nu}$ (`contrasts_mu` and `contrasts_nu`, respectively), as well as a user-contributed input regarding the maximum allowable number of iterations to determine $\boldsymbol{\lambda}$ (`maxlambdaiter`), and a specified range of possible values for $\boldsymbol{\lambda}$ (the lower and upper bounds `lambdalb` and `lambdaub` are defaulted as $1 \times 10^{-10} \leq \boldsymbol{\lambda} \leq 1900$). `DGLMExtPois::glm.CMP` meanwhile includes inputs (1) `maxiter_series` to predetermine the number of iterations used to compute the normalizing constant, (2) `opts` to supply options to the nonlinear optimizer function `nloptr`, and (3) the logical values for `x`, `y`, and `z` to determine whether or not to return the model matrix for $\boldsymbol{\mu}$, the response vector, and the model matrix for $\boldsymbol{\nu}$, respectively.

Beyond creating the respective MCMP1 regression objects, `mpcmp` and `DGLMExtPois` share an array of analogous capabilities, including summarizing the regression results via the `summary` command, obtaining confidence interval information (`confint`), determining fitted values and residuals, producing residual plots of interest, obtaining predicted values (`predict`), conducting a likelihood ratio test, and computing information criterions of interest. Some of these functions (e.g. `summary`, `confint`, and `predict`) operate as usual and do so in conjunction with either MCMP1 regression object. The `cmplrtest` function in `mpcmp` performs a likelihood ratio test between two `glm.cmp` objects, `object1` and `object2`, representing the respective (nested) MCMP1 models considered in association with a dataset of interest. More specifically, the `LRTnu` function performs the likelihood ratio test comparing the Poisson and MCMP1 (based on the `glm.cmp` object) models, thus checking for statistically significant data dispersion. Similarly, the `lrtest` function (`DGLMExtPois`) conducts the likelihood ratio test for two nested models stemming from the `glm.CMP` objects. The `mpcmp::AIC.cmp` and `DGLMExtPois::AIC_CMP` functions each compute an information criterion for the associated object (`glm.cmp` (`mpcmp`) or `glm.CMP` (`DGLMExtPois`), respectively). They both default to compute the AIC (`k = 2`) but can be used to compute (for example) the BIC (setting `k=log(n)`, where $n$ equals the sample size). The `DGLMExtPois` package likewise offers the `BIC` function to directly determine the BIC.

The `plot` command maintains its ability to provide residual plots associated with an MCMP1 regression. When applied to a `glm.cmp` (`mpcmp`) object, the `plot` command produces four figures by default, namely a (1) fitted values versus deviance residuals plot, (2) nonrandomized probability integral transform histogram, (3) scale-location plot associating the square-root of the absolute value of the residuals with the fitted values, and (4) leverage versus Pearson residuals plot. Additional plots accessible to analysts include a Q–Q plot comparing the nonrandomized probability integral transform with the uniform distribution, a histogram and Q–Q plot of the normal randomized residuals, and a Cook's distance plot. Meanwhile, using the `plot` command on a `glm.CMP` object (`DGLMExtPois`) can produce two residual plots, namely a fitted versus residual plot, and a normal Q–Q plot. The default option for the residuals via the `mpcmp` package is to produce deviance residuals, while the `DGLMExtPois` package uses quantile residuals. Both packages, however, also allow for `type` to be updated as `"pearson"` or `"response"`. The `residuals` command is likewise available in both packages in order to extract residuals associated with an MCMP1 regression object. One should, however, be cautious using the `residuals` function in `DGLMExtPois`; be sure to specify `type="quantile"` because the function instead has `"pearson"` as the default type.

*Bayesian Estimation for ACMP Regression*

The `combayes` package (Chanialidis, 2020) conducts the Bayesian ACMP regression via exact samplers. While help pages are not currently available with this package, one can gain a rudimentary understanding of package functionality based on an illustrative example provided in `GitHub`. The supplied codes allow analysts to sample from an ACMP distribution via rejection sampling, and to evaluate bounds for $Z_1(\mu_*, \nu)$, where the MCMC is conducted with `algorithm = "bounds"` or `"exchange"` (the default setting). Selecting `algorithm = "bounds"`, however, causes the MCMC to perform more slowly (Chanialidis, 2020). The `combayes` package particularly contains the `cmpoisreg` function that produces the samples of interest and reports the acceptance probabilities. Along with the provided response and explanatory variable inputs (`y` and `X`, respectively) required for the `cmpoisreg` function, analysts supply the desired number of samples (`num_samples`), burn-in (`burnin`), and the variance–covariance structures for the MVN prior distributions for $\boldsymbol{\beta}$ and $\boldsymbol{\gamma}$, respectively. Note, however, that the `combayes` package denotes the parameter `delta` (instead of, say, `gamma` for consistency with notation provided in this reference) to associate with $\nu$; thus, the respective prior variance–covariance inputs are named `prior_var_beta` and `prior_var_delta`. Given these inputs, `cmpoisreg` generates a list of five results: respective dataframes `posterior_beta` and `posterior_delta` of size `num_samples` representing the generated posterior distributions for $\boldsymbol{\beta}$ and $\boldsymbol{\gamma}$; and `accept_beta`, `accept_delta`, `accept_mixed` providing the respective acceptance probabilities for $\boldsymbol{\beta}$, $\boldsymbol{\gamma}$, and $(\boldsymbol{\beta}, \boldsymbol{\gamma})$. Analysts can use these outputs to conduct MCMC and produce trace plots and caterpillar plots associated with the respective regression coefficients associated with $\boldsymbol{\mu}_*$ and $\boldsymbol{\nu}$.

### 5.2.5 Illustrative Examples

Over- and under-dispersed data examples demonstrate the ability of the COM–Poisson regression to address and account for data dispersion, thus resulting in improved performance relative to the Poisson regression. Over-dispersed examples tend to show that the COM–Poisson regression performs at best as well as a NB regression while under-dispersed examples demonstrate more optimal performance for analyzing such data. Nonetheless, this model flexibility allows for the COM–Poisson regression to provide an additional tool for determining proper model selection because varying levels of dispersion may attribute to the perceived dispersion initially identified by the analyst when, in truth, there could be differing levels or even types of dispersion present (Sellers and Shmueli, 2013). This section considers three data examples where various COM–Poisson regressions are illustrated and compared with other models: two examples contain under-dispersed observed response outcomes and one contains over-dispersed response data. Much of the presented analysis assumes a constant dispersion for ease of discussion and interpretation; however, one can consider varying dispersion for these examples as well.

*Example: Number of Children in a Subset of German Households*

More substantive illustrative examples of COM–Poisson regression are in its ability to address under-dispersed data. One example of such accommodation is with a well-studied dataset regarding the number of children in a subset of German households. As introduced in Chapter 2, Winkelmann (1995) presents data from the 1985 German Socio-Economic Panel where a random sample of 1,243 women over 44 years of age and in their first marriages is studied. Let us consider a regression for the number of children in association with other variables from the dataset, including years of schooling (`edu`), type of training (whether vocational (`voc`) or university (`uni`)), religious denomination (whether Catholic (`cath`), Protestant (`prot`), or Muslim (`musl`)), and age at marriage (`agemarr`). This dataset allows for direct comparison between the Poisson regression and various COM–Poisson regression alternatives in order to assess and compare how different computational techniques impact resulting output and statistical inference.

Table 5.2 reports the resulting regression coefficients and standard errors (in parentheses) from the Poisson, CMP (via `COMPoissonReg::glm.cmp` and `CompGLM::glm.comp`, respectively), and MCMP1 (via `mpcmp::glm.cmp` and `DGLMExtPois::glm.CMP`, respectively) regressions computed in `R`. The table clearly demonstrates the superior model performance of the COM–Poisson regressions (i.e. larger log-likelihood and smaller AIC) than the Poisson model because of the ability of the former to properly account for data dispersion. In particular, the COM–Poisson dispersion estimate (under either parametrization) is approximately $\hat{\gamma} = 0.355 > 0$; thus, $\hat{\nu} = \exp(\hat{\gamma}) \approx 1.4262 > 1$, which denotes the presence of data under-dispersion; this result is consistent with what was initially recognized in Section 2.4.4. Thus, as an aside, it is worth noting that the NB regression produces the same coefficient estimates and standard errors as reported for the Poisson model, while the resulting output regarding the NB dispersion parameter reads

Table 5.2 *Coefficient estimates and standard errors (in parentheses) associated with the number of children from women over 44 years of age and in their first marriage. Respective outputs likewise report the associated log-likelihood and Akaike information criterion (AIC) associated with each model. The* `glm.cmp` *(*`COMPoissonReg`*) and* `glm.comp` *(*`CompGLM`*) functions conduct CMP regression, while the* `glm.cmp` *(*`mpcmp`*) and* `glm.CMP` *(*`DGLMExtPois`*) functions conduct MCMP1 regression. NR = not reported.*

| | Poisson | | CMP | | | | MCMP1 | | | |
|---|---|---|---|---|---|---|---|---|---|---|
| | | | `COMPoissonReg::glm.cmp` | | `CompGLM::glm.comp` | | `mpcmp::glm.cmp` | | `DGLMExtPois::glm.CMP` | |
| Variable | Coeff. | SE | Coeff. | SE | Coeff. | SE | Coeff. | SE | Coeff. | SE |
| Constant | 1.2163 | (0.2945) | 1.8000 | (0.3508) | 1.7971 | (0.0468) | 1.2146 | (0.2564) | 1.2146 | (0.2566) |
| German | −0.1874 | (0.0710) | −0.2521 | (0.0827) | −0.2493 | (0.3498) | −0.1862 | (0.0616) | −0.1862 | (0.0616) |
| Years of schooling | 0.0345 | (0.0325) | 0.0477 | (0.0375) | 0.0457 | (0.0825) | 0.0348 | (0.0283) | 0.0348 | (0.0284) |
| Vocational training | −0.1617 | (0.0430) | −0.2137 | (0.0501) | −0.2131 | (0.0373) | −0.1625 | (0.0374) | −0.1625 | (0.0375) |
| University | −0.1750 | (0.1579) | −0.2873 | (0.1833) | −0.2313 | (0.0500) | −0.1794 | (0.1377) | −0.1794 | (0.1379) |
| Catholic | 0.2176 | (0.0707) | 0.2904 | (0.0822) | 0.2880 | (0.1811) | 0.2164 | (0.0615) | 0.2165 | (0.0616) |
| Protestant | 0.1155 | (0.0762) | 0.1572 | (0.0878) | 0.1530 | (0.0820) | 0.1142 | (0.0664) | 0.1142 | (0.0665) |
| Muslim | 0.5470 | (0.0850) | 0.7479 | (0.1035) | 0.7398 | (0.0876) | 0.5466 | (0.0733) | 0.5466 | (0.0734) |
| Rural | 0.0572 | (0.0381) | 0.0774 | (0.0440) | 0.0759 | (0.1033) | 0.0565 | (0.0330) | 0.0565 | (0.0331) |
| Age at marriage | −0.0286 | (0.0062) | −0.0386 | (0.0074) | −0.0380 | (0.0439) | −0.0286 | (0.0054) | −0.0286 | (0.0054) |
| Dispersion parameter | — | — | 0.3594 | (0.0468) | 0.3551 | (0.0073) | 0.3570 | (NR) | 0.3549 | (0.0468) |
| Log-likelihood | −2101.8 | | −2078.6 | | −2078.6 | | −2078.6 | | −2078.6 | |
| AIC | 4224.6 | | 4179.3 | | 4179.1 | | 4179.2 | | 4179.2 | |

```
              Theta:  39037
          Std. Err.:  183425
Warning while fitting theta: iteration limit reached
```

This report illustrates that, because the response data are actually under-dispersed, the NB regression output mirrors that of the Poisson regression because the NB model is unable to address data under-dispersion; it can only account for data equi- or over-dispersion. The large dispersion estimate $\hat{\theta}$ implies that the variance $V(Y) = \mu + \mu^2/\theta \approx \mu$ because $1/\hat{\theta} = 2.562 \times 10^{-5} \approx 0$, implying that the variance and mean are approximately equal (i.e. the data are equi-dispersed). Accordingly, given the constrained parameter space for optimization, the resulting NB estimates equal the corresponding Poisson estimates, while the dispersion parameter $\theta$ continues to grow toward infinity, thus converging to the Poisson model; see Section 1.2 for discussion. The resulting AIC value for the NB model equals 4226.6 (i.e. two more than the Poisson AIC value), accounting for the (unnecessary) additional dispersion parameter. Meanwhile, the `mpcmp::glm.cmp` function does not allow for varying dispersion; hence, it explicitly reports the estimated dispersion on the original scale ($1.429 \approx \exp(0.3570)$ in this example). While all of the COM–Poisson regressions produce the same approximate log-likelihood, their reported AIC values differ slightly due to roundoff error.

The MCMP1 regression functions produce nearly identical coefficient estimates and associated standard errors. The `mpcmp::glm.cmp` function, however, does not report a standard error associated with the dispersion parameter. The MCMP1 regression coefficient and standard error estimates more closely align with the corresponding Poisson values for a more relatable model comparison, and coefficient interpretation is consistent with that of the usual generalized linear model. The CMP estimates, however, do not allow for an easy apples-to-apples comparison with the Poisson estimates because the CMP model considers an association between $\boldsymbol{\lambda}$ (which is not the mean vector) and the explanatory variables. The `COMPoissonReg::glm.cmp` and `CompGLM::glm.comp` functions produce different coefficient estimates and standard errors; however, these respective estimate results produce the same log-likelihood values; this demonstrates that the respective computations for the gradient function at these values is within the error bound. Taking the associated uncertainty into account as estimated by the reported standard errors likewise shows that the respective coefficients are reasonably similar under either model.

While all of the COM–Poisson regression model functions produce output containing the corresponding coefficient tables,

COMPoissonReg::glm.cmp also reports the estimate and standard error for $\nu$ on the original scale, along with a chi-squared test for equi-dispersion. For this example, we find that statistically significant data dispersion exists in this example ($C_\nu = 47.3513$, df = 1, $p$-value = $5.9338 \times 10^{-12}$), where COMPoissonReg::glm.cmp likewise reports that $\hat{\nu} = 1.4324$ (with standard error 0.067); this infers the presence of statistically significant data under-dispersion. COMPoissonReg::glm.cmp further reports the optimization method, elapsed time for estimation, and convergence status. The mpcmp::glm.cmp meanwhile reports the null and residual deviance, and AIC along with the coefficient table; the corresponding log-likelihood value reported in Table 5.2 was deduced from this function's reported AIC.

### *Example: Airfreight Breakage Study*

Another illustrative example of an under-dispersed dataset is a small airfreight study that considers the association between the amount of damage attained over a series of transit flights. Consider the airfreight breakage dataset provided in Table 5.3 (supplied in Kutner et al. (2003)), where data from 10 air shipments is supplied. In this example, each flight carries 1,000 jars on the flight where broken ($Y_i$) denotes the number of broken jars

Table 5.3 *Airfreight breakage dataset, where* broken *denotes the number of broken jars detected following a flight that involved a number of transfers,* transfers *(Kutner et al., 2003).*

| broken | transfers |
|---|---|
| 16 | 1 |
| 9 | 0 |
| 17 | 2 |
| 12 | 0 |
| 22 | 3 |
| 13 | 1 |
| 8 | 0 |
| 15 | 1 |
| 19 | 2 |
| 11 | 0 |

detected following a flight that involved a number of carton transfers in order for the shipment to reach its destination; `transfers` ($X_i$) denotes the number of times the carton is transferred to subsequent flights (Sellers and Shmueli, 2010).

Table 5.4 reports the resulting regression coefficients and standard errors (in parentheses) from the Poisson, CMP, and MCMP1 regressions computed in `R`. Again, compared to the Poisson model, the COM–Poisson regressions outperform because of their ability to handle data underdispersion. The MCMP1 functions produce nearly identical coefficient estimates to those from the Poisson model. This is to be expected because the MCMP1 functions likewise allow for the traditional representation of a generalized linear model for a loglinear relationship between the distribution mean and the considered explanatory variables. The corresponding standard errors associated with the MCMP1 models, however, are smaller than those reported for the Poisson regression. This is a natural by-product of the MCMP1 model's ability to address the existing underdispersion contained in this dataset. The reported standard errors stemming from the `mpcmp::glm.cmp` and `DGLMExtPois::glm.CMP` functions are likewise similar to each other, when represented on the same scale. The `mpcmp::glm.cmp` function, however, actually reports the dispersion parameter on the original scale; for this example, it produces a reported $\hat{\nu} = 5.306$ (i.e. approximately $\exp(1.6688)$). The `DGLMExtPois::glm.CMP` function meanwhile reports a dispersion parameter (on the log scale) as 1.7466. The difference in estimated dispersion parameters is presumably the cause for the slight difference in reported log-likelihoods ($-18.6943$ and $-18.6799$) and AICs (43.3886 and 43.3597) for `mpcmp::glm.cmp` and `DGLMExtPois::glm.CMP`, respectively. Such minute differences in the respective metrics, however, demonstrate relatively equivalent performance between the two functions.

The coefficient estimates stemming from the CMP regressions cannot readily be compared with those from the MCMP1 regressions because the underlying CMP parametrization directly associates $\boldsymbol{\lambda}$ with the explanatory variables via a loglinear model; yet $\boldsymbol{\lambda}$ does not define the mean of a CMP distribution. What is curious in this illustration, however, is that the respective CMP functions do not produce similar coefficient estimates. While both functions report estimates based on iterative schemes that reportedly attain convergence, it is clear from Table 5.4 that the `CompGLM::glm.comp` estimates do not adequately represent the potential MLEs because the associated log-likelihood is smaller than the log-likelihood reported in association with the `COMPoissonReg::glm.cmp`

Table 5.4 *Coefficient estimates and standard errors (in parentheses) associating the number of broken jars detected following a flight that involved a number of transfers. Respective outputs likewise report the associated log-likelihood and Akaike information criterion (AIC) associated with each model. The* `glm.cmp` *(*`COMPoissonReg`*) and* `glm.comp` *(*`CompGLM`*) functions conduct CMP regression, while the* `glm.cmp` *(*`mpcmp`*) and* `glm.CMP` *(*`DGLMExtPois`*) functions conduct MCMP1 regression. NA = not applicable; NR = not reported.*

| | Poisson | | CMP | | | | MCMP1 | | | |
|---|---|---|---|---|---|---|---|---|---|---|
| | | | `COMPoissonReg::glm.cmp` | | `CompGLM::glm.comp` | | `mpcmp::glm.cmp` | | `DGLMExtPois::glm.CMP` | |
| Variable | Coeff. | SE | Coeff. | SE | Coeff | SE | Coeff. | SE | Coeff. | SE |
| Constant | 2.3530 | (0.1317) | 13.8314 | (6.2643) | 5.3924 | (NA) | 2.3534 | (0.0582) | 2.3533 | (0.0560) |
| transfers | 0.2638 | (0.0792) | 1.4844 | (0.6916) | 0.5560 | (NA) | 0.2635 | (0.0349) | 0.2636 | (0.0336) |
| $\ln(\hat{\nu})$ | — | — | 1.7552 | (0.4509) | 0.8105 | (NA) | 1.6688 | (NR) | 1.7466 | (0.4491) |
| Log-likelihood | −23.1973 | | −18.6449 | | −20.3180 | | −18.6943 | | −18.6799 | |
| AIC | 50.3946 | | 43.2898 | | 46.6361 | | 43.3886 | | 43.3597 | |

function MLEs; clearly, the log-likelihood function has not achieved a maximum via the CompGLM::glm.comp function. Even more interesting, the CompGLM::glm.comp function appears unable to quantify the uncertainty associated with the coefficient estimates, thus reporting NA for the corresponding standard errors, test statistics, and $p$-values (see below); this does not occur with any of the other COM–Poisson regression R functions.

```
> freight.comp <- glm.comp(lamFormula=broken~transfers,
  data = freight)
> summary(freight.comp)

Call:
glm.comp(lamFormula = broken ~ transfers, data = freight)

Beta:
            Estimate Std.Error t.value p.value
(Intercept)  5.39239        NA      NA      NA
transfers    0.55602        NA      NA      NA

Zeta:
            Estimate Std.Error t.value p.value
(Intercept)   0.8105        NA      NA      NA

AIC: 46.63607
Log-Likelihood: -20.31804
```

Modifying the starting values for the optimization introduces different resulting coefficient estimates; however, this does not appear to circumvent the issue of uncertainty quantification. Additional input updates including the consideration of a different optimization method in optim or changing the Winsorization bound for the normalizing constant via sumTo also do not resolve the matter. Upon further inspection of the freight.comp object, freight.comp$hessian reports a $3 \times 3$ matrix whose elements are all NaN. Clearly, this causes the standard errors to be reported as NA; however, it remains unclear what is causing the Hessian matrix to be reported as NaN. This illustration thus demonstrates the superior performance of the COMPoissonReg over the CompGLM, and the need for analysts to use the CompGLM package with caution. Even still, all of the COM–Poisson models outperform the Poisson regression because of their ability to account for data under-dispersion. The COMPoissonReg::glm.cmp object further contains the hypothesis testing results (`$equitest`) that can be used to infer the existence of statistically significant data dispersion. For this example, `$equitest` reports the test statistic (`$teststat`) equaling 9.104771 with the corresponding $p$-value (`$pvalue`) (approximately 0.0025). These results (partnered with either the reported coefficient estimate associated with the regression model for $\nu$ or the direct $\nu$ estimate reported in `$DF.nu`) imply that the data are statistically significantly under-dispersed because $\hat{\nu} = \exp(\hat{\gamma}_0) = \exp(1.7552) = 5.7845 > 1$.

*Example: Number of Faults in Textile Fabrics*

Hinde (1982) provided data regarding the number of faults in rolls of fabric along with the length of the respective rolls. The resulting 32 observations were then analyzed and modeled via a compound Poisson regression model where the number of faults served as the response variable, and the roll length (i.e. the explanatory variable) was considered on the log-scale. While the discussion of the compound Poisson regression is outside of the scope of this reference, it was observed that the textile fabric data were over-dispersed. Subsequent discussion here thus considers various regression models for analyzing these count data, including the Poisson, NB, and MCMP1 (via the `mpcmp` and `DGLMExtPois` packages), where each model considers the loglinear relation

$$\ln(\mu) = \beta_0 + \beta_1 \ln(X), \tag{5.24}$$

with $X$ denoting the roll length of the fabric. The CMP regression (via `COMPoissonReg`) is likewise considered with the model $\ln(\lambda) = \beta_0 + \beta_1 \ln(X)$.

Table 5.5 reports the resulting coefficient estimates and standard errors (in parentheses) for the respective models, along with the corresponding AIC. The dispersion estimates are likewise reported, along with their respective standard error; these measures, however, are reported on differing scales, and thus are not necessarily directly comparable. Of the five considered models, the NB regression is the optimal selection (AIC = 181.38), while the Poisson regression is the worst performer (AIC = 191.82) showing essentially no empirical support ($\Delta_P = 10.44$) relative to the NB model. The COM–Poisson models perform at comparable levels with the CMP model (via `COMPoissonReg`) having the smallest AIC (183.61 versus 183.80 for `mpcmp` and `DGLMExtPois`) among the COM–Poisson models. They all fall just shy of being considered to have substantial empirical support relative to the NB model ($\Delta_{\text{CMP}} = 2.23$ and

Table 5.5 *Coefficient estimates and standard errors (in parentheses) associating the number of faults in rolls of fabric with the logarithm of the corresponding roll length. Respective outputs likewise report the corresponding Akaike information criterion (AIC) for each model. CMP regression was conducted via the* `COMPoissonReg` *package, while MCMP1 regressions were performed via the* `mpcmp` *and* `DGLMExtPois` *packages, respectively. NR = not reported. Dispersion measures are reported on varying scales (with standard errors rounded to two decimal places) as provided in the respective outputs and do not allow for direct comparison.*

| | Constant | | log(RollLength) | | Dispersion | | |
|---|---|---|---|---|---|---|---|
| | Est. | SE | Est. | SE | Est. | SE | AIC |
| Poisson | −4.1741 | (1.1352) | 0.9971 | (0.1759) | — | — | 191.82 |
| Negative binomial | −3.7959 | (1.4576) | 0.9379 | (0.2279) | 8.6713 | (4.17) | 181.38 |
| CMP (`COMPoissonReg`) | −2.2647 | (0.9375) | 0.5052 | (0.1742) | −0.7869 | (0.30) | 183.61 |
| MCMP1 (`mpcmp`) | −4.1306 | (1.5296) | 0.9904 | (0.2375) | 0.4666 | (NR) | 183.80 |
| MCMP1 (`DGLMExtPois`) | −4.0896 | (1.5350) | 0.9841 | (0.2384) | −0.7773 | (0.30) | 183.80 |

$\Delta_{\text{MCMP1}} = 2.42$, respectively); however, they are considerably closer to that designation rather than classifying as having considerably less empirical support relative to the NB; see Table 1.2.

The Poisson, NB, and MCMP1 models allow analysts to compare coefficient estimates for $\beta_0$ and $\beta_1$ directly because each of these regressions has a traditional generalized linear model form, as illustrated in Equation (5.24). Accordingly, the respective estimates for $\beta_0$ and $\beta_1$ all lie within one standard deviation from each other. Their corresponding dispersion estimates are not directly comparable, however; the NB dispersion is reported on the raw scale, while dispersion for the MCMP1 regression (whether `mpcmp` or `DGLMExtPois`) is reported on the logarithmic scale. The `MASS::glm.nb` output for the NB regression reports the dispersion parameter as $\theta = r$; recall the discussion of the NB distribution in Section 1.2. In this example, $\hat{r} = 8.6713 > 1$ infers a less extreme case of over-dispersion than represented by the geometric distribution, while at the same time, the dispersion measure is not substantially large to argue potential convergence toward infinity (and thus considers the Poisson regression). This contributes toward explaining why the NB regression outperforms all of the other presented models with respect to AIC, including the COM–Poisson regressions (irrespective of parametrization and the associated `R` package and code). At the same time, however, it is interesting to note that both MCMP1 regression packages (i.e. `mpcmp` and `DGLMExtPois`) report the same approximate AIC value (to two decimal places), while their respective dispersion estimates have different signs. Their reported dispersion estimates are directly comparable because both models consider the same loglinear model to describe dispersion ($\ln(\hat{\nu}) = \hat{\gamma}_0$), where the dispersion estimate obtained via `mpcmp` implies that the dispersion $\hat{\nu} = \exp(0.4666) > 1$ (i.e. under-dispersion), while $\hat{\nu} = \exp(-0.7773) < 1$ (i.e. over-dispersion) via `DGLMExtPois`. The `COMPoissonReg` package likewise produces dispersion and standard errors that align with those supplied by `DGLMExtPois`; the `COMPoissonReg` $\hat{\nu} = \exp(-0.7869) < 1$ with a reported standard error equal to that from `DGLMExtPois`. The `COMPoissonReg` and `DGLMExtPois` dispersion estimates indicate data over-dispersion, agreeing with the assessment obtained via the NB regression; however, we cannot directly compare the estimated amount of dispersion between the NB and COM–Poisson distributions, respectively, on the basis of their respective reported dispersion estimates. All three models (NB and the COM–Poisson regressions attained via the `COMPoissonReg` and `DGLMExtPois` packages) report estimated standard errors that infer a statistically significant level of data over-dispersion. Unfortunately, however, the `mpcmp` output only reports a dispersion estimate (which, in this case, implies data under-dispersion) without a corresponding standard error, so it is unclear if the quantified variation in their estimate perhaps infers that the dispersion may not be statistically significant (i.e. the standard error may be sufficiently large); this would help to explain the differing sign for the estimate.

The `cmpoisreg` function (`combayes`) performs the Bayesian ACMP regression with the aid of the `mvnfast` package. For this example, the code

```
textile.res <- cmpoisreg(y=textile$NoOfFaults,
                         X=log(textile$RollLength),
                         num_samples=1e4, burnin=1e3,
                         prior_var_beta=diag(2),
                         prior_var_delta=diag(2))
```

considers the model of interest for both $\mu_*$ and $\nu$ (i.e. $\ln(\mu_*) = \beta_0 + \beta_1 \ln(\text{RollLength})$ and $\ln(\nu) = \gamma_0 + \gamma_1 \ln(\text{RollLength})$) with 10,000 generated values for $\boldsymbol{\beta}$ and $\boldsymbol{\gamma}$, respectively, starting from bivariate normal prior distributions each with $2 \times 2$ identity matrices for the respective variance–covariance matrices and an initial burn-in of 1,000 generated

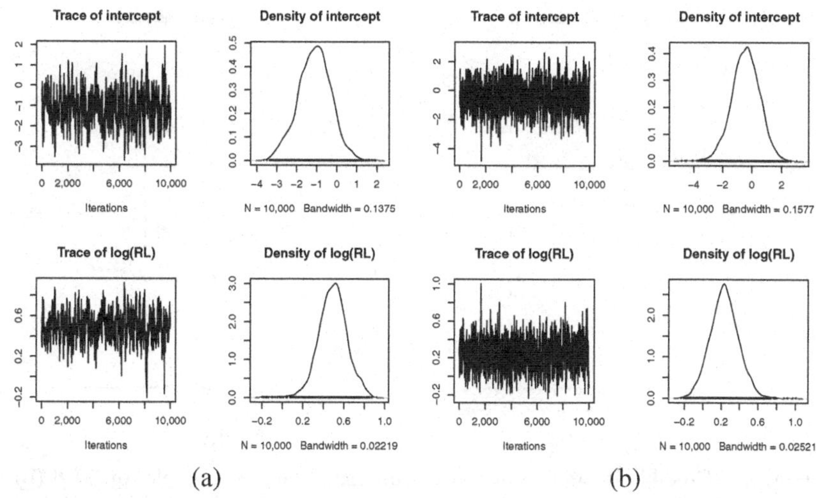


Figure 5.1 Trace and density plots associated with textile fabrics example for (a) $\boldsymbol{\beta}$ (b) $\boldsymbol{\gamma}$; RL = RollLength.

outcomes; recall (as discussed in Section 5.2.4) that the `combayes` package uses `delta` instead of `gamma` to denote the parameters associated with $\nu$. The resulting trace and density plots associated with $\boldsymbol{\beta}$ and $\boldsymbol{\gamma}$, respectively, are contained in Figures 5.1(a)–(b), while the acceptance probabilities are 0.185 for $\boldsymbol{\beta}$, 0.525 for $\boldsymbol{\gamma}$, and 0.183 for $(\boldsymbol{\beta}, \boldsymbol{\gamma})$.

Figures 5.2(a)–(b) supply the respective caterpillar plots that show the estimates, and 68% and 95% credible intervals for $\boldsymbol{\beta}$ and $\boldsymbol{\gamma}$, respectively. The respective means are approximately $\hat{\boldsymbol{\beta}} = (-1.1103, 0.4927)$ and $\hat{\boldsymbol{\gamma}} = (-0.4292, 0.2371)$. Of the four credible intervals displayed, however, only the 95% credible interval for $\beta_1$ does not contain 0. Meanwhile, the 68% credible intervals for $\beta_0$, $\beta_1$, and $\gamma_1$ do not contain 0; only the 68% credible interval for $\gamma_0$ contains 0. Recalling that the underlying model for $\nu$ is $\ln(\boldsymbol{\nu}) = -\boldsymbol{G}\gamma$, we then see that the variable model for $\nu$ implies varying levels of potential data over-dispersion regarding the number of faults observed relative to the roll length of the fabric.

The `cmpoisreg` function seems to currently require the same model form for $\mu_*$ and $\nu$. Naive attempts to modify the variance–covariance structure for $\gamma$ (e.g. to only consider constant dispersion via $\gamma_0$) produced errors noting unequal size in relation to the number of columns for the design matrix. Further, it is currently unclear how to conduct a comparable Bayesian analysis that allows only for a constant dispersion.

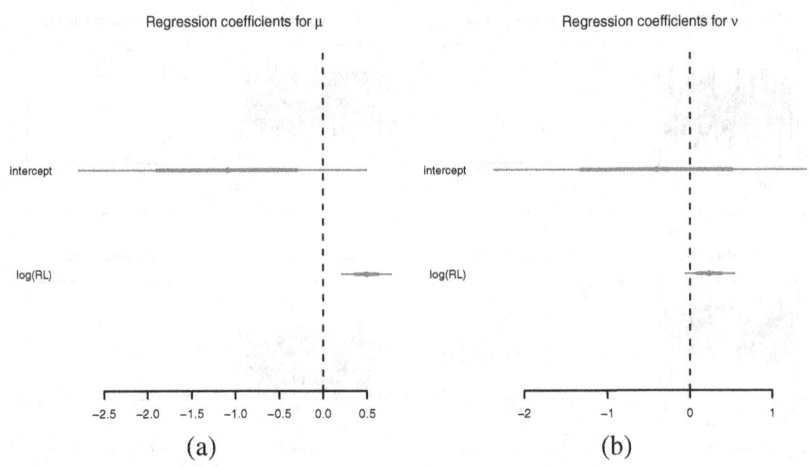


(a) (b)

Figure 5.2 Credible intervals associated with textile fabrics example for (a) $\boldsymbol{\beta}$ (b) $\boldsymbol{\gamma}$; RL = RollLength. Inner and outer intervals, respectively, represent 68% and 95% credible intervals.

## 5.3 Accounting for Excess Zeroes: Zero-inflated COM–Poisson Regression

Excess zeroes relative to a chosen distribution are frequently encountered in count datasets (in particular, those with substantially small sample mean values). A common feature of such data is that the expected number of zeroes decreases as the count distribution's mean increases (Brooks et al., 2017). While excess zeroes give way to apparent data over-dispersion, NB models do not necessarily perform well with such data because they may under-predict the frequency of zeroes (or overestimate nonzero count values). In fact, any sort of analysis that ignores zero-inflation or attempts to compensate for it via a model for over-dispersion runs the risk of producing parameter estimates that are biased. Accordingly, many works (e.g. Famoye and Singh, 2006; Greene, 1994; Lambert, 1992) consider zero-inflation along with various count data models. Several such distributions (including zero-inflated Poisson (ZIP), and zero-inflated negative binomial regression) are available for modeling, for example, in the Vector Generalized Linear and Additive Models (`VGAM`) package in `R` (Yee, 2008).

Excess zeroes are commonly cited as a cause for data over-dispersion; see, for example, Hilbe (2007). This is generally true because, while also affecting the variance, the zero observations contribute to a smaller mean than an analogous dataset where excess zeroes do not exist. The assertion that excess zeroes cause data over-dispersion, however, is not unilaterally accurate. There can exist data structures with underlying data under-dispersion such that the inclusion of excess zeroes can result in perceived data equi- or (still) under-dispersion (Sellers and Shmueli, 2013). Sellers and Raim (2016) argue that analysts should consider a flexible distribution that not only can account for excess zeroes but can also address potential underlying over- or under-dispersion, and develop a zero-inflated Conway–Maxwell–Poisson (ZICMP) regression to model the relationship between response and explanatory variables while accounting for excess zeroes and significant data dispersion. Other models based on ZI COM–Poisson parametrizations can likewise be considered; Section 5.3.1 discusses the various possible model formulations for ZI COM–Poisson regression.

This section focuses solely on addressing excess zeroes; therefore, research that incorporates zero-inflation within a broader analytical context (e.g. clustered data analysis in Section 5.4 (Choo-Wosoba and Datta, 2018; Choo-Wosoba et al., 2016)) is discussed in later sections.

### 5.3.1 Model Formulations

The COM–Poisson distribution (under any parametrization) can be generalized to allow for excess zeroes by considering a random sample $Y_i, i = 1, \ldots, n$, where

$$Y_i \sim \begin{cases} 0 & \text{w.p. } \pi_{i*} \\ \text{COM–Poisson} & \text{w.p. } 1 - \pi_{i*}. \end{cases} \tag{5.25}$$

Under this framework, the resulting probability mass function can be represented as a Bernoulli probability form where the associated probabilities stemming from the respective COM–Poisson parametrization for $y = 1, 2, 3, \ldots$ are multiplied by $1 - \pi_{i*}$, and the zero probability equals the sum of $\pi_{i*}$ and $(1 - \pi_{i*})$ times the corresponding COM–Poisson probability mass at zero. For example, the CMP parametrization described in Equation (2.8) implies that the ZICMP distribution has the form

$$P(Y_i = y_i) = \left(\pi_{i*} + \frac{1 - \pi_{i*}}{Z(\lambda_i, \nu_i)}\right)^{u_i} \left[\frac{(1 - \pi_{i*})\lambda_i^{y_i}}{(y_i!)^{\nu_i} Z(\lambda_i, \nu_i)}\right]^{1-u_i}, \quad y = 0, 1, 2, \ldots, \tag{5.26}$$

for $u_i = 1(0)$ if $y_i = (\neq)0$. The corresponding log-likelihood function is (Sellers and Raim, 2016)

$$\ln L(\boldsymbol{\lambda}, \boldsymbol{\nu}, \boldsymbol{\pi}_*; \mathbf{y}) = \sum_{i=1}^{n} \{u_i \ln(\pi_{i*} Z(\lambda_i, \nu_i) + (1 - \pi_{i*})) \\ + (1 - u_i)[\ln(1 - \pi_{i*}) + y_i \ln(\lambda_i) \\ - \nu_i \ln(y_i!)] - \ln Z(\lambda_i, \nu_i)\}. \quad (5.27)$$

Similarly, the MCMP1 form described in Equation (2.51) produces a zero-inflated MCMP1 (ZIMCMP1) model with an analogous log-likelihood form as displayed in Equation (5.27) where $\lambda(\mu_i, \nu_i)$ replaces $\lambda_i$. For the special case where equi-dispersion holds (i.e. $\nu_i = 1$ for all $i$), ZICMP and ZIMCMP1 modeling reduce to ZIP modeling. Meanwhile, when $\nu_i = 0$ and $\lambda_i < 1$ for all $i$, they reduce to zero-inflated geometric (ZIG) modeling; and when $\nu_i \to \infty$ for all observations, the ZICMP and ZIMCMP1 models reduce to a (model-adjusted) logistic regression.

The ACMP probability mass function (Equation (2.48)) meanwhile implies that zero-inflated ACMP (ZIACMP) has the probability

$$P(Y_i = y_i) = \left(\pi_{i*} + \frac{1 - \pi_{i*}}{Z_1(\mu_{i*}, \nu_i)}\right)^{u_i} \left[(1 - \pi_{i*}) \frac{1}{Z_1(\mu_{i*}, \nu_i)} \left(\frac{\mu_{i*}^{y_i}}{y_i!}\right)^{\nu_i}\right]^{1-u_i}, \\ y = 0, 1, 2, \ldots \quad (5.28)$$

thus producing the associated log-likelihood

$$\ln L(\boldsymbol{\mu}_*, \boldsymbol{\nu}, \boldsymbol{\pi}_*; \mathbf{y}) = \sum_{i=1}^{n} \{u_i \ln(\pi_{i*} Z_1(\mu_{i*}, \nu_i) + (1 - \pi_{i*})) \\ + (1 - u_i)[\ln(1 - \pi_{i*}) \\ + \nu_i y_i \ln(\mu_{i*}) - \nu_i \ln(y_i!)] - \ln Z_1(\mu_{i*}, \nu_i)\}. \quad (5.29)$$

Similarly, the MCMP2 probability mass function (Equation (2.53)) induces a zero-inflated MCMP2 (ZIMCMP2) probability

$$P(Y_i = y_i) = \left(\pi_{i*} + \frac{1 - \pi_{i*}}{Z_2(\mu_i, \phi_i)}\right)^{u_i} \\ \left[(1 - \pi_{i*}) \left(\mu_i + \frac{e^{\phi_i} - 1}{2e^{\phi_i}}\right)^{ye^{\phi_i}} \frac{(y!)^{-e^{\phi_i}}}{Z_2(\mu_i, \phi_i)}\right]^{1-u_i}, \quad y = 0, 1, 2, \ldots \\ (5.30)$$

thus producing a log-likelihood of the form

$$\ln L(\boldsymbol{\mu}, \boldsymbol{\phi}, \boldsymbol{\pi}_*; \mathbf{y}) = \sum_{i=1}^{n} \Bigg\{ u_i \ln\left(\pi_{i*} Z_2(\mu_i, \phi_i) + (1 - \pi_{i*})\right) + (1 - u_i)\Bigg[ \ln(1 - \pi_{i*}) + y_i e^{\phi_i} \ln\left(\mu_i + \frac{e^{\phi_i} - 1}{2e^{\phi_i}}\right) - e^{\phi_i} \ln(y_i!)\Bigg] - \ln Z_2(\mu_i, \phi_i)\Bigg\}. \quad (5.31)$$

Under all of the aforementioned COM–Poisson formulations, the parameters can be modeled via the canonical link-generalized linear models (e.g. a loglinear model associating the intensity parameter (whether $\boldsymbol{\lambda}$, $\boldsymbol{\mu}_*$, $\boldsymbol{\lambda}(\boldsymbol{\mu}, \boldsymbol{\nu})$, or $\boldsymbol{\mu}$) with $\boldsymbol{X\beta}$, where $\boldsymbol{\beta} = (\beta_1, \ldots, \beta_{p_1})^{\mathrm{T}}$; and a logistic model for the zero probability vector, i.e. $\text{logit}(\boldsymbol{\pi}_*) = \boldsymbol{W\zeta}$ where $\boldsymbol{\zeta} = (\zeta_0, \ldots, \zeta_{p_3-1})^{\mathrm{T}}$). The dispersion parameter $\boldsymbol{\nu}$ can either be considered as a constant across observations (i.e., $\nu_i \equiv \nu$ for all $i$) or likewise modeled via the loglinear link, $\ln(\boldsymbol{\nu}) = \boldsymbol{G\gamma}$, where $\boldsymbol{\gamma} = (\gamma_0, \ldots, \gamma_{p_2-1})$, to ensure nonnegativity in the resulting value for $\boldsymbol{\nu}$. The ZIMCMP2 construct, however, already accounts for and reparametrizes the dispersion as $\boldsymbol{\nu} = e^{\boldsymbol{\phi}}$; hence, one can directly consider $\boldsymbol{\phi} = \boldsymbol{G\gamma}$.

*A Further Extension: The ZISCMP Regression*

The sum-of-CMPs (sCMP) distribution introduced in Section 3.2 proves valuable in many contexts. Here, it is helpful in establishing a zero-inflated sCMP (ZISCMP) regression. Constructed analogously to the Lambert (1992) ZIP and Sellers and Raim (2016) ZICMP models, this model is likewise a flexible tool for analyzing count data containing an excess number of zeroes among response counts that also contain significant data dispersion. Defining $\boldsymbol{Y} = (Y_1, \ldots, Y_n)'$ to be the response vector where

$$Y_i \sim \begin{cases} 0 & \text{with probability } \pi_{i*} \\ \text{sCMP}(\lambda_i, \nu_i, m) & \text{with probability } 1 - \pi_{i*}, \end{cases} \quad (5.32)$$

the resulting probability mass function is

$$P(Y_i = y_i) = \begin{cases} \dfrac{\pi_{i*} Z^m(\lambda_i, \nu_i) + (1 - \pi_{i*})}{Z^m(\lambda_i, \nu_i)} & \text{for } y_i = 0 \\ \dfrac{(1 - \pi_{i*})\lambda_i^{y_i}}{(y_i!)^{\nu_i} Z^m(\lambda_i, \nu_i)} \displaystyle\sum_{\substack{c_1, \ldots, c_m = 0 \\ c_1 + \cdots + c_m = y_i}}^{y_i} \binom{y_i}{c_1 \cdots c_m}^{\nu} & \text{for } y_i = 1, 2, 3, \ldots, \end{cases} \quad (5.33)$$

where $m$ is a count value, and the association between the response and covariate terms is modeled via analogous canonical forms as considered for the CMP parameters above, namely $\ln(\boldsymbol{\lambda}) = \boldsymbol{X\beta}$, $\ln(\boldsymbol{\nu}) = \boldsymbol{G\gamma}$, and $\text{logit}(\boldsymbol{\pi}_*) = \boldsymbol{W\zeta}$, where $\boldsymbol{\beta} = (\beta_0, \ldots, \beta_{p_1-1})\text{T}$, $\boldsymbol{\gamma} = (\gamma_0, \ldots, \gamma_{p_2-1})$, and $\boldsymbol{\zeta} = (\zeta_0, \ldots, \zeta_{p_3-1})\text{T}$ are the unknown parameters (Sellers and Young, 2019). The special case where $m = 1$ simplifies to the ZICMP model. The ZISCMP model has the resulting log-likelihood

$$\ln L(\boldsymbol{\lambda}, \boldsymbol{\nu}, \boldsymbol{\pi}_* \mid \boldsymbol{X}, \boldsymbol{G}, \boldsymbol{W}, m)) = \sum_{i=1}^{n} \Bigg\{ u_i \ln\left(\pi_{i*} Z^m(\lambda_i, \nu_i) + (1 - \pi_{i*})\right)$$
$$+ (1 + u_i) \Bigg[ \ln(1 - \pi_{i*}) + y_i \ln(\lambda_i) - \nu_i \ln(y_i!)$$
$$+ \ln\left( \sum_{\substack{c_1, \ldots, c_m = 0 \\ c_1 + \cdots + c_m = y_i}}^{y_i} \binom{y_i}{c_1 \ \cdots \ c_m}^{\nu} \right) \Bigg] - m \ln(Z(\lambda_i, \nu_i)) \Bigg\}, \quad (5.34)$$

where $u_i$ is the indicator value noting when $Y_i = 0$ (i.e., $u_i = 1$) or not ($u_i = 0$).

### 5.3.2 Parameter Estimation

*Frequentist Approach*

The method of maximum likelihood is a natural approach for parameter estimation, determining those parameter estimates $\hat{\boldsymbol{\beta}}, \hat{\boldsymbol{\gamma}}, \hat{\boldsymbol{\zeta}}$ that maximize the log-likelihood of interest (i.e. Equations (5.27), (5.29), (5.31), or (5.34) for ZICMP or ZIMCMP1, ZIACMP, ZIMCMP2, or ZISCMP regression, respectively). Analysts interested in only considering a constant dispersion can constrain $\boldsymbol{\gamma} = \boldsymbol{\gamma}_0$. Assuming $m$ to be discrete in the ZISCMP regression allows for a "profile likelihood" approach toward estimating the other parameters. This method simplifies the overall model structure's complex computational nature.

As described in Section 2.4.2, a closed-form solution does not exist for the MLEs; thus, statistical computing (e.g. via R) is required in order to conduct a Newton-type optimization procedure that identifies the estimates that maximize the log-likelihood; see Section 5.5 for details. Natural starting values for any such Newton-type algorithm are the MLEs attained assuming a ZIP model. Under this construct, the initial dispersion parameter value is $\nu^{(0)} = 1$ (i.e. $\boldsymbol{\gamma}^{(0)} = \mathbf{0}$). Standard errors associated with the parameter estimates are obtained by deriving the Fisher information matrix, attainable from some optimization functions via the approximate corresponding Hessian matrix that can retained as output. The Fisher information matrix for the ZISCMP model is a nonorthogonal, block symmetric matrix whose components can likewise be determined. In all cases, the corresponding standard errors can be obtained by isolating the diagonal components of the inverted resulting Fisher information matrix.

*Bayesian Formulation*

Alternatively, one can consider a full Bayesian approach for parameter estimation and inference where $L(\boldsymbol{\theta}; \boldsymbol{D})$ denotes the likelihood associated with a ZI COM–Poisson-inspired model (whose log-likelihoods are supplied within Equations (5.27)–(5.31) for the various ZI COM–Poisson parametrizations and Equation (5.34) for ZISCMP) that is a function of $\boldsymbol{\theta} = (\boldsymbol{\beta}', \boldsymbol{\gamma}', \boldsymbol{\zeta}')'$ given $n$ independent observations $\boldsymbol{D} = ((y_1, u_1, \boldsymbol{x}_1, \boldsymbol{g}_1, \boldsymbol{w}_1), \ldots, (y_n, u_n, \boldsymbol{x}_n, \boldsymbol{g}_n, \boldsymbol{w}_n))$, and $u_i = 1(0)$ if $y_i = 0$ (otherwise) for $i = 1, \ldots, n$. Barriga and Louzada (2014), for example, further assume the parameters $\boldsymbol{\beta}$, $\boldsymbol{\gamma}$, and $\boldsymbol{\zeta}$ to be independent a priori (i.e. $\pi(\boldsymbol{\theta}) = \prod_{i=0}^{p_1-1} \pi(\beta_i) \prod_{i=0}^{p_2-1} \pi(\gamma_i) \prod_{i=0}^{p_3-1} \pi(\zeta_i)$) with respective normal prior distributions $\beta_j \sim N(0, \sigma^2_{\beta_j}), j = 0, \ldots, p_1 - 1$; $\gamma_k \sim N(0, \sigma^2_{\gamma_k}), k = 0, \ldots, p_2 - 1$; and $\zeta_l \sim N(0, \sigma^2_{\zeta_l})$, $l = 0, \ldots, p_3 - 1$, where the hyperparameters have noninformative priors. While these assumptions are proposed only in the consideration of the ZIACMP model, they are reasonable for any of the ZI COM–Poisson-inspired models described in Section 5.3.1.

The joint posterior distribution for $\boldsymbol{\theta}$ determined by $\pi(\boldsymbol{\theta}|\boldsymbol{D}) \propto L(\boldsymbol{\theta}; \boldsymbol{D})\pi(\boldsymbol{\theta})$ is difficult to analyze directly; however, Gibbs sampling with the Metropolis–Hastings algorithm offers some understanding of the joint posterior form. The expectation–maximization (EM) algorithm is another popular technique for parameter estimation in a zero-inflated model; however, no published works to date utilize this approach for the ZI COM–Poisson regressions under any parametrization.

### 5.3.3 Hypothesis Testing

Analogous to the discussion in Section 5.2.3, one can conduct a hypothesis test to assess whether a statistically significant amount of data dispersion exists such that a ZI COM–Poisson model (under any parametrization) is considered more appropriate than a ZIP model. For example, Sellers and

Raim (2016) consider the constant dispersion ZICMP model and hypotheses, $H_0: \nu = 1$ versus $H_1: \nu \neq 1$, because the concern does not center around the type of potential data dispersion but rather the existence of any such form (whether under- or over-dispersion). The likelihood ratio test statistic is

$$C_\nu = -2 \ln \Lambda_\nu = 2 \ln L(\hat{\boldsymbol{\beta}}, \hat{\nu}, \hat{\boldsymbol{\zeta}}) - 2 \ln L(\hat{\boldsymbol{\beta}}_0, \hat{\nu}_0 = 1, \hat{\boldsymbol{\zeta}}_0), \tag{5.35}$$

where $\ln L(\hat{\boldsymbol{\beta}}, \hat{\nu}, \hat{\boldsymbol{\zeta}})$ and $\ln L(\hat{\boldsymbol{\beta}}_0, \hat{\nu}_0 = 1, \hat{\boldsymbol{\zeta}}_0)$ denote the respective log-likelihood values associated with the ZICMP and ZIP MLEs, respectively. Computing $C_\nu$ is straightforward because the respective log-likelihood values are easily extracted as outputs from computational tools in R. Under the null hypothesis, $C_\nu$ has an approximate $\chi^2$ distribution with one degree of freedom. Generalizing to the variable model is straightforward and conducted in a manner analogous to that described in Section 5.2.3.

One can also conduct a test for zero-inflation, e.g. $H_0: \pi_* = 0$ versus $H_1: \pi_* > 0$ for the case where $\pi_*$ does not depend on covariates. The test is measured through the likelihood ratio test statistic,

$$C_{\pi_*} = -2 \ln \Lambda_{\pi_*} = 2 \ln L(\hat{\boldsymbol{\beta}}, \hat{\nu}, \hat{\boldsymbol{\zeta}}) - 2 \ln L(\hat{\boldsymbol{\beta}}_0, \hat{\nu}_0, \hat{\boldsymbol{\zeta}}_0 = 0) \xrightarrow{d} 0.5 p_0 + 0.5 \chi_1^2, \tag{5.36}$$

where $p_0$ denotes the probability at zero. Numerical issues can arise, however, when applying this test in practice; thus, an alternative approach is to conduct a null bootstrapping procedure to obtain $p$-values associated with zero-inflation (Choo-Wosoba and Datta, 2018). While these tests are described for the ZICMP regression, they can likewise be pursued assuming any of the other parametrizations described in Section 5.3.1.

### *5.3.4 A Word of Caution*

While zero-inflated models such as the ZIP, ZINB, or any of the parametrized ZI COM–Poisson models are useful for describing data containing excess zeroes, analysts should be mindful of the underlying process that generated the data and not use such models where the data do not exhibit two distinct generating processes (Kadane et al., 2006b; Lord et al., 2008). At the same time, while the COM–Poisson can predict more zeroes than the NB model for the same mean value, neither should be used as a direct substitute to zero-inflated models when such formulations are appropriate (Kadane et al., 2006b).

### 5.3.5 Alternative Approach: Hurdle Model

Hurdle models serve as an alternative approach to account for excess zeroes, where the random variable of interest has zero and nonzero probability components (e.g. Choo-Wosoba et al. (2018)). Let $\boldsymbol{Y} = (Y_1, \ldots, Y_n)'$ be a response vector such that

$$Y_i \sim \begin{cases} 0 & \text{with probability } \pi_{i*} \\ \text{zero-truncated CMP}(\lambda_i, \nu_i) & \text{with probability } 1 - \pi_{i*}, \end{cases} \tag{5.37}$$

where a zero-truncated CMP (ZTCMP) distribution has the probability mass function

$$P(Y_i = y \mid Y_i > 0) = \frac{\lambda_i^y}{(y!)^{\nu_i}[Z(\lambda_i, \nu_i) - 1]}, \quad y = 1, 2, \ldots. \tag{5.38}$$

A common approach in hurdle modeling is to account for the excess zeroes via a probit model, in which case, $Y_i$ has the distribution

$$P(Y_i = y) = \begin{cases} 1 - \Phi(\boldsymbol{W}_i\boldsymbol{\zeta}) & y = 0 \\ \Phi(\boldsymbol{W}_i\boldsymbol{\zeta}) \cdot \frac{\lambda_i^y}{(y!)^{\nu_i}[Z(\lambda_i, \nu_i) - 1]} & y = 1, 2, \ldots, \end{cases} \tag{5.39}$$

where $\boldsymbol{\lambda}$ and $\boldsymbol{\nu}$ can both be modeled via loglinear links, i.e. $\ln(\boldsymbol{\lambda}) = \boldsymbol{X\beta}$ and $\ln(\boldsymbol{\nu}) = \boldsymbol{G\gamma}$, respectively. While the above represents a hurdle CMP regression, analogous hurdle COM–Poisson models can be formulated based on the alternate COM–Poisson parametrizations described in Section 2.6. As is the case with its zero-inflated analog, hurdle COM–Poisson regression models outperform their corresponding hurdle Poisson regression models when significant dispersion is present and performs comparably otherwise.

## 5.4 Clustered Data Analysis

Clustered data often result from situations that introduce some measures of within-cluster correlation. Common examples include repeated measures, hierarchical constructs, and longitudinal studies. Count data containing such interdependence naturally violate the usual random sample assumption. This violation usually leads to observed over-dispersion if the data are assumed to be independent and identically distributed. Longitudinal count data (i.e. discrete data collected from repeated measurements on a subject) serve as an example of clustered data in that they contain a within-subject correlation. Ways to address the dependence in this scenario range from directly accounting for the correlation within clusters (e.g. Choo-Wosoba et al., 2016), to considering a mixed-effects Poisson model that relaxes the equi-dispersion assumption by parametrizing subject-level variability. The

Poisson longitudinal model, however, assumes that the underlying count mechanism (minus the longitudinal structure) exhibits equi-dispersion. Alternatively, a longitudinal COM–Poisson regression model allows analysts to model additional variability due to the correlation of repeated measures as well as the over- or under-dispersion of the underlying count process. Various works have considered the analysis of clustered count data via a (zero-inflated) COM–Poisson model. All of the constructs considered in this section assume a constant dispersion for their (ZI)CMP model. This assumption is arguably sufficient for most applications; a varying dispersion can lead to an identifiability issue or other issues regarding modeling, computation, and/or convergence (Choo-Wosoba and Datta, 2018).

Choo-Wosoba et al. (2016) consider a general ZICMP regression for clustered data that incorporates correlation and propose two methods for analysis: (1) a modified expectation–solution (MES) approach that contains a cluster bootstrap-based variance estimator and (2) a maximum pseudo-likelihood (MPL) approach that includes an adjusted variance estimator to address the interdependence in clusters. This construct builds on the Sellers and Raim (2016) ZICMP parametrization (Equation (5.25)) where clustered responses $\{Y_{ij}\}$ are assumed for the $j$th observation ($1 \leq j \leq m_i$) in the $i$th cluster ($1 \leq i \leq M$) with a constant dispersion $\nu$ for all subjects, while the shape parameters $\lambda_{ij}$ and zero-inflation parameters $p_{ij}$ vary with each subject. As defined, correlation is assumed within the cluster but not across clusters. The EM algorithm is not appropriate for clustered data and the expectation–solution (ES) algorithm of Rosen et al. (2000) only applies to exponential dispersion families (of which the CMP is not; see Section 2.3); thus, Choo-Wosoba et al. (2016) instead create an MES to estimate the coefficients associated with $\boldsymbol{\lambda}$ (i.e. $\boldsymbol{\beta}$ via $\ln(\boldsymbol{\lambda}) = \boldsymbol{X\beta}$ with correlation coefficient $\rho$) and $\boldsymbol{p}$ (i.e. $\boldsymbol{\zeta}$ via the model, $\text{logit}(\boldsymbol{p}) = \boldsymbol{W\zeta}$ with correlation coefficient $\eta$).

This MES algorithm assumes a complete pseudo-log-likelihood

$$\begin{aligned}
\ln L_c(\boldsymbol{\beta}, \boldsymbol{\zeta}, \nu; y_{ij}, w_{ij}) &= \sum_{i=1}^{M} \sum_{j=1}^{m_i} \ln L_{ij}(\boldsymbol{\beta}, \boldsymbol{\zeta}, \nu; y_{ij}, u_{ij}) \\
&= \sum_{i=1}^{M} \sum_{j=1}^{m_i} w_{ij} \ln (p_{ij}) + \sum_{i=1}^{M} \sum_{j=1}^{m_i} (1 - w_{ij}) \ln (1 - p_{ij}) \\
&\quad + \sum_{i=1}^{M} \sum_{j=1}^{m_i} (1-w_{ij})(y_{ij} \ln (\lambda_{ij}) - \nu \ln (y_{ij}!) - \ln (Z(\lambda_{ij}, \nu))),
\end{aligned} \tag{5.40}$$

where $w_{ij} = 1(0)$ if $y_{ij} = 0$ ( $\neq 0$) is a latent indicator function accounting for zeroes. The authors refer to Equation (5.40) as a "pseudo-likelihood" because the individual terms are assumed independent of each other. The MES algorithm takes this equation and alternates between an expectation step and a solution step until convergence is reached. The expectation step calculates the expectation $E(\ln L_c(\boldsymbol{\beta}, \boldsymbol{\zeta}, \nu; \boldsymbol{y}, \boldsymbol{w})) = \sum_{i=1}^{M} \sum_{j=1}^{m_i} \ln L(\boldsymbol{\beta}, \boldsymbol{\zeta}, \nu; y_{ij}, E(w_{ij}))$ as defined in Equation (5.40), where $\boldsymbol{w}^{(k)}$ denotes the result from

$$E(w_{ij}) = \frac{p_{ij}}{p_{ij} + (1 - p_{ij})/Z(\lambda_{ij}, \nu)}$$

for all $i,j$ in the $k$th iteration. The solution step applies this value to estimate $\boldsymbol{\beta}$ and $\boldsymbol{\zeta}$ via generalized estimating equations (GEE) to update from $\boldsymbol{\beta}^{(k)}, \boldsymbol{\zeta}^{(k)}$ to $\boldsymbol{\beta}^{(k+1)}, \boldsymbol{\zeta}^{(k+1)}$, while $\nu$ is estimated and thus updated via a pseudo-likelihood equation to obtain $\nu^{(k+1)}$. The `pscl` (Jackman et al., 2017) ZIP estimates provide starting values for the MES algorithm. The MPL estimation method is an alternative approach to estimate $\boldsymbol{\beta}, \boldsymbol{\zeta}$, and $\nu$ where the observed log-transformed pseudo-likelihood

$$\begin{aligned}\ln L_o(\boldsymbol{\beta}, \boldsymbol{\zeta}, \nu; y_{ij}) = & \sum_{i=1}^{M} \sum_{j=1}^{m_i} I_{\{y_{ij} \geq 1\}} [\ln(1 - p_{ij}) + y_{ij} \ln(\boldsymbol{\lambda}_{ij}) - \nu \ln(y_{ij}!) \\ & - \ln(Z(\boldsymbol{\lambda}_{ij}, \nu))] \\ & + \sum_{i=1}^{M} \sum_{j=1}^{m_i} I_{\{y_{ij}=0\}} \ln[p_{ij} + (1 - p_{ij})/Z(\boldsymbol{\lambda}_{ij}, \nu)] \end{aligned} \quad (5.41)$$

is considered with a sandwich variance to address the interdependency within clusters. The standard errors for the coefficient estimates obtained via MPL are derived via an adjusted sandwich variance method, while the standard errors corresponding to the MES-attained estimators are obtained via a cluster bootstrapping scheme. The cluster bootstrap method is actually a reasonable approach for obtaining variance estimates associated with the estimated coefficients from either approach. Some `R` packages that can compute the sandwich variances, however, will not produce estimates comparably to those attained using a bootstrap estimate (Choo-Wosoba et al., 2016).

The MES method produces slightly more efficient estimators than those attained via the MPL method; Choo-Wosoba et al. (2016) attribute the added efficiency to using a working variance–covariance matrix (e.g. GEE); however, this occurs at a computational cost. The MPL approach, meanwhile, produces a closed-form variance estimator. This approach (like all) is not immune to convergence issues; for datasets where such issues arise, analysts can try changing the initial values, consider a different optimization method, etc.

Instead of a generally defined (ZI)CMP regression that allows for clustering, one can construct a mixed-effects model that assumes normal random intercepts to handle within-cluster interdependence (Choo-Wosoba and Datta, 2018; Morris and Sellers, 2022). Let $Y_{ij}$ (conditioned on the random effect $u_i$ associated with Cluster $i$) have a (ZI)CMP distribution with parameters $\lambda_{ij}$ and $\nu$, where

$$\ln(\lambda_{ij}) = \beta_0 + \beta_1 x_{ij1} + \cdots + \beta_p x_{ijp} + u_i \doteq \boldsymbol{X}_{ij}\boldsymbol{\beta} + u_i, \tag{5.42}$$

with $u_i$ normally distributed with mean 0 and variance $\sigma^2$. This model produces the conditional log-likelihood $\ln L(\lambda_{ij}, \nu \mid y_{ij})$ that has a form analogous to Equation (2.9), where $\lambda_{ij}$ is defined in Equation (5.42), thus producing a marginal log-likelihood

$$\begin{aligned}\ln L(\boldsymbol{\beta}, \nu, \sigma^2 \mid \boldsymbol{y}) = {} & \sum_{i=1}^{M}\sum_{j=1}^{m_i} y_{ij}\ln(\lambda_{ij}) - \nu\sum_{i=1}^{M}\sum_{j=1}^{m_i}\ln(y_{ij}!) - \sum_{i=1}^{M}\ln(\sigma\sqrt{2\pi}) \\ & + \sum_{i=1}^{M}\ln\left(\int_{u_i} e^{u_i\sum_{j=1}^{m_i} y_{ij} - u_i^2/2\sigma^2}\left(\prod_{j=1}^{m_i} Z(\lambda_{ij}e^{u_i}, \nu)\right)^{-1} du_i\right).\end{aligned} \tag{5.43}$$

Incorporating excess zeroes is further possible via a logistic model $\text{logit}(\boldsymbol{p}) = \boldsymbol{W}\boldsymbol{\zeta}$ such that the likelihood function for the $i$th cluster is

$$\begin{aligned}L_i(\boldsymbol{\beta}, \nu, \sigma^2 \mid \boldsymbol{y}) = \int_{-\infty}^{\infty}\Bigg[\prod_{j=1}^{m_i}\Bigg\{ & I_{[Y_{ij}=0]}\left(p_{ij} + \frac{1-p_{ij}}{Z(\lambda_{ij}, \nu)}\right) \\ & + I_{[Y_{ij}\geq 1]}\left((1-p_{ij}) \cdot \frac{\lambda_{ij}^{y_{ij}}}{(y_{ij}!)^{\nu} Z(\lambda_{ij}, \nu)}\right)\Bigg\}\Bigg] p(u_i)du_i.\end{aligned} \tag{5.44}$$

Neither Equation (5.43) nor (5.44) has a closed form for easy integration. One can, however, obtain MLEs for $\boldsymbol{\beta}$, $\nu$, and $\sigma^2$ in `R` by numerically integrating the equation to obtain an approximated marginal log-likelihood for Cluster $i$ (i.e. $\ln(\tilde{L}_i)$), say via the `integrate` function or approximating the integral via Gaussian–Hermite quadrature (the `fastGHQuad` (Blocker, 2018) package in `R`) to obtain the approximated log-likelihood across clusters, $\ln(\tilde{L}) = \sum_{i=1}^{M}\ln(\tilde{L}_i)$. Analysts, however, should be mindful when utilizing the Gaussian–Hermite quadrature method because it relies heavily on choosing a good number of quadrature points. Nonetheless, $\ln(\tilde{L})$ can then be maximized to obtain the MLEs $\hat{\boldsymbol{\beta}}$, $\hat{\nu}$, and $\hat{\sigma}^2$, either via `nlminb` in order to take into account the constrained space for $\nu$, or one can first transform the parametrization from $\nu$ to $\ln(\nu)$ to allow for unconstrained optimization and utilize the alternative optimization functions such as `nlm` or `optim`. Morris et al. (2017) meanwhile describe how to conduct this analysis via `SAS`; see Section 5.7.2. The ZICMP mixed-effects model can further consider incorporating a random intercept into the logistic model for $\boldsymbol{p}$ and conduct parameter estimation as described. Variation can be estimated via the usual inverse of the approximate Fisher information matrix (achieved by obtaining the approximate Hessian matrix associated with the MLEs from `optim`) or, for added robustness in the case of model misspecification, an estimated sandwich variance–covariance matrix via the Gaussian–Hermite quadrature.

This mixed-effects approach cannot incorporate a large number of random effects because the approximate log-likelihood is obtained via computationally expensive numerical procedures. The Gaussian–Hermite quadrature procedure particularly faces difficulties estimating more elaborate correlation structures beyond equi-correlation (Choo-Wosoba et al., 2018) and can produce biased results in the approximate log-likelihood function. A Bayesian ZICMP regression for clustered data is a viable alternative that can allow for broader correlation structures and dependencies. Choo-Wosoba et al. (2018) combine the Bayesian ZICMP analytic approach with the hurdle CMP model (both described in Section 5.3) while incorporating random effects. The zero model assumes a probit regression with mixed effects to determine the probability of a nonzero outcome $P(Y_{ij} > 0) = \Phi\left(\boldsymbol{W}_{ij}\boldsymbol{\zeta} + \boldsymbol{U}_{ij}\boldsymbol{\delta}_i^{(1)}\right)$ while $\ln(\lambda_{ij}) = \boldsymbol{X}_{ij}\boldsymbol{\beta} + \boldsymbol{U}_{ij}\boldsymbol{\delta}_i^{(2)}$, and $\nu$ remains constant. Accordingly, $Y_{ij}$ now has the distribution

$$P(Y_{ij} = y) = \begin{cases} 1 - \Phi(\boldsymbol{W}_{ij}\boldsymbol{\zeta} + \boldsymbol{U}_{ij}\boldsymbol{\delta}_i^{(1)}) & y = 0 \\ \Phi(\boldsymbol{W}_{ij}\boldsymbol{\zeta} + \boldsymbol{U}_{ij}\boldsymbol{\delta}_i^{(1)}) \cdot \frac{\lambda_{ij}^y}{(y!)^\nu [Z(\lambda_{ij},\nu)-1]} & y = 1, 2, \ldots, \end{cases} \tag{5.45}$$

where the random effects associated with the $i$th cluster have an MVN distribution with a zero mean and variance–covariance matrix, $\Sigma$ (i.e. $(\boldsymbol{\delta}_i^{(1)}, \boldsymbol{\delta}_i^{(2)})' \sim \text{MVN}(0, \Sigma)$), and $\boldsymbol{\beta}$ and $\boldsymbol{\zeta}$ are likewise MVN distributed with zero means and variance–covariance matrices, $\Omega_\beta$ and $\Omega_\zeta$, respectively. Meanwhile, $\Sigma$ assumes an inverse Wishart prior, and $\nu$ has a lognormal prior with median at 1 and 95% probability between 0.38 and 2.66. The hyperparameters can be determined either by subject-matter experts or assumed as proper, weakly-informative priors. An iterative MCMC sampling scheme can aid with inference since the zero-inflated/hurdle CMP does not have a closed form; see Choo-Wosoba et al. (2018) for an example algorithm. Analogous to Section 5.3.5, hurdle CMP mixed-effects models are shown to perform at least comparably to hurdle Poisson mixed-effects models and outperform such models when significant data-dispersion exists. One should note however that, where a frequentist approach can face difficulties approximating the likelihood, the Bayesian scheme circumvents such issues producing a more flexible random effects design matrix. This requires a considerable computation time, however, to perform the updates.

## 5.5 `R` Computing for Excess Zeroes and/or Clustered Data

Of the four regression functions described in Section 5.2.4, only the `COMPoissonReg:glm.cmp` and `glmmTMB` functions can conduct ZI COM–Poisson regression. The `glm.cmp` function conducts ZICMP regression, estimating the regression parameters via the method of maximum likelihood. The function defaults to consider the constant dispersion zero-inflated model (i.e. `formula.nu = 1`) but allows for variable dispersion, while the analyst provides the model description for the loglinear model $\ln(\boldsymbol{\lambda}) = \boldsymbol{X\beta}$ and logistic model $\text{logit}(\boldsymbol{\pi}_*) = \boldsymbol{W\zeta}$ in the `formula.lambda` and `formula.p`, respectively.

The `glm.cmp` utilizes the `optim` procedure in `R` to estimate the (ZI)CMP parameters; however, the MLEs for any of the respective zero-inflated models can be determined numerically via other computational tools (e.g. `nlminb`, `nlm`). These procedures require specifying an objective function to be minimized and a starting point from which to apply one of the aforementioned Newton-type algorithms. These `R` optimization functions are set to minimize the objective function; hence, the objective function must be defined to be the negated log-likelihood function (Equation (5.27) for ZICMP or its analog for ZIMCMP1, Equation (5.29) for ZIACMP, or Equation (5.31) for ZIMCMP2); accordingly, the resulting values that minimize the objective function maximize the likelihood function (i.e. the resulting values are the MLEs). The ZIP estimates obtained from the `zeroinfl` function in the `pscl` package (Jackman et al., 2017) serve as natural starting values for the respective algorithms. Accordingly, analysts can set `beta.init` equal to the ZIP estimates and `gamma.init = 0`, although other starting values can be supplied as well. While any of the Newton-type algorithms can optimize the log-likelihood of interest, Sellers and

Young (2019) found `nlminb` to be most efficient for ZISCMP parameter estimation and that the computing time significantly increases with $m$.

The `glmmTMB` function (contained in the `glmmTMB` package) conducts generalized linear mixed modeling that allows for data containing excess zeroes (Magnusson et al., 2020). It can perform ZIMCMP1 regressions of the form

$$\ln(\boldsymbol{\lambda}(\boldsymbol{\mu}, \boldsymbol{\nu})) = \boldsymbol{X\beta} + \boldsymbol{Z\alpha} \tag{5.46}$$

$$\ln(\boldsymbol{\nu}) = \boldsymbol{G\gamma} \tag{5.47}$$

$$\text{logit}(\boldsymbol{\pi}_*) = \boldsymbol{W\zeta}, \tag{5.48}$$

accounting for excess zeroes by specifying a zero-inflated (`family = compois`) or hurdle (`family = truncated_compois`) MCMP1 construct; the default `ziformula = 0` assumes no excess zeroes. The `glmmTMB` function assumes a generalized linear mixed model for $\boldsymbol{\lambda}$ (Equation (5.46)) and allows for a loglinear fixed-effects model for the dispersion (Equation (5.47)); however, the dispersion function is assumed constant (i.e. `dispformula = ~ 1`). Analysts should note that the reported dispersion parameter in the output equals $1/\hat{\nu}$. The `glmmTMB` interface is comparable to `lme4` (Bates et al., 2015); yet, it is faster with its computation of the Laplace approximation to integrate over the random effects (Brooks et al., 2017). Analysts are encouraged against using the same covariates in both the conditional and dispersion models for threats of algorithm nonconvergence. Likewise, when little information exists regarding the levels of the random effects, analysts are warned of a potentially poor Laplacian approximation (Brooks et al., 2017). Additional `glmmTMB` function inputs allow for weights, contrasts, an offset, and the ability to specify how to handle missing data. Given the `glmmTMB` function's ability to handle more broad generalized linear mixed models along with excess zeroes, it can likewise consider generalized linear models that only contain fixed effects and excess zeroes, either via zero-inflation or hurdle models. Accordingly, this function can perform ZIMCMP1 or hurdle MCMP1 (HMCMP1) regression involving fixed and/or random effects.

### *5.5.1 Examples*

The following examples illustrate the `COMPoissonReg:glm.cmp` and `glmmTMB` functions and their respective capabilities. Both of these functions can conduct generalized linear modeling that allows for zero-inflation; however, the `glmmTMB` function can further perform analogous generalized linear mixed modeling. Accordingly, it is utilized for both considered examples: the first example illustrates functionality in R to handle excess zeroes in a generalized linear model setting, and the second example illustrates computational abilities to conduct mixed-effects regressions that may or may not account for excess zeroes. The `combayes` package is also reported to have the ability to perform generalized linear mixed modeling for clustered data. The current package, however, does not provide help pages to aid with associated discussion and illustration.

### *Example: Unwanted Pursuit Behavior Perpetrations*

Loeys et al. (2012) presented a nice illustration of a dataset containing excess zeroes; this dataset has since been well considered for analysis under various count data models. The study investigates the association between the number of unwanted pursuit behavior perpetrations in the context of couple separation trajectories and one's education level (an indicator where 1 (0) denotes having at least a bachelor's degree (otherwise)) and anxiety level (a continuous measure) regarding attachment in the relationship. Of the 387 subjects considered in the sample, 246 of them have zero unwanted pursuit behavior perpetrations. Previous works have considered Poisson, ZIP, NB, ZINB, CMP, and ZICMP regressions to analyze these data (Loeys et al., 2012; Sellers and Raim, 2016). Here, we focus our attention on ZICMP and ZIMCMP1 regressions as we revisit these data and broaden the analysis.

Table 5.6 reports the coefficient estimates and standard errors (in parentheses) associated with the model

$$\ln(\theta) = \beta_0 + \beta_1 \text{Education} + \beta_2 \text{Anxiety} \tag{5.49}$$

$$\text{logit}(\pi_*) = \zeta_0 + \zeta_1 \text{Education} + \zeta_2 \text{Anxiety}, \tag{5.50}$$

where $\theta = \lambda$ for the ZIP and ZICMP models, $\theta = \mu$ for the ZINB and ZIG models, and $\theta = \lambda(\mu, \nu)$ for the ZIMCMP1 model. All models assume a constant dispersion where necessary; $\hat{\gamma}$ denotes the ZINB-estimated dispersion as described in Section 1.2, while $\hat{\nu}$ estimates the COM–Poisson dispersion. Along with the coefficient estimates and standard errors, the table reports the negated log-likelihood and AIC associated with the respective models. The respective AIC measures show that the ZIP model is least appropriate for modeling these data as it produces the largest AIC (1616.9) among the considered zero-inflation models. This makes sense because a preliminary exploratory analysis of the data finds the variable regarding the number of unwanted pursuit behavior perpetrations to

Table 5.6 *Estimated coefficients and standard errors (in parentheses), negated log-likelihood, and Akaike information criterion (AIC) for various zero-inflated regressions associating the number of unwanted pursuit behavior perpetrations in the context of couple separation trajectories with the levels of education (an indicator function where 1 (0) denotes having at least bachelor's degree (otherwise)) and anxious attachment (a continuous measure) among 387 participants. Considered models are zero-inflated Poisson (ZIP), negative binomial (ZINB), CMP (ZICMP), Huang (2017) mean-parametrized COM–Poisson (ZIMCMP1), and geometric (ZIG), as well as a hurdle MCMP1 (HMCMP1) model. NR = not reported.*

| | ZIP | | ZINB | | ZICMP | | ZIMCMP1 | | HMCMP1 | | ZIG | |
|---|---|---|---|---|---|---|---|---|---|---|---|---|
| Count comp. | | | | | | | | | | | | |
| Intercept | 1.921 | (0.044) | 1.723 | (0.150) | −0.160 | (0.077) | 1.770 | (0.122) | 1.769 | (0.122) | 1.770 | (0.122) |
| Education | −0.350 | (0.071) | −0.490 | (0.206) | −0.068 | (0.034) | −0.476 | (0.191) | −0.474 | (0.191) | −0.476 | (0.191) |
| Anxiety | 0.133 | (0.034) | 0.205 | (0.108) | 0.023 | (0.015) | 0.199 | (0.100) | 0.201 | (0.099) | 0.199 | (0.100) |
| Zero comp. | | | | | | | | | | | | |
| Intercept | 0.673 | (0.142) | 0.340 | (0.210) | 0.418 | (0.167) | 0.422 | (0.159) | 0.675 | (0.142) | 0.422 | (0.159) |
| Education | −0.232 | (0.222) | −0.459 | (0.297) | −0.388 | (0.268) | −0.416 | (0.271) | −0.220 | (0.221) | −0.416 | (0.271) |
| Anxiety | −0.483 | (0.111) | −0.520 | (0.147) | −0.524 | (0.133) | −0.503 | (0.135) | −0.486 | (0.111) | −0.503 | (0.135) |
| $\hat{\gamma}$ | – | | 0.821 | (0.226) | – | | – | | – | | – | |
| $\hat{\nu}$ | – | | – | | 0.000 | (0.031) | 0.000 | (NR) | 0.000 | (NR) | – | |
| $-\ln L$ | 802.45 | | 626.14 | | 627.17 | | 626.42 | | 626.52 | | 626.42 | |
| AIC | 1616.9 | | 1266.3 | | 1268.3 | | 1266.8 | | 1267.0 | | 1264.8 | |

be over-dispersed (mean and variance approximately equaling 2.284 and 23.302, respectively).

The COMPoissonReg or `glmmTMB` packages conduct a COM–Poisson regression that accounts for excess zeroes on these data; `COMPoissonReg::glm.cmp` conducts ZICMP regression, while the `glmmTMB` function can perform either ZIMCMP1 or hurdle MCMP1 (HM-CMP1) regression. The respective codes to run these models are provided for the data frame `couple` which contains the variables `UPB` (containing the number of unwanted pursuit behavior perpetrations), `EDUCATION`, and `ANXIETY`.

```
# Using COMPoissonReg for ZICMP regression
couples.zicmp <- glm.cmp(UPB ~ EDUCATION + ANXIETY,
formula.p = ~EDUCATION + ANXIETY, data=couple)
summary(couples.zicmp)
couples.zicmp

#Using glmmTMB for ZIMCMP1 regression
couples.zicmpglmm <- glmmTMB(UPB ~ EDUCATION + ANXIETY,
ziformula = ~EDUCATION + ANXIETY, data=couple,
family=compois)
summary(couples.zicmpglmm)

#Using glmmTMB for HMCMP1 regression
couples.hcmp <- glmmTMB(UPB ~ EDUCATION + ANXIETY,
ziformula = ~EDUCATION + ANXIETY, data=couple,
family=truncated_compois)
summary(couples.hcmp)
```

Among the three COM–Poisson models for excess zeroes, the ZIMCMP1 regression produces the smallest AIC (1266.8); however, substantial support holds for the HMCMP1 and ZICMP models, as discussed in Table 1.2 ($\Delta_{\text{HMCMP1}} = 0.2$ and $\Delta_{\text{ZICMP}} = 1.5$, respectively). While the respective coefficient and standard error estimates are not directly comparable across the three models, there exist pairwise comparisons that deserve recognition. The ZICMP and ZIMCMP1 models produce similar coefficient estimates and standard errors for their respective zero-component models; this makes sense because both approaches address excess zeroes in similar fashion, while the respective count model components have different structural representations. The count model coefficient estimates and standard errors are meanwhile essentially identical for the ZIMCMP1 and HMCMP1 models, while their respective zero-component estimates are within one standard error of each other. This too makes sense because the MCMP1 parametrization serves as the underlying structure for both frameworks; what differs here is how the excess zeroes are modeled.

Most interesting is that all three COM–Poisson regressions addressing excess zeroes produce a dispersion estimate $\hat{\nu} = 0.000$, indicating consideration to analyze the dataset via a geometric distribution that can account for excess zeroes. Sellers and Raim (2016) report the resulting coefficient estimates, standard errors, log-likelihood and AIC values stemming from a zero-inflated geometric (ZIG) model; those results are likewise provided in Table 5.6. The resulting coefficient estimates, standard errors, and negated log-likelihood are identical to those reported for the ZIMCMP1 model, validating that the ZIG regression is a special case of the ZIMCMP1 where $\nu = 0$. Considering the reduced model of the ZIG thus decreases the AIC by a value of 2 because the dispersion parameter is no longer considered in the analysis. Meanwhile, the ZIG model not only captures the ZIMCMP1 representation but likewise contains similar output to either the count or zero components from the ZICMP and HMCMP1 models.

Table 5.6 overall thus reports that the ZIG attains the minimal AIC (1264.8), while the ZINB (1266.3) and ZIMCMP1 (1266.8) show substantial support. The strong comparison between the ZIG and ZINB models likewise yields a natural result, recognizing that the ZIG is also a special case of a ZINB model where $\gamma = 1$. In fact, while the ZINB model produces the smallest negated log-likelihood, it does so at the expense of an additional (dispersion) parameter where the resulting difference between the two estimated log-likelihoods equals 0.28. Further, the reported ZINB dispersion estimate $\hat{\gamma} = 0.821$ has a corresponding standard error (0.226) such that one can see that it is not unreasonable to believe $\gamma = 1$ as a plausible true dispersion parameter under the ZINB model, again implying the consideration of the ZIG model.

*Example: Epilepsy and Progabide*

To illustrate the functionality of the `glmmTMB` function for the MCMP1 mixed model regression, consider the longitudinal data analysis of an epilepsy dataset that examines the number of seizures experienced by 59 patients in an eight-week baseline period, followed by four consecutive two-week periods where the patients are treated with progabide. This dataset has been well studied and analyzed via various count data models (Booth et al., 2003; Breslow, 1984; Diggle and Milne, 1983; Molenberghs et al., 2007; Morris and Sellers, 2022; Thall and Vail, 1990); here, we compare the Poisson and MCMP1 regressions as baseline, zero-inflated, and hurdle mixed models for analysis. In all cases, let the Poisson ($\lambda_{ij}$) or MCMP1 ($(\lambda(\mu, \nu))_{ij}$) mean associate with explanatory variables via the loglinear model whose right-hand side of the equation equals

$$\beta_0 + \beta_1 x_{ij1} + \beta_2 x_{ij2} + \beta_3 x_{ij1} x_{ij2} + \ln(T_{ij}) + \alpha_i, \tag{5.51}$$

where, for Subject $i$, $T_{ij}$ denotes the length (in weeks) of the time period $j$, $x_{ij1}$ is an indicator function of a period after the baseline (i.e. weeks 8 through 16), $x_{ij2}$ is an indicator function noting whether or not the progabide medication is administered, and $\alpha_i$ denotes the random intercept. This model is consistent with that considered in Diggle and Milne (1983); accordingly, $\ln(T_{ij})$ serves as an offset in this model.

Table 5.7 reports the resulting coefficient estimates and standard errors (in parentheses) stemming from various generalized linear mixed models assuming baseline Poisson or MCMP1 regressions, as well as their zero-inflated and hurdle-regression counterparts. Along with these estimates, the table further reports the associated negative log-likelihood, AIC, and deviance values, respectively. These results show that, irrespective of the type of regression (i.e. baseline, zero-inflation, or hurdle), the MCMP1 model outperforms the corresponding Poisson regression, reporting a smaller negated log-likelihood, AIC, and deviance. Overall, the MCMP1 baseline mixed model is optimal with a negated log-likelihood (905.3), AIC (1822.5), and deviance (1810.5). In fact, the zero-inflated and baseline MCMP1 regressions report identical coefficient estimates and standard errors, with a considerably large coefficient standard error (3408.77) associated with $\hat{\zeta}_0 = -20.24$, demonstrating that statistically significant zero-inflation does not exist in this dataset when accounting for the random effect and dispersion. The zero-inflated and hurdle Poisson models, however, detect statistical significance when accounting for excess zeroes, apparently accounting for the additional variation through the zero-component model because it is constrained from compensating for the variation beyond that accounted for via the random intercept.

The hurdle Poisson and MCMP1 regressions report coefficients and corresponding standard errors that reflect similar contributions, associating the respective components to the average number of seizures among these epileptic patients. What is more interesting is that they further produce identical coefficients and standard errors for $\zeta_0$; the coefficient is further statistically significant ($p$-value $< 0.001$). This makes sense in that both

Table 5.7 *Estimated coefficients and standard errors (in parentheses), log-likelihood, Akaike information criterion (AIC), and deviance for various epilepsy longitudinal data analyses associating the number of seizures experienced by 59 patients in an eight-week baseline period, followed by four consecutive two-week periods where the patients are treated with progabide. Baseline Poisson and mean-parametrized COM–Poisson (MCMP1) regressions (along with their zero-inflated and hurdle analog models) are considered for constructing generalized linear mixed models, where, for Subject* $\mathrm{i}$, $\mathrm{T_{ij}}$ *denotes the length (in weeks) of the time period* $\mathrm{j}$, $\mathrm{x_{ij1}}$ *is an indicator function of a period after the baseline (i.e. weeks 8 through 16),* $\mathrm{x_{ij2}}$ *is an indicator function noting whether or not the progabide medication is administered, and* $\sigma^2$ *is the variance associated with the random intercept. Zero-inflation and hurdle regressions are performed assuming a constant model (i.e. Equation (5.48) reduces to* $logit(\pi_*) = \zeta_0$*). The parameter* $\nu$ *denotes the associated MCMP1 dispersion component under each respective model.*

| | Baseline model | | | | Zero-inflation model | | | | Hurdle model | | | |
|---|---|---|---|---|---|---|---|---|---|---|---|---|
| | Poisson | | MCMP | | Poisson | | MCMP | | Poisson | | MCMP | |
| Intercept | 1.0326 | (0.1526) | 1.0530 | (0.1632) | 1.0433 | (0.1493) | 1.0530 | (0.1632) | 1.0411 | (0.1479) | 1.0352 | (0.1640) |
| $x_{ij1}$ | 0.1108 | (0.0469) | 0.0996 | (0.0890) | 0.1575 | (0.0473) | 0.0996 | (0.0890) | 0.1567 | (0.0473) | 0.1165 | (0.0892) |
| $x_{ij2}$ | −0.0239 | (0.2106) | −0.0285 | (0.2248) | −0.0126 | (0.2060) | −0.0285 | (0.2248) | 0.0033 | (0.2040) | −0.0004 | (0.2257) |
| $x_{ij1}x_{ij2}$ | −0.1037 | (0.0651) | −0.1388 | (0.1235) | −0.1178 | (0.0658) | −0.1388 | (0.1235) | −0.0994 | (0.0657) | −0.1227 | (0.1239) |
| $\sigma^2$ | 0.6083 | | 0.5944 | | 0.5800 | | 0.5944 | | 0.5684 | | 0.6026 | |
| $\nu$ | – | | 0.2433 | | – | | 0.2433 | | – | | 0.2577 | |
| ZI-intercept | – | | – | | −2.9388 | (0.3255) | −20.2400 | (3408.77) | −2.4703 | (0.2172) | −2.4703 | (0.2172) |
| -logLik | 1010.7 | | 905.3 | | 988.5 | | 905.3 | | 992.1 | | 915.8 | |
| AIC | 2031.4 | | 1822.5 | | 1988.9 | | 1824.5 | | 1996.2 | | 1845.7 | |
| deviance | 2021.4 | | 1810.5 | | 1976.9 | | 1810.5 | | 1984.2 | | 1831.7 | |

models account for the zeroes in the dataset in identical fashion, thus obtaining the same coefficient results to account for excess zeroes. The larger variance and added accountability via the MCMP1 dispersion parameter estimate, however, presumably contribute toward the better model performance for the hurdle MCMP1 versus the hurdle Poisson mixed-effects model (respective AICs are 1996.2 for the Poisson versus 1845.7 for the MCMP1). This reduced dispersion estimate, however, may contribute toward the less than optimal model performance for the hurdle MCMP1 mixed-effects model, in comparison to its baseline MCMP1 counterpart, since the hurdle MCMP1 model is based on a truncated MCMP1 distribution. The hurdle MCMP1 requires more parameters and produces a smaller log-likelihood than the baseline MCMP1 mixed-effects model.

All of the presented MCMP1 mixed-effects models produce dispersion estimates $\hat{\nu} < 1$, thus recognizing that, while accounting for additional dispersion via the random effect, the epilepsy dataset is over-dispersed. This motivates one to consider a negative binomial mixed-effects model. Running this model via `glmer.nb` (`lme4` package in `R`) produces the output presented in Code 5.1; running the model via `glmmTMB` produces a similar output. Note that this model achieves a smaller AIC (1789.5) than the baseline MCMP1-generalized linear mixed model (1822.5). While the NB model achieves a better AIC, utilizing the MCMP1 regression proves helpful nonetheless in aiding us toward the NB GLMM by informing the analyst of the type of dispersion inherent in the epilepsy data, when accounting for additional dispersion caused by the random intercept and/or excess zeroes.

Code 5.1 Negative binomial GLMM regression output for the epilepsy and progabide data example. Output produced via the `glmer.nb` function, contained in the `lme4` package in `R`.

```
> prog.glmernb <- glmer.nb(NumSeizures ~ visitne0 +
TX + pbide.int + (1|ID), offset=log(Weeks), data=pbide)
> summary(prog.glmernb)
Generalized linear mixed model fit by maximum
likelihood (Laplace Approximation)
['glmerMod']
 Family: Negative Binomial(6.7803)  ( log )
Formula: NumSeizures ~ visitne0 + TX + pbide.int +
         (1 | ID)
   Data: pbide
 Offset: log(Weeks)

     AIC      BIC   logLik deviance df.resid
  1789.5   1811.6   -888.7   1777.5      289
```

```
Scaled residuals:
    Min      1Q  Median      3Q     Max
-1.8269 -0.5474 -0.0971  0.4245  3.9026

Random effects:
 Groups Name        Variance Std.Dev.
 ID     (Intercept) 0.661    0.813
Number of obs: 295, groups:  ID, 59

Fixed effects:
            Estimate Std. Error z value Pr(>|z|)
(Intercept)  1.08031    0.17542   6.159 7.34e-10 ***
visitne0     0.02349    0.10029   0.234   0.8149
TX           0.07256    0.24205   0.300   0.7644
pbide.int   -0.31042    0.14052  -2.209   0.0272 *
---

Correlation of Fixed Effects:
          (Intr) vistn0 TX
visitne0  -0.402
TX        -0.725  0.291
pbide.int  0.287 -0.712 -0.399
```

## 5.6 Generalized Additive Model

COM–Poisson generalized additive model (GAM) regression can likewise be established (via any of the considered parametrizations) to associate dispersed count data with nonlinear relationships to covariates of interest. As an illustration, the CMP-GAM considers a CMP distribution as defined in Equation (2.8) with a model of the form

$$\ln(\lambda_i) = \boldsymbol{x}_i^* \boldsymbol{\beta}^* + \sum_{j=1}^{p_1} f_{1j}(x_{ij}), \tag{5.52}$$

$$\ln(\nu_i) = \boldsymbol{g}_i^* \boldsymbol{\gamma}^* + \sum_{j=1}^{p_2} f_{2j}(g_{ij}), \tag{5.53}$$

for $i = 1, \ldots, n$, where the first components of the respective equations account for the parametric part of $\ln(\lambda_i)$ and $\ln(\nu_i)$, respectively, via the $i$th row of the respective model matrices ($X^*$ and $G^*$) and the associated parameter vectors $\boldsymbol{\beta}^*$ and $\boldsymbol{\gamma}^*$; and $f_{\ell j}$ ($\ell = 1, 2$) are smooth functions of covariates $\boldsymbol{x}_j$ and $\boldsymbol{g}_j$, respectively, that are subject to identifiability constraints (e.g. $\sum_{i=1}^{n} f_{\ell j}(x_{ij}) = 0$ for all $j, \ell$) (Chatla and Shmueli, 2018).

A penalized splines approach is a popular estimation method for estimating $f_{\ell j}$. Here, the model is estimated via a penalized maximum likelihood, where the maximization can be conducted via a penalized iterative reweighted least squares that minimizes the objective functions

$$\|\sqrt{\boldsymbol{W}_1^{(k)}}(\boldsymbol{T}_1^{(k)} - \boldsymbol{X}\boldsymbol{\beta})\|^2 + \boldsymbol{\beta}'\boldsymbol{A}_1\boldsymbol{\beta} + \sum_j \phi_{1j}\boldsymbol{\beta}'\boldsymbol{S}_{1j}\boldsymbol{\beta} \quad \text{w.r.t.} \quad \boldsymbol{\beta}, \tag{5.54}$$

$$\|\sqrt{\boldsymbol{W}_2^{(k)}}(\boldsymbol{T}_2^{(k)} - \boldsymbol{G}\boldsymbol{\gamma})\|^2 + \boldsymbol{\gamma}'\boldsymbol{A}_2\boldsymbol{\gamma} + \sum_j \phi_{2j}\boldsymbol{\gamma}'\boldsymbol{S}_{2j}\boldsymbol{\gamma} \quad \text{w.r.t.} \quad \boldsymbol{\gamma}, \tag{5.55}$$

where for $\ell = 1, 2$, $\boldsymbol{W}_\ell^{(k)}$ are the $k$th iteration weight matrices, $\phi_{\ell j}$ are the smooth parameters for the models, respectively, regressing $\ln(\lambda_i)$ and $\ln(\nu_i)$, $\boldsymbol{A}_\ell$ are the fixed positive semidefinite penalty matrices, and $\boldsymbol{T}_\ell^{(k)}$ are as described in Algorithm 4. A penalized least-squares method is used to solve Equations (5.54) and (5.55) for given smoothing parameters $\phi_{\ell j}$ that are themselves first estimated via generalized cross validation. Statistical inference for the CMP-GAM model spline regression coefficients is conducted via a Bayesian perspective such that the posterior distribution of the coefficients is

$$\boldsymbol{\beta}|\boldsymbol{\phi}_1, Y \sim N(\hat{\boldsymbol{\beta}}, \boldsymbol{\Sigma}_\beta), \text{ where } \boldsymbol{\Sigma}_\beta = \left(\boldsymbol{X}'\boldsymbol{W}_1\boldsymbol{X} + \sum_{j=1}^{p_1} \phi_{1j}\boldsymbol{S}_{1j}\right)^{-1} \eta_1, \text{ and} \tag{5.56}$$

$$\boldsymbol{\gamma}|\boldsymbol{\phi}_2, Y \sim N(\hat{\boldsymbol{\gamma}}, \boldsymbol{\Sigma}_\gamma), \text{ where } \boldsymbol{\Sigma}_\gamma = \left(\boldsymbol{G}'\boldsymbol{W}_2\boldsymbol{G} + \sum_{j=1}^{p_2} \phi_{2j}\boldsymbol{S}_{2j}\right)^{-1} \eta_2, \tag{5.57}$$

where, for $\ell = 1, 2$, $\eta_\ell$ are the estimated scale parameters and $\boldsymbol{W}_\ell$ are the weight matrices incorporated to achieve the convergence of the penalized iterative reweighted least-squares algorithm (Chatla and Shmueli, 2018).

A hypothesis test can be considered that asks whether a considered smooth function is a significant contributor to the model or not. For such a test, the null hypothesis is $H_0 \colon f_{\ell j} = 0$ for any $j$ where, for $\ell = 1, 2$, it addresses the smoothing function associated with $\ln(\lambda)$ or $\ln(\nu)$, respectively. The Wood (2012) Wald test statistic is an appropriate measure to conduct inference as it is based on the marginal likelihood to ensure that the pseudo-inverse remains optimal and the covariance matrix remains positive definite. See Chatla and Shmueli (2018) for detailed discussion.

## 5.7 Computing via Alternative Softwares

While this reference focuses on statistical computing via R, one would be remiss to not acknowledge the popularity of the COM–Poisson distribution (particularly with regard to regression analysis) and its influence on statistical computation developments via other softwares. Functionality is available to conduct COM–Poisson regression via MATLAB (Section 5.7.1) and SAS (Section 5.7.2).

### *5.7.1* MATLAB *Computing*

MATLAB software for the MCMP1 regression is provided in the online supplement associated with the Huang (2017) manuscript. This routine reports the coefficient estimates $\hat{\boldsymbol{\beta}}$, their corresponding standard errors, and the resulting log-likelihood if those outputs are requested by the analyst. The reported standard errors are obtained via a plug-in estimator,

$$V(\hat{\beta}) = \left[ \sum_{i=1}^{n} \frac{\left( \frac{\partial \mu(\boldsymbol{X}_i\hat{\boldsymbol{\beta}})}{\partial(\boldsymbol{X}_i\hat{\boldsymbol{\beta}})} \right)^2 \boldsymbol{X}_i'\boldsymbol{X}_i}{V(\mu(\boldsymbol{X}_i\hat{\boldsymbol{\beta}}), \hat{\nu})} \right]^{-1}.$$

The codes apply Newton–Raphson algorithms to the MCMP1 regression score equations (Equations (5.5) and (5.6)) in a computationally efficient manner such that the routine is up to one order of magnitude faster than the glm.cmp function in COMPoissonReg (Huang, 2017). While these codes are available, advancements to the mpcmp R package supersede these contributions; analysts are encouraged to utilize the mpcmp package accordingly.

### *5.7.2* SAS *Computing*

The SAS/ETS procedure COUNTREG (SAS Institute, 2014) uses the ACMP parametrization as the default (when not accounting for excess zeroes) but allows for the CMP parametrization by specifying the option, PARAMETER = LAMBDA. Meanwhile, specifying DISP=CMPOISSON in the MODEL statement while omitting the DISPMODEL statement will perform ACMP regression where $\nu$ is a constant. Loglinear links are assumed for both the COM–Poisson location and dispersion parameters, i.e. $\ln(\lambda_i) = \boldsymbol{x}_i\boldsymbol{\beta}$ and $\ln(\nu_i) = -\boldsymbol{g}_i\boldsymbol{\gamma}$. The negated loglinear relationship for the dispersion allows analysts to more naturally interpret the coefficients in relation to the direction of dispersion; $\gamma_j > (<)0$ associated with a quantitative covariate

$g_{ij}$, $j = 1, \ldots, p$, indicates a multiplicative impact $e^{-\gamma_j}$ in data over-(under-)dispersion (holding the other terms constant). ZI COM–Poisson regression in SAS meanwhile assumes an underlying CMP parametrization. ZICMP regression in COUNTREG is performed by including a ZEROMODEL statement, where LINK = LOGISTIC tells SAS to supply a logistic link function to account for zero-inflation; alternatively, LINK = NORMAL considers a probit link function to address the excess zeroes. The COUNTREG procedure conducts parameter estimation via the method of maximum likelihood where the Newton–Raphson method serves as the default approach for maximizing the log-likelihood. Other optimization schemes, however, can be considered by inserting/updating the METHOD option in the MODEL statement. Further, COUNTREG can likewise consider a Bayesian approach for analysis by inputting a BAYES statement in the model call. The resulting output "Model Fit Summary" provides detailed information regarding the model of interest, including the maximum log-likelihood, the number of iterations required to reach convergence, the performed optimization method, the AIC, and Schwarz's BIC. The usual coefficient table follows the Model-Fit Summary and provides the respective coefficient estimates, standard errors, $t$ values, and corresponding $p$-values used to assess the statistical significance of the coefficients in question.

SAS Institute (2014) reports that the "Lord et al. (2008) specification makes the model comparable to the negative binomial model because it has only one parameter." The reader, however, is advised to use the reparametrized regression with caution. Recall from Chapter 2 that $\mu_* = \lambda^{1/\nu}$ is a further approximation of the approximated COM–Poisson mean $\lambda^{1/\nu} - \frac{1-\nu}{2\nu}$, where this approximation is reasonable when $\nu \leq 1$ (i.e. the data are equi- or over-dispersed) or $\lambda > 10^{\nu}$ (i.e. $\mu_* > 10$). First, be advised that the Guikema and Coffelt (2008) and Lord et al. (2008) references denote this reparametrization as "$\mu$," which already can lead readers to a false sense of confidence such that analysts can presumably interpret the reparametrized form to easily identify the mean. It remains unknown, however, under what circumstances such further approximation is reasonable. Assuming the constraint $\nu \leq 1$ to hold (implying that the response variable is equi- or over-dispersed), an NB regression is naturally considered to model such data; hence, the comparison to an NB model is direct in that both models assume a constant dispersion parameter that is constrained such that only equi- or over-dispersion is feasible. However, for under-dispersed response data, even $\lambda > 20^{\nu}$ shows some loss in precision (Chanialidis et al., 2017); thus, the COUNTREG procedure should be used with caution when conducting ACMP regression since it relies on all $\boldsymbol{\lambda}$, $\boldsymbol{\nu}$, satisfying the appropriate constraints; such a restriction is not guaranteed if the data are under-dispersed.

Generalized linear mixed modeling can likewise be performed in SAS via the NLMIXED procedure. In particular, COM–Poisson mixed-effects modeling can be considered under any parametrization because NLMIXED allows the option of specifying a user-defined likelihood function. The generalized linear mixed model assuming a random intercept with a CMP parametrization, for example, has been considered via the NLMIXED procedure (Morris et al., 2017). For a given dataset, the user supplies starting values for the model coefficients `parms` and provides the log-likelihood in the `model` statement and underlying distributional form of the random component `random`. Given the model formulation described in Equation (5.42), users can directly supply code to compute the normalizing function $Z(\lambda_{it}, \nu)$ and the log-likelihood

$$\ln L(\lambda_{it}, \nu \mid y_{it}) = y_{it} \ln(\lambda_{it}) - \nu \ln(y_{it}!) - \ln(Z(\lambda_{it}, \nu));$$

see Morris et al. (2017) for a detailed illustration. While Morris et al. (2017) focus on the CMP parametrization, the other parametrizations (i.e. ACMP, MCMP1, and MCMP2) can likewise be considered for analogous mixed-effects modeling.

### *Revisiting Example: Airfreight Breakage*

Revisiting the airfreight breakage example described in Section 5.2.5, we can instead consider conducting COM–Poisson regression analysis via SAS. Assuming that the dataset provided in Table 5.3 has been read into SAS and given the data object name `freight`, the commands

```
proc countreg data=freight;
   model broken=transfers / dist=compoisson;
run;
```

perform an ACMP regression while updating the `model` command to include `parameter=lambda` conducts CMP regression. Resulting coefficient tables are supplied in Tables 5.8 and 5.9 for the CMP and ACMP regressions, respectively. The only estimate that is comparable between the two approaches is that for the dispersion because its representation remains the same under either of the parametrizations. Here, both models produce $\ln(\hat{\nu}) \approx -1.7568 < 0$ where, given the modified representation that estimates the dispersion parameter, the negative coefficient implies data under-dispersion in this example. More broadly, however, as discussed in

Table 5.8 COUNTREG *output from the airfreight breakage example with CMP regression.*

| Parameter | DF | Estimate | Std. error | $t$ value | Approx Pr >\| $t$ \| |
|---|---|---|---|---|---|
| Intercept | 1 | 13.8542 | 6.2941 | 2.20 | 0.0277 |
| transfers | 1 | 1.4868 | 0.6947 | 2.14 | 0.0323 |
| _lnNu | 1 | −1.7568 | 0.4523 | −3.88 | 0.0001 |

Table 5.9 COUNTREG *output from the airfreight breakage example with approximate COM–Poisson (ACMP) regression.*

| Parameter | DF | Estimate | Std. error | $t$ value | Approx Pr >\| $t$ \| |
|---|---|---|---|---|---|
| Intercept | 1 | 2.3911 | 0.0539 | 44.38 | <.0001 |
| transfers | 1 | 0.2566 | 0.0325 | 7.90 | <.0001 |
| _lnNu | 1 | −1.7568 | 0.4491 | −3.91 | <.0001 |

Section 5.2.5, the other coefficient results are not immediately comparable to each other, except for the fact that both approaches produce statistically significant results. Instead, one can better compare the CMP output achieved via SAS to its counterpart obtained in R. Another potential comparison can consider how the ACMP results obtained via SAS compare with (say, for example) the MCMP1 parametrization obtained via R; this allows analysts to better assess how well the ACMP regression effectively captures a traditional generalized linear model structure between the mean and covariates for a given dataset.

Comparing the outputs supplied in Tables 5.4 and 5.8, we obtain comparable estimates for the coefficients and associated standard errors; thus, while not displayed, one can recognize that the coefficient table produced via R would likewise contain similar $t$ statistics and $p$-values. Meanwhile, the respective $\ln(\nu)$ estimates differ primarily in sign (positive in R and negative in SAS), while their respective absolute values match to two decimal places; recall that the difference in sign stems from the altered model formulation $\ln(\boldsymbol{\nu}) = -\boldsymbol{G\gamma}$ in SAS, while the R form maintains $\ln(\boldsymbol{\nu}) = \boldsymbol{G\gamma}$. Tables 5.4 and 5.9 meanwhile likewise display relatively comparable results for the MCMP1 regressions obtained via the mpcmp and DGLMExtPois packages in R and the ACMP regression attained via the COUNTREG procedure in SAS. The difference in estimated coefficients is greater here than the difference between CMP estimates because the latter comparison stems from models utilizing the same parametrization. This example, however, is one for which the ACMP reparametrization appears to reasonably approximate the mean.

## 5.8 Summary

Regression analysis via the COM–Poisson distribution has been the most studied aspect of related statistical methodology considered, outside of the research conducted regarding the distribution itself. Whether considering the generalized linear model, or analog models that allow for excess zeroes, clustered data analysis, or a GAM, all of these constructs can be represented for a COM–Poisson model assuming any of the parametrizations discussed in this reference (i.e. CMP, ACMP, MCMP1, and MCMP2).

Along with the substantive theoretical work done in this area, there has likewise been a great deal of computational development achieved regarding COM–Poisson regression models. Many of the aforementioned regression models have `R` packages and functions associated with them for applied data analysis, and statistical computing for some regressions are likewise attainable via `MATLAB` and `SAS`. The fact that COM–Poisson regression analysis is readily available via multiple computational tools demonstrates the importance of this distribution and its abilities for flexible discrete data analysis when dispersion is present.

# 6

# COM–Poisson Control Charts

While the Poisson model motivated much of the classical control chart theory for count data (e.g. the Shewhart $c$- and $u$-charts), several works (e.g. Albers, 2011; Chen et al., 2008; Mohammed and Laney, 2006; Spiegelhalter, 2005) note the constraining equi-dispersion assumption. Dispersion must be addressed because over-dispersed data can produce false out-of-control detections when using Poisson limits, while under-dispersed data will produce Poisson limits that are too broad, resulting in potential false negatives. In the latter case, out-of-control states would require a longer study period for detection. An alternative statistical control chart based on a shifted generalized Poisson distribution of Consul and Jain (1973) can address data over- or under-dispersion; however, it gets truncated under certain conditions regarding the dispersion parameter, and thus its underlying distributional form is not a true probability model (Famoye, 2007).

Sellers (2012b) established a seminal work, introducing a flexible COM–Poisson-based Shewhart control chart for count data expressing data dispersion. Section 6.1 describes this work in greater detail, demonstrating its flexibility in assessing in- or out-of-control status, along with advancements made to this control chart. These initial works have further led to a wellspring of flexible control chart development motivated by the COM–Poisson distribution. Section 6.2 describes a generalized exponentially weighted moving average (EWMA) control chart, and Section 6.3 describes the cumulative sum (CUSUM) charts for monitoring COM–Poisson processes. Meanwhile, Section 6.4 introduces generally weighted moving average (GWMA) charts based on the COM–Poisson, and Section 6.5 presents the Conway–Maxwell–Poisson chart via the progressive mean statistic. Finally, Section 6.6 concludes the chapter with summary and discussion. All of the works described in this chapter stem from the

CMP parametrization of the COM–Poisson distribution as described in Section 2.2; however, analogous results can be determined based on the reparametrizations presented in Section 2.6.

## 6.1 CMP-Shewhart Charts

Sellers (2012b) developed CMP-Shewhart charts (i.e. *cmpc*- and *cmpu*-charts) to generalize and bridge the classical Shewhart charts for count data, namely the *c*- and *u*-charts developed via the Poisson model, the *np*- and *p*-charts motivated by the Bernoulli distribution, and the *g*- and *h*-charts described in Kaminsky et al. (1992) stemming from the geometric distribution. Its underlying flexibility implies that the CMP-Shewhart charts are more effective than the Poisson *c*- and *u*-charts in detecting large shifts. Assume a process that generates a random sample of $n$ events $X_1, X_2, \ldots, X_n$ according to a shifted CMP($\lambda$, $\nu$) distribution with probability

$$P(X = x \mid \lambda, \nu) = \frac{\lambda^{x-a}}{[(x-a)!]^{\nu} Z(\lambda, \nu)}, \quad x = a, a+1, a+2, \ldots, \tag{6.1}$$

where $\lambda > 0$, $\nu \geq 0$, $a \in \mathbb{N} = \{0, 1, 2, \ldots\}$, and $T = \sum_{i=1}^{n} X_i$ and $\bar{X} = \frac{T}{n}$, respectively, denote the total and average number of events. Their respective means and standard deviations can be used to construct the Shewhart $\mu \pm k\sigma$ control bounds for the *cmpc*- and *cmpu*-charts, respectively; Table 6.1 provides the resulting centerline and Shewhart $k\sigma$ upper/lower control limits.

One can alternatively determine upper and lower control limits given a predetermined Type I error $\alpha$ so that the probability of being beyond those limits equals $\alpha$. This approach is particularly favorable for small counts, given the skewness of the CMP distribution; see Section 6.1.1 for further

Table 6.1 *Centerline and Shewhart* k$\sigma$ *upper/lower control limits for* cmpc- *and* cmpu-*charts (Sellers, 2012b).*

| | *cmpc*-chart | *cmpu*-chart |
|---|---|---|
| Centerline | $n\left(\lambda \frac{\partial \ln(Z(\lambda, \nu))}{\partial \lambda} + a\right)$ | $\lambda \frac{\partial \ln(Z(\lambda, \nu))}{\partial \lambda} + a$ |
| Upper/lower bounds | $n\left(\lambda \frac{\partial \ln(Z(\lambda, \nu))}{\partial \lambda} + a\right)$ | $\lambda \frac{\partial \ln(Z(\lambda, \nu))}{\partial \lambda} + a$ |
| | $\pm k \sqrt{n \left(\frac{\partial E(X)}{\partial \ln(\lambda)}\right)}$ | $\pm k \sqrt{\frac{1}{n} \left(\frac{\partial E(X)}{\partial \ln(\lambda)}\right)}$ |

discussion. Accordingly, the upper and lower control limits (UCL and LCL, respectively) for the total number of events $T$ are determined such that

$$P(T \leq \mathrm{LCL}_{\alpha/2}) = \sum_{t=0}^{\mathrm{LCL}_{\alpha/2}} P(T = t) \leq \frac{\alpha}{2} \text{ and} \tag{6.2}$$

$$P(T \geq \mathrm{UCL}_{\alpha/2}) = \sum_{\mathrm{UCL}_{\alpha/2}}^{\infty} P(T = t) \leq \frac{\alpha}{2}, \tag{6.3}$$

where $P(T = t)$ is the probability mass function from a sum-of-COM-Poissons (sCMP) distribution with parameters $\lambda$, $\nu$, and $n$; see Section 3.2 for details regarding the sCMP distribution. Similarly, we can determine the probability limits associated with the average number of events $\bar{X}$. The *cmpc*- and *cmpu*-charts are flexible control charts that contain three established control charts as special cases. The *cmpc*- and *cmpu*-charts include the *c*-chart and *u*-chart derived from the Poisson distribution (for $\nu = 1$), the Kaminsky et al. (1992) *g*-chart and *h*-chart derived from the geometric distribution (when $\nu = 0$ and $\lambda < 1$), and the *np*- and *p*-charts obtained from the Bernoulli distribution (as $\nu \rightarrow \infty$ with success probability $\frac{\lambda}{\lambda+1}$).

### 6.1.1 CMP Control Chart Probability Limits

Using the traditional Shewhart limits or the more precise probability limits can be debated based on several factors, e.g. the size of the parameters and sample size, and their impact on resulting computations. The below discussion holds for either the $T$ or the $\bar{X}$ chart.

The Shewhart chart assumes that the underlying distribution is symmetric; however, the COM–Poisson distribution contains skewness; this can be problematic under certain circumstances. The empirical rule, for example, infers that the Shewhart $3\sigma$ bounds will contain 99.7% of the data; yet, simulation studies illustrate that the CMP-Shewhart construct requires $k > 3$ in order to achieve the desired statistical significance level. Further, the difference between the Shewhart and probability-based LCL bounds is considerable, resulting in a smaller power associated with detecting a parameter shift (Saghir et al., 2013). These issues arise for all of the traditional count-based Shewhart charts, however, so they are not particular to the CMP-Shewhart chart. Holding any two parameters among $\{\lambda, \nu, n\}$ fixed, however, results in the desired value for $k$ decreasing toward 3 as the third parameter increases, making $k = 3$ reasonable for sufficiently large $\lambda$, $\nu$, or $n$ (Saghir et al., 2013).

Saghir et al. (2013) further report that the Shewhart limit scheme produces a biased power function when the sample size is small given either type of data dispersion, but particularly so when the data are under-dispersed; this bias, however, decreases as $n$ increases. Nonetheless, to detect a downward shift, the probability limit bounds produce a better power function than using Shewhart limits. Both approaches perform comparably for detecting an upward shift in a parameter, and the power increases with the sample size under either control chart construct. When the data are under-dispersed, the Shewhart chart produces a larger power for upward parameter shifts in $\lambda$ than the probability limits approach for small $n$; however, the disparity between the two methods diminishes as $n$ increases. Probability limits likewise circumvent any need for Winsorization, thus providing another benefit over Shewhart limits. The Shewhart LCL may require Winsorization to zero if the resulting computation is negative. The probability LCL, however, avoids any need to Winsorize because the CMP support space only contains nonnegative integers; hence, the probability-based LCL cannot be negative.

### 6.1.2 `R` Computing

The `R` package `CMPControl` (Sellers and Costa, 2014) provides the user with the ability to produce the Shewhart control charts based on the traditional Poisson or CMP distribution; further, the produced plot provides both the $3\sigma$ bounds and those bounds based on the probability limits. The function `ControlCharts` only requires the desired dataset for analysis (`data`), while the other inputs aid in designing the desired control chart figure. `xlabel` and `ylabel` allow the analyst to input the labels for the respective axes. `CMP`, `P`, and `CMPProb` are each logical inputs that allow the user to choose which control limits they desire to have appear in the figure, where `CMP` identifies the $3\sigma$ CMP-Shewhart control bounds, `P` the $3\sigma$ Poisson-Shewhart control bounds, and `CMPProb` the CMP probability limits. In all cases, the respective logical option is `TRUE` such that all of the respective upper and lower bounds are defaulted to appear in any produced control chart image. This provides the analyst with a more well-rounded understanding of unit performance.

This package uses the `compoisson` and `MASS` packages to perform the various Poisson and CMP calculations, including parameter estimation, computing the mean and standard deviation, and probabilities of the resulting estimated distributions. In particular, the estimates for $\lambda$, $\nu$, and $a$ are attained via maximum likelihood estimation with the likelihood function

$$L(\lambda, \nu, a \mid \boldsymbol{x}) = \prod_{i=1}^{n} \frac{\lambda^{x_i - a}}{[(x_i - a)!]^{\nu} Z(\lambda, \nu)} I_a(\boldsymbol{x}), \tag{6.4}$$

where $I_a(\boldsymbol{x}) = 1(0)$ when $x_i \geq a$ for all $i$ (otherwise). Accordingly, $\hat{a} = X_{(1)}$, the minimum order statistic, while $\hat{\lambda}$ and $\hat{\nu}$ are determined by optimizing Equation (6.4) via `nlminb` in R (Sellers, 2012b; Sellers and Costa, 2014). All of these calculations aid in determining the respective CMP $3\sigma$ and probability control limits. Along with the resulting figure displaying the control bounds of interest, the function further outputs the CMP parameter estimates $\hat{\lambda}$ and $\hat{\nu}$ for the provided data, along with the resulting estimates for the mean and standard deviation, and the Poisson mean and standard deviation.

Code 6.1 R code and corresponding output stemming from the `nonconformities` dataset. The Poisson- and CMP-Shewhart upper and lower control limits are computed as "$\mu \pm 3\sigma$"; accordingly, negative bounds are Winsorized at zero. "Upper Out of Control Observations" and "Lower Out of Control Observations" list the observation/point/sample number(s) that are out of control because they are greater than the upper bound or less than the lower bound, respectively.

```
> ControlCharts(nonconformities,''Sample Number'',
  ''Number of nonconformities'')
$'CMP Lambda Hat and Nu Hat'
[1] 1.9370111 0.2554058

$'CMP Mean and Standard Deviation'
[1] 19.838957  7.193244

$'CMP Shewhart Upper and Lower Bounds'
[1] 41.418689 -1.740774

$'Poisson Mean and Standard Deviation'
[1] 19.846154  4.454902

$'Poisson Shewhart Upper and Lower Bounds'
[1] 33.210861  6.481447

$'Upper Out of Control Observations'
[1] 20

$'Lower Out of Control Observations'
[1] 6

$'CMP Probability Limits'
[1] 39  0
```

### 6.1.3 Example: Nonconformities in Circuit Boards

Montgomery (2001) tracks the number of nonconformities in 26 samples of circuit boards; the resulting data are supplied in the `CMPControl` package (named `nonconformities`). The `ControlCharts` function (also contained in the `CMPControl` package) can be used to determine the existence and number of any out-of-control samples. The command and resulting output are shown in Code 6.1, while the resulting control chart is displayed in Figure 6.1.

Montgomery (2001) assumes a *c*-chart; `ControlCharts` output reports that the Poisson mean and standard deviation are 19.846154 and 4.454902, respectively. As a result, the centerline value for the *c*-chart is 19.846154, and the $\pm 3\sigma$ control limits are approximately 33.2109 and 6.4814. Code 6.1 and Figure 6.1 further show that two points (namely Samples 6 and 20) fall beyond the control limits. Applying the *cmpc*-chart model to the nonconformities data, however, `ControlCharts` reports the CMP parameter estimates $\hat{\lambda} = 1.9370111$ and $\hat{\nu} = 0.2554058$, thus recognizing the

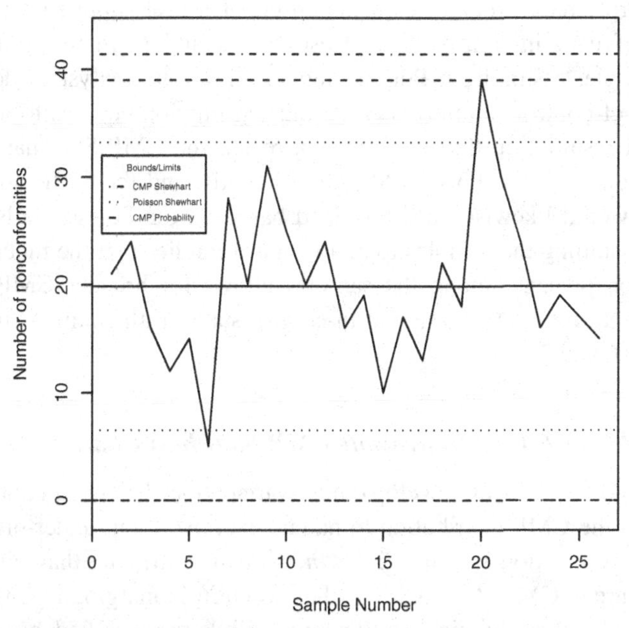


Figure 6.1 Control chart associated with `nonconformities` data analysis via `CMPControl` package: corresponding `R` code and output supplied in Code 6.1. Lower CMP-Shewhart bound is Winsorized to 0, while the lower CMP probability bound equals 0.

dataset to be over-dispersed. Assuming these samples to be in control, the CMP distribution better models the data distribution of the samples. The `ControlCharts` function outputs that the CMP mean and standard deviation are 19.838957 and 7.193244, respectively; thus, the CMP-Shewhart centerline is 19.838957, and the upper bound is 41.418689; the computed CMP-Shewhart lower bound (−1.740774) reported in Code 6.1 is Winsorized to zero; see Figure 6.1. Meanwhile, `ControlCharts` determines the upper and lower CMP probability limits to be 39 and 0, respectively.

The centerlines based on the Poisson and CMP models are approximately equal; however, their upper and lower Shewhart bounds differ because of their respective underlying data dispersion assumption. The amount of variation in the data determines the size of the Shewhart bounds via the "$\mu \pm 3\sigma$" computation. Meanwhile, the CMP-Shewhart versus probability bounds utilize an approximate shape based on assumed relative symmetry versus a more precise representation of the CMP distributional form. The core concern with all of these models and, more broadly, the flexible model approach in general is that the proper determination of the control chart structure depends on either a user-supplied a priori estimate of $\nu$ or in-control data to estimate $\lambda$ and $\nu$. In this illustration, for example, assuming a Poisson model allows an analyst to detect the two out-of-control samples because this assumption equates to an expert supplied assumption that $\hat{\nu} = 1$; as a result, the CMP-Shewhart bounds would equate to the Poisson-Shewhart bounds, and the CMP probability bounds would likewise be determined based on the Poisson tails. Meanwhile, assuming the samples to be in control implies that the inherent data dispersion present among the samples allows for broader CMP control bounds that reflect the assumed in-control state of all samples, including Samples 6 and 20.

### *6.1.4 Multivariate CMP-Shewhart Chart*

Saghir and Lin (2014a) develop a multivariate Shewhart-type control chart based on the CMP distribution to handle data over- and under-dispersion. Assume a collection $X_i$, $i = 1, 2, \ldots, n$, of quality attributes that jointly have a multivariate CMP (MultiCMP) where each $X_i$ is marginally CMP($\lambda_i$, $\nu_i$) distributed with correlation coefficient $r_{ij}$. While not specified, based on the research to date, the required distributional assumptions imply considering a MultiCMP distribution derived either via the Sarmanov family or via copulas. See Chapter 4 for discussion.

Consider the statistic $T = \sum_{i=1}^{n} X_i$ that studies the sum of the $n$ attributes. A MultiCMP Shewhart chart has bounds $\mu_T \pm k\sigma_T$ and centerline $\mu_T$, where

$$\mu_T = \sum_{i=1}^{n} \lambda_i \frac{\partial \ln(Z(\lambda_i, \nu_i))}{\partial \lambda_i} \approx \sum_{i=1}^{n} \left[ \lambda_i^{1/\nu_i} - \frac{\nu_i - 1}{2\nu_i} \right] \tag{6.5}$$

$$\sigma_T^2 = \sum_{i=1}^{n} \frac{\partial E(X_i)}{\partial \ln(\lambda_i)} + 2\sum_{i<j} r_{ij} \sqrt{\frac{\partial E(X_i)}{\partial \ln(\lambda_i)} \cdot \frac{\partial E(X_j)}{\partial \ln(\lambda_j)}} \tag{6.6}$$

$$\approx \sum_{i=1}^{n} \frac{1}{\nu_i} \lambda_i^{1/\nu_i} + 2\sum_{i<j} r_{ij} \sqrt{\frac{1}{\nu_i \nu_j} \lambda_i^{1/\nu_i} \lambda_j^{1/\nu_j}} \tag{6.7}$$

denote the mean and variance of $T$, and the respective approximations hold when $\nu_i \leq 1$ or $\lambda_i > 10^{1/\nu_i}$. If $T$ goes beyond the control limits thus marking an out-of-control signal, analysts should determine the minimum and maximum score statistics, $S_{(1)}$ and $S_{(n)}$, respectively, where $S_i = X_i - E(X_i)$, $i = 1, 2, \ldots, n$. For $T >$ UCL, $S_{(n)}$ identifies which attribute most contributes toward the upward shift, while for $T <$ LCL, $S_{(1)}$ identifies which attribute most contributes toward the downward shift.

The MultiCMP Shewhart chart simplifies to the multivariate Poisson chart when the $n$-length vector, $\boldsymbol{\nu} = \mathbf{1}$. For the special case when $\boldsymbol{\nu} \to \infty$, the transformed statistic $T^* = \sum_{i=1}^{k} X_i/\sqrt{p_i}$ produces control bounds that represent the multivariate $np$ chart where $n = 1$ with "equal nonconformance severity of all $p$ quality characteristics" (Saghir and Lin, 2014a). Through simulated and real-data examples, the MultiCMP Shewhart chart demonstrates itself as a flexible multivariate extension of the Sellers (2012b) CMP-Shewhart chart. Saghir and Lin (2014a) evaluate chart performance via average run length (ARL) to study location shifts in the control process. Letting $\boldsymbol{X}$ have a MultiCMP($\boldsymbol{\lambda}$, $\boldsymbol{\nu}$, $\boldsymbol{\rho}$) distribution with a shift in the location parameter implies that the distribution is updated to MultiCMP($\boldsymbol{\lambda}_s$, $\boldsymbol{\nu}$, $\boldsymbol{\rho}$). Data simulations studying the in-control ARL show that $k$ varies, often not equaling approximately 3. This is because of the skewed nature of the CMP distribution; thus, a normal approximation is not appropriate for small means. To study the out-of-control ARL, data simulations on a bivariate Shewhart-CMP chart find that the $\text{ARL}_1$ values have a concave parabolic relationship as the correlation between the two attributes increases. Meanwhile, as the dispersion level deviates from 1, the $\text{ARL}_1$ values increase; i.e., $\text{ARL}_1$ has a convex parabolic relationship as $\nu$ increases.

## 6.2 CMP-inspired EWMA Control Charts

While discrete Shewhart charts are a useful tool for quality control, they are critiqued for only being able to detect large process shifts, ignoring historical data and only focusing on the most recent sample information. EWMA charts are an attractive alternative to Shewhart charts because EWMA charts take both historical and current sample information into account. These "memory-type control charts" use the larger collective of past and current sample data to gain more information, thus making them more capable than their Shewhart counterparts to detect small or moderate control shifts (Alevizakos and Koukouvinos, 2022). Gan (1990) introduced three modified EWMA charts that monitor Poisson observations through the mean and round the EWMA statistic to an integer value. The three charts are based on the relation

$$Q_t = (1-\omega)Q_{t-1} + \omega R_t, \qquad t = 1, 2, \ldots, \tag{6.8}$$

i.e. a weighted average of $Q_{t-1}$ (the EWMA value at time $t-1$) and $R_t$ (the $t$th observation from a Poisson($\lambda$) distribution); $0 < \omega \leq 1$ is the associated weighting constant, and $Q_0$ is the initial EWMA value. The Shewhart control chart is a special case of EWMA where $\omega = 1$. A lower-sided (upper-sided) EWMA control chart detects shifts in the process when $Q_t <$ LCL ($Q_t >$ UCL) for some lower (upper) control limit, and a two-sided EWMA chart takes both limits into account. The modifications to Equation (6.8) force $Q_t$ to be integer-valued, either through the ceiling ($\lceil \cdot \rceil$), round, or floor ($\lfloor \cdot \rfloor$) operators and hence are referred to as the ceiling, round, and floor-EWMA control charts, respectively. The center-EWMA and floor-EWMA charts are respectively designed to detect movement toward the upper and lower limits, while the round-EWMA approximates the unmodified EWMA chart defined in Equation (6.8). The averages, probability functions, and percentiles of the run lengths associated with these modified EWMAs are computable via a Markov-chain approach; see Gan (1990) for details. These discrete EWMA charts do not necessarily allow for exact design equal to a stated in-control ARL; however, these charts can get close to a desired in-control ARL. Meanwhile, they become more effective as $\omega$ decreases.

Borror et al. (1998) introduced a Poisson EWMA (PEWMA) chart to monitor count data, where the procedure is likewise evaluated via the Markov-chain approximation. This method, however, does not lose information because the precise EWMA statistic (as defined in Equation (6.8)) is retained. Their model is likewise defined by Equation (6.8) where, when the process is in control, $E(Q_t) = \lambda_0$ with exact variance $V(Q_t) = \frac{\omega}{2-\omega}[1 - (1-\omega)^{2t}]\lambda_0$ and asymptotic variance $V(Q_t) \approx \frac{\omega\lambda_0}{2-\omega}$. This PEWMA chart thus detects an out-of-control signal when $Q_t < \text{LCL} = \lambda_0 - B_L\sqrt{V(Q_t)}$ or $Q_t > \text{UCL} = \lambda_0 + B_U\sqrt{V(Q_t)}$, where $V(Q_t)$ is either the exact or asymptotic variance under the in-control process assumption, and $B_L$ and $B_U$ can be assumed equal (i.e. $B$) or not, depending on the desired performance of the EWMA chart.

The PEWMA chart is a natural choice of EWMA design for monitoring the number of nonconformities or defects; however, the underlying equi-dispersion assumption associated with the motivating Poisson distribution can produce unreliable inferences. Several alternative EWMA charts are presented below, each motivated by the CMP distribution. Section 6.2.1 introduces the basic CMP-EWMA chart, while Sections 6.2.2 and 6.2.3 introduce CMP-EWMA charts based on multiple dependent state sampling and repetitive sampling, respectively. Section 6.2.4 presents a modified CMP-EWMA chart, while Section 6.2.5 discusses a double EWMA chart that assumes CMP attribute data, and Section 6.2.6 concludes this section introducing a hybrid EWMA. All of these works assume the CMP parametrization as described in Section 2.2, while some works (e.g. Aslam et al. 2016a; 2016b) further assume that its mean and variance approximations (Equations (2.23) and (2.24)) hold.

### *6.2.1 COM–Poisson EWMA (CMP-EWMA) Chart*

Saghir and Lin (2014c)[1] developed a flexible EWMA chart for discrete attribute data, motivated by the COM–Poisson distribution. Their CMP-motivated exponentially weighted moving average (CMP-EWMA) control chart (referred in their original work as a generalized EWMA (GEWMA) chart) demonstrates robustness when working with dispersed data and contains three EWMA charts as special cases for count data measuring the number of nonconformities. Let $X_t$ be an independent and identically CMP-distributed sequence of quality measurements from a production process. Assuming the process to be in control implies that the process mean is $\lambda_0 \frac{\partial \ln(Z(\lambda_0,\nu))}{\partial \lambda_0}$ (see Equation (2.14)). Suppose one wishes to monitor changes in $\lambda$ for a fixed dispersion level $\nu$ (say $\nu_0$). The CMP-EWMA control chart statistic is defined as

[1] While Saghir and Lin (2014c) refer to $\lambda$ as the average number of nonconformities, this description holds only for $\nu = 1$, i.e. the PEWMA chart.

$$C_t = (1 - \omega)C_{t-1} + \omega X_t, \tag{6.9}$$

i.e. a weighted average of the statistic at time $t - 1$ and the $t$th observation. While the weighting constant is $0 < \omega < 1$ by definition, Saghir and Lin (2014c) maintain the Montgomery (2001) recommendation for $0.05 \leq \omega \leq 0.25$.

Given the recursion defined in Equation (6.9), the CMP-EWMA statistic can be represented as $C_t = \omega \left[ \sum_{i=0}^{t-1} (1 - \omega)^t X_{t-i} \right] + (1 - \omega)^t C_0$, where $C_0 = \lambda_0 \frac{\partial \ln (Z(\lambda_0, \nu))}{\partial \lambda_0}$. The mean and variance of $C_t$ are thus

$$E(C_t) = \lambda_0 \frac{\partial \ln (Z(\lambda_0, \nu))}{\partial \lambda_0} \text{ and} \tag{6.10}$$

$$V(C_t) = \frac{\omega}{2 - \omega} \left[ 1 - (1 - \omega)^{2t} \right] \frac{\partial E(X_t)}{\partial \ln (\lambda_0)}. \tag{6.11}$$

Accordingly, the upper and lower control limits are

$$\begin{aligned} \text{UCL} &= \lambda_0 \frac{\partial \ln Z(\lambda_0, \nu)}{\partial \lambda_0} + B_U \sqrt{\frac{\omega}{2 - \omega} \left( \frac{\partial E(X)}{\partial \ln (\lambda_0)} [1 - (1 - \omega)^{2t}] \right)} \\ &\rightarrow \lambda_0 \frac{\partial \ln Z(\lambda_0, \nu)}{\partial \lambda_0} + B_U \sqrt{\frac{\omega}{2 - \omega} \left( \frac{\partial E(X)}{\partial \ln (\lambda_0)} \right)} \end{aligned} \tag{6.12}$$

$$\text{CL} = \lambda_0 \frac{\partial \ln Z(\lambda_0, \nu)}{\partial \lambda_0} \tag{6.13}$$

$$\begin{aligned} \text{LCL} &= \lambda_0 \frac{\partial \ln Z(\lambda_0, \nu)}{\partial \lambda_0} - B_L \sqrt{\frac{\omega}{2 - \omega} \left( \frac{\partial E(X)}{\partial \ln (\lambda_0)} [1 - (1 - \omega)^{2t}] \right)} \\ &\rightarrow \lambda_0 \frac{\partial \ln Z(\lambda_0, \nu)}{\partial \lambda_0} - B_L \sqrt{\frac{\omega}{2 - \omega} \left( \frac{\partial E(X)}{\partial \ln (\lambda_0)} \right)}, \end{aligned} \tag{6.14}$$

where $B_L$ and $B_U$ can be set equal to each other (say, some value $B$), and the asymptotic bounds for UCL and LCL occur as $t \rightarrow \infty$. As usual, the LCL is Winsorized to zero whenever computations lead to a negative value. By design, the CMP-EWMA generalizes the Poisson, Bernoulli, and geometric EWMA charts. The CMP-EWMA chart equals the Poisson EWMA (PEWMA) when $\nu = 1$, the geometric EWMA when $\nu = 0$ and $\lambda_0 < 1$, and the Bernoulli EWMA when $\nu \rightarrow \infty$ with $p = \frac{\lambda_0}{1+\lambda_0}$. The CMP-EWMA likewise reduces to the Sellers (2012b) Shewhart chart for $\omega = 1$ (Alevizakos and Koukouvinos, 2019).

The CMP-EWMA chart is more effective than the CMP-Shewhart chart for detecting small and large positive shifts. The CMP-EWMA chart becomes more effective at detecting small positive shifts as $\omega$ decreases; however, the CMP-Shewhart chart is better than the CMP-EWMA chart at detecting small negative shifts. This result is common – control charts with symmetric control limits are less effective at detecting small shifts toward smaller values when they are more effective at detecting small shifts toward larger values (Gan, 1990). Meanwhile, for equi- or over-dispersed data, the CMP-EWMA chart gives unbiased results for upward and downward shifts when the weighting constant $\omega$ is small. The results are biased, however, for over-dispersed data when $\omega$ is moderate to large for upward or downward shifts, and for downward shifts in equi-dispersed data. Finally, when the data are under-dispersed, the CMP-EWMA chart performs well only for detecting upward shifts (Saghir and Lin, 2014c).

### 6.2.2 CMP-EWMA Chart with Multiple Dependent State Sampling

Aslam et al. (2016a) broaden the CMP-EWMA chart definition to develop a CMP-EWMA chart with multiple dependent state sampling (i.e. CMP-MDSEWMA chart). Assume a CMP parametrization where its mean and variance approximations (i.e. Equations (2.23) and (2.24)) hold such that the CMP-MDSEWMA statistic $C_t$ is approximately normally distributed for large $t$ with mean $\lambda_0^{1/\nu} - \frac{\nu-1}{2\nu}$ and variance $\frac{\omega}{2-\omega}\left(\frac{\lambda_0^{1/\nu}}{\nu}\right)$ when the process is in control. Accordingly, the control limits are defined as

$$\text{LCL}_i = \left(\lambda_0^{1/\nu} - \frac{\nu - 1}{2\nu}\right) - B_i\sqrt{\frac{\omega}{2-\omega}\left(\frac{\lambda_0^{1/\nu}}{\nu}\right)} \text{ and} \tag{6.15}$$

$$\text{UCL}_i = \left(\lambda_0^{1/\nu} - \frac{\nu - 1}{2\nu}\right) + B_i\sqrt{\frac{\omega}{2-\omega}\left(\frac{\lambda_0^{1/\nu}}{\nu}\right)} \tag{6.16}$$

for $i = 1, 2$, where $i = 1$ (2) produces the outer (inner) limits, respectively; Algorithm 7 provides the steps to produce the chart. The CMP-MDSEWMA chart reduces to the CMP-EWMA chart when $g = 0$ and $B_1 = B_2 = B$.

For $\nu$ fixed and a process shift multiple $d$ such that $\lambda = d\lambda_0$ reflecting a substantive shift, the CMP-MDSEWMA chart has an in-control ARL,

$$\begin{aligned}\text{ARL}_0 = [1 - [(\Phi(B_2) - \Phi(-B_2)) + \{(\Phi(-B_2) - \Phi(-B_1)\\ + (\Phi(B_1) - \Phi(B_2))\}\{\Phi(B_2) - \Phi(-B_2)\}^g]]^{-1},\end{aligned}$$

1. Randomly select an observation at time $t$ and determine the number of nonconformities $X_t$ associated with this observation.
2. Calculate the EWMA statistic at time $t$, $C_t = (1 - \omega)C_{t-1} + \omega X_t$, for a weighting constant $\omega$.
3. The process is considered out of (in) control if $C_t \geq \text{UCL}_1$ or $C_t \leq \text{LCL}_1$ ($\text{LCL}_2 \leq C_t \leq \text{UCL}_2$). If $C_t$ falls outside of these ranges, go to Step 4.
4. Check if $g$-preceding subgroups are in control. If yes, then declare the process as in control; otherwise, declare the process as out of control.

**Algorithm 7** Steps to establish the CMP-MDSEWMA chart.

and out-of-control ARL, $\text{ARL}_1 = \left[1 - A_1^1 + (A_2^1 + A_3^1)(A_1^1)^g\right]^{-1}$, where

$$A_1^1 = \Phi\left(\frac{(1 - d^{1/\nu})\lambda_0^{1/\nu} + B_2\sqrt{\frac{\omega}{2-\omega}\left(\frac{\lambda_0^{1/\nu}}{\nu}\right)}}{\sqrt{\frac{\omega}{2-\omega}\left(\frac{(d\lambda_0)^{1/\nu}}{\nu}\right)}}\right)$$

$$- \Phi\left(\frac{(1 - d^{1/\nu})\lambda_0^{1/\nu} - B_2\sqrt{\frac{\omega}{2-\omega}\left(\frac{\lambda_0^{1/\nu}}{\nu}\right)}}{\sqrt{\frac{\omega}{2-\omega}\left(\frac{(d\lambda_0)^{1/\nu}}{\nu}\right)}}\right)$$

$$A_2^1 = \Phi\left(\frac{(1 - d^{1/\nu})\lambda_0^{1/\nu} - B_2\sqrt{\frac{\omega}{2-\omega}\left(\frac{\lambda_0^{1/\nu}}{\nu}\right)}}{\sqrt{\frac{\omega}{2-\omega}\left(\frac{(d\lambda_0)^{1/\nu}}{\nu}\right)}}\right)$$

$$- \Phi\left(\frac{(1 - d^{1/\nu})\lambda_0^{1/\nu} - B_1\sqrt{\frac{\omega}{2-\omega}\left(\frac{\lambda_0^{1/\nu}}{\nu}\right)}}{\sqrt{\frac{\omega}{2-\omega}\left(\frac{(d\lambda_0)^{1/\nu}}{\nu}\right)}}\right)$$

$$A_3^1 = \Phi\left(\frac{(1-d^{1/\nu})\lambda_0^{1/\nu} + B_1\sqrt{\frac{\omega}{2-\omega}\left(\frac{\lambda_0^{1/\nu}}{\nu}\right)}}{\sqrt{\frac{\omega}{2-\omega}\left(\frac{(d\lambda_0)^{1/\nu}}{\nu}\right)}}\right) - \Phi\left(\frac{(1-d^{1/\nu})\lambda_0^{1/\nu} + B_2\sqrt{\frac{\omega}{2-\omega}\left(\frac{\lambda_0^{1/\nu}}{\nu}\right)}}{\sqrt{\frac{\omega}{2-\omega}\left(\frac{(d\lambda_0)^{1/\nu}}{\nu}\right)}}\right),$$

where $\Phi(\cdot)$ denotes the usual cumulative distribution function of the standardized normal distribution. Simulation studies show that the CMP-MDSEWMA chart not only outperforms the CMP-EWMA chart producing smaller $\text{ARL}_1$ values for all shifts but also increases in efficiency as $g$ increases.

### 6.2.3 CMP-EWMA Chart with Repetitive Sampling

Aslam et al. (2016b) advance the CMP-EWMA chart to allow for repetitive sampling, thus developing what is hereafter referred to as a CMP-RSEWMA chart. Again assuming a CMP parametrization where the mean and variance approximations hold, the CMP-RSEWMA chart maintains the same outer and inner control limits as described for the CMP-MDSEWMA chart (see Equations (6.15) and (6.16)) while the associated decision process is modified; see Algorithm 8.

Maintaining the process shift design as described in Section 6.2.2 (i.e. process shift $\lambda = d\lambda_0$ with stationary $\nu$), the control bounds are determined based on the in-control ARL, $\text{ARL}_0 = \frac{1-2(\Phi(B_1)-\Phi(B_2))}{2(1-\Phi(B_1))}$, where $\Phi(\cdot)$

1. Randomly select an observation at time $t$ and determine the number of non conformities $X_t$ associated with this observation.
2. Calculate the EWMA statistic at time $t$, $C_t = (1-\omega)C_{t-1} + \omega X_t$, for a weighting constant $\omega$.
3. The process is considered out of (in) control if $C_t \geq \text{UCL}_1$ or $C_t \leq \text{LCL}_1$ ($\text{LCL}_2 \leq C_t \leq \text{UCL}_2$). If $C_t$ falls outside of these ranges, repeat the process selecting a new observation.

**Algorithm 8** Steps to establish the CMP-RSEWMA chart.

denotes the standardized normal cumulative distribution function; meanwhile, $\mathrm{ARL}_1 = \frac{1-P^1_{\mathrm{rep}}}{P^1_{\mathrm{out},1}}$, where

$$P^1_{\mathrm{out},1} = \Phi\left(\frac{(1-d^{1/\nu})\lambda_0^{1/\nu} - B_1\sqrt{\frac{\omega}{2-\omega}\left(\frac{\lambda_0^{1/\nu}}{\nu}\right)}}{\sqrt{\frac{\omega}{2-\omega}\left(\frac{(d\lambda_0)^{1/\nu}}{\nu}\right)}}\right) + 1 - \Phi\left(\frac{(1-d^{1/\nu})\lambda_0^{1/\nu} + B_1\sqrt{\frac{\omega}{2-\omega}\left(\frac{\lambda_0^{1/\nu}}{\nu}\right)}}{\sqrt{\frac{\omega}{2-\omega}\left(\frac{(d\lambda_0)^{1/\nu}}{\nu}\right)}}\right)$$

denotes the probability of declaring the process out of control for the shifted process $\lambda = d\lambda_0$ based on one sample, and

$$P^1_{\mathrm{rep}} = \Phi\left(\frac{(1-d^{1/\nu})\lambda_0^{1/\nu} + B_1\sqrt{\frac{\omega}{2-\omega}\left(\frac{\lambda_0^{1/\nu}}{\nu}\right)}}{\sqrt{\frac{\omega}{2-\omega}\left(\frac{(d\lambda_0)^{1/\nu}}{\nu}\right)}}\right) - \Phi\left(\frac{(1-d^{1/\nu})\lambda_0^{1/\nu} + B_2\sqrt{\frac{\omega}{2-\omega}\left(\frac{\lambda_0^{1/\nu}}{\nu}\right)}}{\sqrt{\frac{\omega}{2-\omega}\left(\frac{(d\lambda_0)^{1/\nu}}{\nu}\right)}}\right) + \Phi\left(\frac{(1-d^{1/\nu})\lambda_0^{1/\nu} - B_2\sqrt{\frac{\omega}{2-\omega}\left(\frac{\lambda_0^{1/\nu}}{\nu}\right)}}{\sqrt{\frac{\omega}{2-\omega}\left(\frac{(d\lambda_0)^{1/\nu}}{\nu}\right)}}\right) - \Phi\left(\frac{(1-d^{1/\nu})\lambda_0^{1/\nu} - B_1\sqrt{\frac{\omega}{2-\omega}\left(\frac{\lambda_0^{1/\nu}}{\nu}\right)}}{\sqrt{\frac{\omega}{2-\omega}\left(\frac{(d\lambda_0)^{1/\nu}}{\nu}\right)}}\right)$$

is the probability of resampling when the shifted process relation holds.

Aslam et al. (2016b) suggest minimizing the in-control process average sample size

$$\mathrm{ASS}_0 = \frac{1}{1 - 2(\Phi(B_1) - \Phi(B_2))}$$

with respect to $B_1$ and $B_2$ so that $\text{ARL}_0$ is no less than a predefined target $\text{ARL}_0$ value. The determined values for $B_1, B_2$, and $d$ are then used to determine $\text{ARL}_1$. Illustrative simulations show that the CMP-RSEWMA $\text{ARL}_1$ decreases as $|d-1|$ increases. A decrease in $\text{ARL}_1$ occurs more slowly as $\nu$ increases; however, this result is derived for $\nu \leq 1$ (i.e. cases where data are equi- to over-dispersed); thus, it remains unknown what happens in the case of data under-dispersion. At the same time, however, maintaining this restriction is reasonable, given the reliance of the CMP-RSEWMA chart on the CMP mean and variance approximations. Finally, $\text{ARL}_1$ for the CMP-RSEWMA chart increases with $\omega$ (holding the other terms fixed). Thus, the CMP-RSEWMA chart is found to be more efficient than the CMP-EWMA chart in notifying analysts that a process is going out of control because the CMP-RSEWMA construct produces smaller ARLs associated with a shifted process than does CMP-EWMA (Aslam et al., 2016b).

### 6.2.4 Modified CMP-EWMA Chart

Aslam et al. (2017) develop a modified EWMA chart based on the CMP distribution (i.e. a CMP-MEWMA chart), which likewise claims a more efficient ability to detect changes in $\lambda$. Given a sample of size $n$ at time $t$ and the associated measure of quality number $X_t$ that is CMP($\lambda$, $\nu$) distributed, the CMP-MEWMA statistic $M_t$ has the form

$$M_t = (1-\omega)M_{t-1} + \omega X_t + k(X_t - X_{t-1}), \tag{6.17}$$

where $0 < \omega < 1$ remains the weighting constant and $k$ is another constant such that the process is deemed out of control if $M_t > \text{UCL}$ or $M_t < \text{LCL}$. The CMP-MEWMA reduces to the CMP-EWMA statistic for $k = 0$; Aslam et al. (2017) define $k = -\omega/2$ as optimal and recommend $0.05 \leq \omega \leq 0.25$. The cascading form of $M_t$ implies that its mean and variance are $E(M_t) = \lambda^{1/\nu} - \frac{\nu-1}{2\nu}$ and $V(M_t) = \frac{\omega+2\omega k+2k^2}{2-\omega}\left(\frac{\lambda^{1/\nu}}{\nu}\right)$; thus, the control limits are now

$$\text{LCL} = \lambda_0^{1/\nu} - \frac{\nu-1}{2\nu} - B\sqrt{\frac{\omega+2\omega k+2k^2}{n(2-\omega)}\left(\frac{\lambda^{1/\nu}}{\nu}\right)}, \tag{6.18}$$

$$\text{UCL} = \lambda_0^{1/\nu} - \frac{\nu-1}{2\nu} + B\sqrt{\frac{\omega+2\omega k+2k^2}{n(2-\omega)}\left(\frac{\lambda^{1/\nu}}{\nu}\right)}, \tag{6.19}$$

where $k = 0$ represents the special case of the CMP-EWMA chart. Simulation studies show that, as $\omega$ increases, $B$ increases, while the first quartile of the run-length distribution decreases; further, the CMP-MEWMA chart is more efficient than the CMP-EWMA chart. As is the case with the CMP-MDSEWMA and CMP-RSEWMA charts, the CMP-MEWMA chart likewise assumes the CMP mean and variance approximations to hold; thus, analysts should use these charts with caution or at least after first confirming that their data are equi- to over-dispersed.

### 6.2.5 Double EWMA Chart for CMP Attributes

Assuming that $R_t$ are independent and identically Poisson($\lambda$)-distributed random variables denoting the number of nonconformities at time $t$, a Poisson-double EWMA (PDEWMA) chart can monitor Poisson data via the PDEWMA statistic $S_t$, defined as

$$\begin{cases} Q_t = (1-\omega)Q_{t-1} + \omega R_t \\ S_t = (1-\omega)S_{t-1} + \omega Q_t \\ S_0 = Q_0 = \lambda_0, \end{cases} \tag{6.20}$$

where $0 < \omega < 1$ denotes the usual weighting constant, and $Q_t$ is the PEWMA statistic in Equation (6.8) (Zhang et al., 2003). This chart produces the control limits $\lambda_0 \pm B\sqrt{V(S_t)}$, where

$$V(S_t) = \lambda_0 \frac{\omega^4[1+(1-\omega)^2 - (t+1)^2(1-\omega)^{2t} + (2t^2+2t-1)(1-\omega)^{2t+2} - t^2(1-\omega)^{2t+4}]}{[1-(1-\omega)]^3}. \tag{6.21}$$

Alevizakos and Koukouvinos (2019) generalize the PDEWMA chart, developing a double EWMA control chart to monitor CMP($\lambda, \nu$)-distributed attribute data; accordingly, this flexible chart can detect shifts in the respective parameters individually or simultaneous shifts in both parameters. Here, the number of nonconformities is now denoted $X_t$ at time $t$ and is CMP($\lambda, \nu$) distributed, $t = 1, 2, \ldots$, such that this CMP-double EWMA (CMP-DEWMA) process is analogously represented as

$$\begin{cases} C_t = (1-\omega)C_{t-1} + \omega X_t \\ D_t = (1-\omega)D_{t-1} + \omega C_t \\ D_0 = C_0 = \lambda_0 \dfrac{\partial \ln(Z(\lambda_0, \nu_0))}{\partial \lambda_0}; \end{cases} \tag{6.22}$$

the process is in control when $\lambda = \lambda_0$ and $\nu = \nu_0$. The control limits and centerline for the CMP-DEWMA chart are

$$\mathrm{UCL}_t = \lambda_0 \frac{\partial \ln(Z(\lambda_0, \nu_0))}{\partial \lambda_0} + B\sqrt{V_0(D_t)} \tag{6.23}$$

$$\mathrm{CL} = \lambda_0 \frac{\partial \ln(Z(\lambda_0, \nu_0))}{\partial \lambda_0} \tag{6.24}$$

$$\mathrm{LCL}_t = \lambda_0 \frac{\partial \ln(Z(\lambda_0, \nu_0))}{\partial \lambda_0} - B\sqrt{V_0(D_t)}, \tag{6.25}$$

where

$$V_0(D_t) = \frac{V(S_t)}{\lambda_0} \cdot \frac{\partial E(X)}{\partial \ln(\lambda_0)}$$

denotes the variance of $D_t$ assuming that the process is in control, and $V(S_t)$ is defined in Equation (6.21). These bounds converge to

$$\mathrm{UCL} = \lambda_0 \frac{\partial \ln(Z(\lambda_0, \nu_0))}{\partial \lambda_0} + B\sqrt{\frac{\omega(2 - 2\omega + \omega^2)}{(2-\omega)^3} \frac{\partial E(X)}{\partial \ln(\lambda_0)}} \tag{6.26}$$

$$\mathrm{CL} = \lambda_0 \frac{\partial \ln(Z(\lambda_0, \nu_0))}{\partial \lambda_0} \tag{6.27}$$

$$\mathrm{LCL} = \lambda_0 \frac{\partial \ln(Z(\lambda_0, \nu_0))}{\partial \lambda_0} - B\sqrt{\frac{\omega(2 - 2\omega + \omega^2)}{(2-\omega)^3} \frac{\partial E(X)}{\partial \ln(\lambda_0)}} \tag{6.28}$$

as $t$ gets large. The CMP-DEWMA simplifies to the PDEWMA chart for the special case where $\nu_0 = 1$; given the underlying CMP distributional qualities, the CMP-DEWMA likewise contains DEWMA charts based on geometric and Bernoulli distributional random variables.

Simulation studies show that, for a fixed $\nu$, (1) $B$ increases with $\omega$ in order to obtain similar $\mathrm{ARL}_0$ values, (2) the standard deviation run length (SDRL) likewise increases with $\omega$ while always respecting the SDRL $\leq$ ARL constraint, and (3) the in-control run-length distribution is right-skewed, and its median run length (MRL) decreases slightly as $\omega$ increases. Meanwhile, for fixed $\omega$, $B$ likewise increases with $\nu$, however at a slower rate. Studying location shift (say, $\lambda = d_\lambda \lambda_0$) for a fixed dispersion $\nu \leq 1$ ($\nu > 1$), the CMP-DEWMA chart is unbiased for small $\omega$ but grows in potential bias for small downward (upward) shifts as $\omega$ increases. Under these circumstances, the CMP-DEWMA chart detects upward (downward) shifts faster than downward (upward) shifts; in particular, a larger $\omega$ is required to detect a larger downward (upward) shift. It is unclear what causes the association between dispersion type and shift detection. Regarding the dispersion shift ($\nu = d_\nu \nu_0$) for a fixed location value $\lambda$, the mean and variance of the CMP distribution increase (decrease) as $\nu$ decreases (increases), implying that a downward (upward) shift in $\nu$ causes the CMP-DEWMA chart to detect an upward (downward) shift in the process mean. The CMP-DEWMA chart is unbiased when the data are over-dispersed in an in-control state (i.e. $\nu_0 < 1$), and larger $\omega$ is required to detect larger shifts in $\nu$. Meanwhile, when $\nu_0 = 1$, the CMP-DEWMA chart is unbiased for small $\omega$ but shows bias for small to moderate shifts and more strongly so as $\omega$ increases. For under-dispersed $\nu_0$, the CMP-DEWMA chart is biased for upward shifts in $\nu$, and the amount of bias only strengthens as $\omega$ increases. In all cases, the CMP-DEWMA chart detects downward shifts in $\nu$ more quickly than upward shifts. Studying either type of shift where the other parameter is held fixed, $\text{ARL}_1$ increases with $\nu$, implying that the CMP-DEWMA chart has a more difficult time with shift detection as the variance associated with a dataset decreases in size (i.e. becomes more under-dispersed). Likewise, the MRL values move closer to their corresponding ARL values as either parameter shift increases; thus, the skewness in the run-length distribution decreases.

The CMP-DEWMA performs at least equally efficiently as (if not better than) the CMP-EWMA chart and is more effective than the latter at detecting downward parameter shifts that associate with negative process mean shifts. Simulation studies demonstrate the greater sensitivity of the CMP-DEWMA chart to detect small to moderate shifts in the location and dispersion parameters, respectively (downward for the location parameter and upward for the dispersion parameter) with small $\omega$ values, when data are equi- or over-dispersed. Meanwhile, when the data are under-dispersed, the CMP-DEWMA demonstrates greater sensitivity than the CMP-EWMA with at least upward shifts and perhaps shifts in both directions for the location parameter, depending on $\omega$. With regard to the dispersion parameter, the CMP-DEWMA demonstrates greater sensitivity to identifying upward shifts with regard to the dispersion parameter than the CMP-EWMA chart. Given the joint influence between $\lambda$ and $\nu$ on the CMP mean, the CMP-DEWMA's ability to outperform the CMP-EWMA chart with regard to detecting upward shifts in $\nu$ is equivalent to the CMP-DEWMA's greater ability to detect downward shifts in the process mean. The broader scope regarding sensitivity comparisons between the CMP-DEWMA and CMP-EWMA charts has some dependence likewise on $\omega$, although the CMP-DEWMA is generally more efficient than CMP-EWMA; see Alevizakos and Koukouvinos (2019a) for details. Nonetheless, the CMP-DEWMA can outperform the CMP-EWMA chart because the

CMP-DEWMA more efficiently detects (1) downward shifts in $\lambda$ for a fixed $\nu$, (2) upward shifts in $\nu$ for fixed $\lambda$, and (3) simultaneous upward or downward shifts in $\lambda$ and $\nu$ that (in either scenario) produce a downward shift in the process mean.

### 6.2.6 Hybrid EWMA Chart

Aslam et al. (2018) design a hybrid EWMA control chart to study attribute data having an underlying CMP($\lambda, \nu$) distribution; we refer to this as a CMP-HEWMA chart. Let $X_t$ denote the number of nonconformities at time $t$ and have a CMP($\lambda, \nu$) distribution. This random variable along with two weighting constants $\omega_1$ and $\omega_2 \in [0, 1]$ designs the HEWMA statistic

$$H_t = (1 - \omega_1)H_{t-1} + \omega_1 E_t, \text{ where} \tag{6.29}$$

$$E_t = (1 - \omega_2)E_{t-1} + \omega_2 X_t. \tag{6.30}$$

The resulting CMP-hybrid EWMA (CMP-HEWMA) chart reduces to the CMP-DEWMA chart when $\omega_1 = \omega_2$, the CMP-EWMA chart when $\omega_1 = 1$, and the CMP-Shewhart chart when $\omega_1 = \omega_2 = 1$. Sufficient conditions are assumed to hold such that $E_t$ and $H_t$, respectively, have approximate normal distributions with mean $\mu_E = \mu_H = \lambda^{1/\nu} - \frac{\nu-1}{2\nu}$ and respective variances $\sigma_E^2 = \frac{\omega_2}{2-\omega_2}\left(\frac{\lambda^{1/\nu}}{\nu}\right)$ and

$$\sigma_H^2 = \frac{\omega_1^2\omega_2}{2-\omega_2}\left(\frac{\lambda^{1/\nu}}{\nu}\right)$$

$$\left[\frac{1-(1-\omega_1)^{2t}}{\omega_1(2-\omega_1)} - \frac{(1-\omega_2)^2\{(1-\omega_2)^{2t} - (1-\omega_1)^{2t}\}}{(1-\omega_2)^2 - (1-\omega_1)^2}\right] \tag{6.31}$$

$$\approx \frac{\omega_1\omega_2}{(2-\omega_1)(2-\omega_2)}\left(\frac{\lambda^{1/\nu}}{\nu}\right) \tag{6.32}$$

when $t$ is sufficiently large. Accordingly, the CMP-HEWMA control limit bounds are

$$\text{LCL} = \left(\lambda_0^{1/\nu} - \frac{\nu-1}{2\nu}\right) - B\sqrt{\frac{\omega_1\omega_2}{(2-\omega_1)(2-\omega_2)}\left(\frac{\lambda^{1/\nu}}{\nu}\right)} \tag{6.33}$$

$$\text{UCL} = \left(\lambda_0^{1/\nu} - \frac{\nu-1}{2\nu}\right) + B\sqrt{\frac{\omega_1\omega_2}{(2-\omega_1)(2-\omega_2)}\left(\frac{\lambda^{1/\nu}}{\nu}\right)} \tag{6.34}$$

for the in-control location parameter $\lambda_0$ and control bound $B$ determined to satisfy the desired $ARL_0$. Note that Aslam et al. (2018) do not assume an in-control dispersion parameter (say $\nu_0$); in fact, they assume that a shifted process is only detected through a shift in $\lambda$; hence, they appear to assume a constant dispersion $\nu$.

Monte Carlo simulations are used assuming an over-dispersed (i.e. $\nu < 1$) structure to show that, for detecting $\lambda = d\lambda_0$ for some shift parameter $d$, the $\text{ARL}_1$ values decrease as $d$ moves away from 1 with this decrease occurring more rapidly for $d > 1$ than $d < 1$. This implies that the CMP-HEWMA chart is more efficient with detecting upward rather than downward shifts in the location parameter. Meanwhile, the $\text{ARL}_1$ values decrease with $\omega_2$. A comparative study likewise demonstrates that the CMP-HEWMA outperforms the CMP-EWMA and CMP-Shewhart charts for over-dispersed data. Aslam et al. (2018) infer that these results likewise hold when data under-dispersion exists; however, the limits depend on the CMP mean and variance approximations being satisfactory (i.e. with the associated constraint space for the respective parameters being satisfied, namely $\nu \leq 1$ (i.e. the data are equi- or over-dispersed) or $\lambda > 10^{\nu}$; see Chapter 2).

## 6.3 COM–Poisson Cumulative Sum (CUSUM) Charts

CUSUM charts are additional "memory-type control charts" serving as an attractive alternative to Shewhart charts because CUSUM charts are more sensitive to smaller process shifts; the development of CUSUM charts likewise takes historical sample data into account along with current observations. Accordingly, CUSUM charts have a faster detection rate on average than other charts containing the same in-control performance. Such charts also have a minimax optimality property.

The Poisson CUSUM (PCUSUM) chart is utilized to detect significant changes (whether an increase or decrease) in a count measure (e.g. the number of nonconformities). The PCUSUM statistic $S_t = \max(0, Y_t - K + S_{t-1})$ (alternatively, $S_t = \max(0, K - Y_t + S_{t-1})$) at time $t$ is used to detect an increase (decrease) in counts, where $Y_t$ denotes an observed value, and $K$ is a reference value; $S_0 = 0$ is used for standard CUSUMs, while $S_0 \approx L/2 > 0$ is recommended (for some decision interval value $L$) for fast initial response CUSUMs (Brook and Evans, 1972; Lucas, 1985). An out-of-control status is detected when $S_t \geq L$ inferring that a statistically significant change in the count rate has occurred. Brook and Evans (1972) study a PCUSUM chart via a Markov-chain approach whose transition probability matrix is used to determine various chart properties, including the ARL, and moments and exact probabilities from the run-length distribution. Lucas (1985) meanwhile notes that the ARL of a standard (or fast initial response)

CUSUM is nearly geometric, however, it either has a lower (or larger) probability of short run lengths. The optimal reference value is determined to be $K = \frac{\lambda_d - \lambda_a}{\ln(\lambda_d) - \ln(\lambda_a)}$, where $\lambda_a$ and $\lambda_d$, respectively, denote the acceptable and detectable mean number of counts per sampling interval. Given the value of $K$, a decision interval value $r$ can then be determined such that the ARL is sufficiently large when the counts are at an acceptable level and sufficiently small when the process is considered out-of-control; see Lucas (1985) for details. A two-sided CUSUM control scheme is encouraged because its properties can easily be obtained by combining the results of the respective one-sided schemes used to detect significant increase or decrease, while this approach allows for flexibility in detecting change.

The PCUSUM control chart is a natural (and hence the most popular) CUSUM chart for tracking the average number of nonconformities in order to detect small to moderate performance shifts. Such charts, however, rely on an underlying assumption of equi-dispersion, which is constraining because real count data usually contain some measure of dispersion. Thus, a flexible alternative to the PCUSUM chart is a COM–Poisson CUSUM (CMP-CUSUM) chart that allows for data over- or under-dispersion; see Section 6.3.1 for details.

### *6.3.1 CMP-CUSUM charts*

Saghir and Lin (2014b) assume a CMP($\lambda$, $\nu$) parametrization where the approximations for the mean and variance (Equations (2.23) and (2.24)) hold; this implies that they further assume that the associated parameter constraints (i.e. $\nu \leq 1$ or $\lambda > 10^{\nu}$) likewise hold. Note that $\lambda$ is not the distribution mean except for the special case where $\nu = 1$. With these assumptions in place, three versions of CMP-CUSUM control charts are developed (namely the $\lambda$-CUSUM, $\nu$-CUSUM, and $s$-CUSUM charts) to detect shifts in the intensity and/or dispersion parameters; the $s$-CUSUM chart detects shifts in both parameters. The $\lambda$-CUSUM chart (used to detect shifts in the intensity parameter) contains three CUSUM charts as special cases: PCUSUM (when $\nu = 1$), geometric CUSUM ($\nu = 0$), and Bernoulli ($\nu \to \infty$) CUSUM charts. In all cases, the aforementioned charts are designed to monitor for any change (either increase or decrease) in the nonconformities rate. In order to develop these CMP-CUSUM control charts, it is assumed that all items are inspected from which the data are attained and that the sequence of random variables $X_t$, $t = 1, 2, 3, \ldots$, is independent and identically CMP($\lambda$, $\nu$) distributed with probability $P(X_t = x_t)$ as defined in Equation (2.8).

*The λ-CUSUM chart*

Assume that the CMP in-control parameter values are known to be $\lambda_0$ and $\nu_0$. Saghir and Lin (2014b) consider a hypothesis test $H_0: \lambda = \lambda_0$ versus $H_1: \lambda = \lambda_1$, where $\lambda_1 > \lambda_0$ ($\lambda_1 < \lambda_0$) infers a rate increase (decrease) in the number of nonconformities. Using a likelihood ratio test approach, the respective one-sided λ-CUSUM statistics have the form $C_t^+ = \max(0, K_{1t} + C_{t-1}^+)$ that gets sufficiently large if $H_1: \lambda = \lambda_1 > \lambda_0$, and $C_t^- = \min(0, K_{1t} + C_{t-1}^-)$ becoming sufficiently small if $H_1: \lambda = \lambda_1 < \lambda_0$ for $i = 1, 2, \ldots$, where

$$K_{1t} = x_t[\ln(\lambda_1) - \ln(\lambda_0)] + \ln(Z(\lambda_0, \nu_0)) - \ln(Z(\lambda_1, \nu_0)); \quad (6.35)$$

the starting value $C_0$ can be zero or nonzero as proposed in Lucas and Crosier (1982). The λ-CUSUM chart rejects $H_0$ when $C_t^+ > B_C^+$ or $C_t^- < B_C^-$, where $B_C^+$ and $B_C^-$, respectively, are the upper and lower control limits, determined via analyst-required in-control performance. Alternatively, one can develop a λ-CUSUM construct based on the sequential probability ratio test. Here,

$$C_t = \frac{\prod_{i=1}^{t} f(X = x_i; \lambda_0, \nu_0)}{\prod_{i=1}^{t} f(X = x_i; \lambda_1, \nu_0)} \text{ or} \quad (6.36)$$

$$\ln(C_t) = \ln(C_{t-1}) + X_t - K_{2t}, \quad (6.37)$$

for $t = 1, 2, 3, \ldots$, where $K_{2t} = \frac{\ln[Z(\lambda_1, \nu_0)/Z(\lambda_0, \nu_0)]}{\ln(\lambda_1/\lambda_0)}$. Under the sequential probability ratio test analysis, the λ-CUSUM statistics are now $C_t^+ = \max(0, C_{t-1}^+ + X_t - K_{2t})$ and $C_t^- = \min(0, K_{2t} - X_t + C_{t-1}^-)$, where $C_t^+ > B_C^+$ and $C_t^- < B_C^-$ infer rejecting $H_0$. Under either construct, the critical values for the respective one-sided tests are determined based on analyst prespecified in-control performance requirements. A two-sided CUSUM chart can be considered by simultaneously performing the respective one-sided CUSUM charts.

The λ-CUSUM chart simplifies to the PCUSUM and the geometric and Bernoulli CUSUM charts under the special cases for $\nu_0$. For $\nu_0 = 1$, $K_{2t} = \frac{\lambda_1 - \lambda_0}{\ln(\lambda_1) - \ln(\lambda_0)}$ is precisely the reference value of a Poisson($\lambda_0$) CUSUM chart of Lucas (1985). Meanwhile, for $\nu_0 = 0$ and $\lambda < 1$, $K_{2t} = \frac{\ln(1-\lambda_0) - \ln(1-\lambda_1)}{\ln(\lambda_1/\lambda_0)}$ is the reciprocal of the reference value of a Bourke (2001) geometric($1 - \lambda_0$) CUSUM chart, and for $\nu_0 \to \infty$, $K_{2t} = \frac{\ln(1-p_0) - \ln(1-p_1)}{\ln(p_1(1-p_0)) - \ln(p_0(1-p_1))}$ is the reference value of a Bernoulli$\left(p_0 = \frac{\lambda_0}{1+\lambda_0}\right)$ CUSUM chart (Lee et al., 2013; Reynolds, 2013).

*The ν-CUSUM Chart*

For a fixed intensity value $\lambda$, the CMP distribution mean and variance decrease as the dispersion value $\nu$ increases; meanwhile, for a fixed dispersion level $\nu$, its mean and variance increase as the intensity value $\lambda$ increases (see Section 2.3). Saghir and Lin (2014b) use these relationships to develop a $\nu$-CUSUM chart in order to detect a significant shift in the dispersion parameter from $\nu_0$ to $\nu_1 \neq \nu_0$. Accordingly, for $H_0: \nu = \nu_0$ versus respective alternate hypotheses $H_1: \nu = \nu_1 > \nu_0$ and $H_1: \nu = \nu_1 < \nu_0$, their associated upper and lower $\nu$-CUSUM control statistics are $D_t^+ = \max(0, G_t + D_{t-1}^+)$ and $D_t^- = \min(0, G_t + D_{t-1}^-)$, respectively, for $t = 1, 2, \ldots$, where $D_0 = 0$ and $G_t$ is the log-likelihood ratio for a shift satisfying

$$G_t = \ln(x_t!)(\nu_0 - \nu_1) + \ln\left(\frac{Z(\lambda_0, \nu_0)}{Z(\lambda_0, \nu_1)}\right). \tag{6.38}$$

Thus, for upper and lower control bounds, $B_D^+$ and $B_D^-$, respectively, $D_t^+ > B_D^+$ serves to detect statistically significant positive change in $\nu$, while $D_t^- > B_D^-$ detects significant negative change.

*The s-CUSUM Chart*

While the previous charts allow analysts to consider isolated charts for only $\lambda$ or $\nu$ Saghir and Lin (2014b) propose the $s$-CUSUM chart to simultaneously detect significant change (increase or decrease) in both $\lambda$ and $\nu$ via a single chart. Denoting in- and out-of-control values of $\lambda$ as $\lambda_0$ and $\lambda_1$, respectively (and similarly $\nu_0$ and $\nu_1$ for $\nu$), the likelihood ratio method calls for the $s$-CUSUM control statistics to be $S_t^+ = \max(0, Q_t + S_{t-1}^+)$ and $S_t^- = \max(0, Q_t + S_{t-1}^-)$ for $t = 1, 2, 3, \ldots$, where $S_0 = 0$ and

$$Q_t = x_t \ln(\lambda_1/\lambda_0) + \ln(x_t!)[\nu_0 - \nu_1] + \ln[Z(\lambda_0, \nu_0)/Z(\lambda_1, \nu_1)]. \tag{6.39}$$

Accordingly, for upper and lower control thresholds $B_S^+$ and $B_S^-$, the $s$-CUSUM chart detects an out-of-control state when $S_t^+ > B_S^+$ or $S_t^- > B_S^-$. Note that the sequential probability ratio test analysis is not a viable approach here because it does not produce unique expressions for the CUSUM control statistics.

*CUSUM Chart Design, and Comparisons*

Instead of using ARL, Saghir and Lin (2014b) assess chart performance for the $\lambda$-, $\nu$-, and $s$-CUSUM charts via two average number of observations to signal (ANOS) values, namely the respective expected number of products until a signal is given when $\lambda$ and $\nu$ or both are in and out of control, i.e. $\text{ANOS}_0$ and $\text{ANOS}_1$. As is the case with ARLs, analysts desire large $\text{ANOS}_0$ when the process is in control and a small $\text{ANOS}_1$ when the process is out of control. Recognizing that the ANOS measures can be determined under the assumption of a zero-state or steady-state framework, the authors explain that ANOS values attained under a zero-state scenario assume that the parameter(s) shifted at time zero, thus implying that the shift occurred under the initial setup of the monitoring process or given the initial framework of the associated chart. Meanwhile, ANOS values attained under steady-state conditions imply that the process operated in control for a steady period of time before shifting out of control.

Saghir and Lin (2014b) use a Markov-chain approach to approximate ANOS where the in-control region $[0, r]$ for the CUSUM statistic is divided into $N$ subintervals $[\ell_j, u_j] = \left[\frac{(j-1)r}{N}, \frac{jr}{N}\right]$ with respective midpoints $m_j = \frac{(2j-1)r}{2N}$ for $j = 1, 2, \ldots, N$. Under this construct, the $(N+1)$st state is an absorbing state because the CUSUM statistic then falls beyond the $[0, r]$ interval and the process is out of control. ANOS can be thought of as the number of moves made from an initial state to an absorbing state; thus, attention focuses on the $(N+1) \times (N+1)$ transition probability matrix

$$\boldsymbol{P} = \begin{bmatrix} \boldsymbol{A} & (\boldsymbol{I}-\boldsymbol{A})\boldsymbol{1} \\ \boldsymbol{0} & 1 \end{bmatrix} \tag{6.40}$$

comprised of an $N \times N$ identity and partial matrices, $\boldsymbol{I}$ and $\boldsymbol{A}$, respectively. These components aid in determining the approximate ANOS $= \boldsymbol{P}_0'(\boldsymbol{I} - \boldsymbol{A})^{-1}\boldsymbol{1}$, where $\boldsymbol{P}_0'$ is an $N$-length vector of initial probabilities.

Specifying the values for $(\lambda_0, \nu_0)$ and $(\lambda_1, \nu_1)$ determine the respective reference values for the appropriate CUSUM chart – $K_{1t}$ or $K_{2t}$ via Equation (6.35) for the $\lambda$-CUSUM chart, $G_t$ in (6.38) for the $\nu$-CUSUM chart, or $Q_t$ in (6.39) for the $s$-CUSUM chart. Given the appropriate CUSUM chart, the control limit can be determined for a given reference value and $\text{ANOS}_0$, albeit not necessarily uniquely; see Saghir and Lin (2014b) for details. This is due to the discrete form of the CMP distribution.

The $\lambda$-CUSUM chart is more efficient than the $\nu$- or $s$-CUSUM charts when it comes to detecting a change in $\lambda$ with $\nu$ fixed. By design, the $\lambda$-CUSUM chart detects out-of-control states more quickly than the CMP-Shewhart chart in the presence of small to moderate changes in $\lambda$, while the CMP-Shewhart chart remains useful for detecting large shifts. The $s$-CUSUM chart meanwhile demonstrates bias under such a scenario because, by design, it seeks to detect shifts in both parameters simultaneously.

For situations where the potential shift is only in the dispersion parameter (i.e. maintaining an in-control rate parameter, $\lambda$), the $\nu$-CUSUM outperforms the other CMP-CUSUM charts (as expected, by design). The $s$-CUSUM chart ranks second in performance because, by allowing for the detection of shifts in both parameters, this chart allows for the possible detection of a significant shift in $\nu$; this, of course, is not possible in the $\lambda$-CUSUM chart, given its framework that assumes a fixed dispersion level. Finally, when both parameters shift, the $s$-CUSUM offers mixed performance results. The chart performs better when detecting shifts of both parameters when the direction of change is the same for both parameters (i.e. both parameters increase or decrease); however, the chart performs poorly in cases where one parameter increases while the other decreases. In particular, a dispersion shift that changes the dispersion type (e.g. over- to under-dispersion) results in a large $s$-CUSUM chart ANOS.

### 6.3.2 Mixed EWMA-CUSUM for CMP Attribute Data

Rao et al. (2020) develop a CMP-based mixed EWMA-CUSUM (i.e. CMP-EWMA-CUSUM) chart as a special case of the Abbas et al. (2013) EWMA-CUSUM chart. This chart combines the respective strengths of a CMP-EWMA chart and a $\lambda$-CUSUM chart to monitor attribute data. Assume that process attributes $X_t, t = 1, 2, \ldots$, are independent and identically CMP($\lambda, \nu$) distributed where the approximations for the mean and variance (Equations (2.23) and (2.24)) hold. The CMP-EWMA-CUSUM approach relies on the upper and lower CUSUM statistics $M_t^+ = \max[0, (C_t - \mu_0) - K_t + M_{t-1}^+]$ and $M_t^- = \max[0, (\mu_0 - C_t) - K_t + M_{t-1}^-]$ that (in turn) rely on the CMP-EWMA statistic $C_t = \omega X_t + (1 - \omega)C_{t-1}$ defined in Equation (6.9); $\omega \in [0, 1]$ remains the weighting constant, and let $C_0 = \mu_0 = \lambda^{1/\nu} - \frac{\nu-1}{2\nu}$ (the approximated target mean) be the starting value for $C_t$. Meanwhile, $M_t^+$ and $M_t^-$, respectively, denote the upper and lower CUSUM statistics (initially set at zero), where $K_*$ and $K_t = K_*\sqrt{\frac{\omega}{2-\omega}[1 - (1 - \omega)^{2t}]\frac{1}{\nu}\lambda^{1/\nu}}$ are CUSUM chart reference values. The resulting CMP-EWMA-CUSUM control chart further relies on the CUSUM chart control limits, $L_*$ and $L_t = L_*\sqrt{\frac{\omega}{2-\omega}[1 - (1 - \omega)^{2t}]\frac{1}{\nu}\lambda^{1/\nu}}$; $K_*$ and $L_*$ are (e.g. classical CUSUM reference value and limit) constants. Accordingly, $M_t^+$ and $M_t^-$ are compared with $L_t$ such that the process is in control if the CUSUM statistics $M_t^+$ and $M_t^-$ are less than $L_t$; otherwise, the process is out of control (i.e. the process has shifted either above or below the target value). The CMP-EWMA-CUSUM approach reduces to the $\mu_0$-shifted $\lambda$-CUSUM scheme for $\omega = 1$.

Rao et al. (2020) consider various Monte Carlo simulations where the resulting data are assumed to be equi- or over-dispersed (i.e. $\nu \leq 1$, thus allowing for the mean and variance approximations to hold) to study the ARL. These studies show that $\text{ARL}_1$ decreases with $\omega$ (thus supporting a small value for $\omega$), while, given $\omega$, the $\text{ARL}_1$ decreases with $\lambda$. The more precise representations,

$$C_0 = \mu_0 = \frac{\partial \ln Z(\lambda, \nu)}{\partial \ln(\lambda)} = \lambda \frac{\partial \ln Z(\lambda, \nu)}{\partial \lambda}$$

$$K_t = K_* \sqrt{\frac{\omega}{2-\omega}[1-(1-\omega)^{2t}]\frac{\partial E(X)}{\partial \ln(\lambda)}}$$

$$L_t = L_* \sqrt{\frac{\omega}{2-\omega}[1-(1-\omega)^{2t}]\frac{\partial E(X)}{\partial \ln(\lambda)}},$$

however would presumably allow for greater flexibility of use, thus not restricting control chart performance based on dispersion type.

## 6.4 Generally Weighted Moving Average

The Poisson generally weighted moving average (PGWMA) chart was designed to better incorporate historical data and advance the PEWMA chart in an effort to more efficiently detect process shifts (Chiu and Sheu, 2008). The PGWMA chart statistic is defined as

$$G_t^* = \sum_{j=1}^{t} (q^{(j-1)^\alpha} - q^{j^\alpha}) X_{t-j+1}^* + q^{t^\alpha} G_0^* \tag{6.41}$$

for a Poisson($\lambda$) distributed number of detected nonconformities $X_t^*$ at time $t$ ($t = 1, 2, 3, \ldots$), where $0 \leq q < 1$ is a design parameter, $\alpha > 0$ an adjustment parameter, and $G_0^* = \lambda_0$ is the PGWMA initial value. The resulting PGWMA chart control limits and centerline are

$$\begin{aligned}
\text{UCL}_t &= \lambda_0 + B\sqrt{\lambda_0 \sum_{j=1}^{t} \left(q^{(j-1)^\alpha} - q^{j^\alpha}\right)^2} \\
\text{CL} &= \lambda_0 \\
\text{LCL}_t &= \max\left(0,\ \lambda_0 - B\sqrt{\lambda_0 \sum_{j=1}^{t} \left(q^{(j-1)^\alpha} - q^{j^\alpha}\right)^2}\right),
\end{aligned} \tag{6.42}$$

where $B$ is the PGWMA control bound that attains the desired $\text{ARL}_0$. The PEWMA chart is a special case of the PGWMA chart where $\alpha = 1$ and

$q = 1 - \omega$, while the Shewhart $c$-chart is a special case of the PGWMA where $\alpha = 1$ and $\omega = 1$ (Sheu and Chiu, 2007).

Chen (2020) generalizes the PGWMA chart in order to analyze CMP attribute data (hereafter referred to as the CMP-GWMA chart). Letting $X_t$, $t = 1, 2, 3, \ldots$, denote the number of nonconformities with an underlying CMP($\lambda$, $\nu$) distribution, assume that the CMP mean and variance approximations (Equations (2.23) and (2.24)) are permissible and an in-control status exists, i.e. $\lambda = \lambda_0$ and $\nu = \nu_0$. The CMP-GWMA chart statistic is defined as

$$G_t = \sum_{j=1}^{t} (q_1^{(j-1)^\alpha} - q_1^{j^\alpha}) X_{t-j+1} + q_1^{t^\alpha} G_0, \tag{6.43}$$

for the design parameter $0 \leq q_1 < 1$, the adjustment parameter $\alpha > 0$, and $G_0 \doteq \lambda_0^{1/\nu_0} - \frac{\nu_0 - 1}{2\nu_0}$. The resulting control limits and centerline for the CMP-GWMA chart are

$$\begin{aligned} \mathrm{UCL}_t &= \lambda_0^{1/\nu_0} - \frac{\nu_0 - 1}{2\nu_0} + B_1 \sqrt{\frac{\lambda_0^{1/\nu_0}}{\nu_0} \sum_{j=1}^{t} \left( q_1^{(j-1)^\alpha} - q_1^{j^\alpha} \right)^2} \\ \mathrm{CL} &= \lambda_0^{1/\nu_0} - \frac{\nu_0 - 1}{2\nu_0} \\ \mathrm{LCL}_t &= \max \left( 0,\ \lambda_0^{1/\nu_0} - \frac{\nu_0 - 1}{2\nu_0} - B_1 \sqrt{\frac{\lambda_0^{1/\nu_0}}{\nu_0} \sum_{j=1}^{t} \left( q_1^{(j-1)^\alpha} - q_1^{j^\alpha} \right)^2} \right), \end{aligned} \tag{6.44}$$

given the CMP-GWMA bound $B_1$ that attains the desired $\mathrm{ARL}_0$. These limits are determined based on the approximations to the CMP mean and variance; hence, analysts should be mindful to ensure that the associated constraints hold for the in-control process ($\nu_0 \leq 1$ or $\lambda_0 > 10^{\nu_0}$). The CMP-GWMA simplifies to the CMP-EWMA chart when $\alpha = 1$ and $q_1 = 1 - \omega$, while the CMP-GWMA reduces to the CMP Shewhart chart when $\alpha = 1$ and $\omega = 1$.

Similarly, Chiu and Lu (2015) design the Poisson-double GWMA (PDGWMA) chart to study Poisson-distributed attribute data. The PDGWMA statistic has the form

$$D_t^* = \sum_{j=1}^{t} V_j X_{t-j+1}^* + \left( 1 - \sum_{j=1}^{t} V_j \right) G_0^*, \quad t = 1, 2, 3, \ldots,$$

where $X^*$ remains Poisson($\lambda$) distributed, $V_j = \sum_{i=1}^{j} (q^{(i-1)^\alpha} - q^{i^\alpha})(q^{(j-i)^\alpha} - q^{(j-i+1)^\alpha})$ for the design parameter $0 \leq q < 1$ and the adjustment parameter

$0 < \alpha \leq 1$, and $G_0^* = \lambda_0$ denotes the initial PGWMA statistic. The resulting PDGWMA chart has the limits

$$\text{UCL}_t = \lambda_0 + B^* \sqrt{\lambda_0 \sum_{j=1}^{t} V_j^2} \tag{6.45}$$

$$\text{CL} = \lambda_0 \tag{6.46}$$

$$\text{LCL}_t = \max\left(0,\ \lambda_0 - B^* \sqrt{\lambda_0 \sum_{j=1}^{t} V_j^2}\right). \tag{6.47}$$

This chart contains the PDEWMA as a special case when $\alpha = 1$; thus (as expected), the PDGWMA outperforms the PDEWMA for detecting process shifts because of its added flexibility.

Chen (2020) likewise generalizes the PDGWMA to create a CMP-based double GWMA (i.e. CMP-DGWMA) chart, designing a double GWMA chart for CMP data in order to supersede the efficiency in shift detection demonstrated by the CMP-DEWMA chart. The CMP-DGWMA statistic is

$$D_t = \sum_{j=1}^{t} W_j X_{t-j+1} + \left(1 - \sum_{j=1}^{t} W_j\right) G_0, \quad t = 1, 2, 3, \ldots, \tag{6.48}$$

where $W_j = \sum_{i=1}^{j} \left(q_1^{(i-1)^\alpha} - q_1^{i^\alpha}\right)\left(q_2^{(j-i)^\beta} - q_2^{(j-i+1)^\beta}\right)$, $0 \leq q_1, q_2 < 1$ are the design parameters, and $0 < \alpha, \beta \leq 1$ the adjustment parameters. Accordingly, the CMP-DGWMA centerline and control limits are

$$\begin{aligned}
\text{UCL}_t &= \lambda_0^{1/\nu_0} - \frac{\nu_0 - 1}{2\nu_0} + B_2 \sqrt{\frac{\lambda_0^{1/\nu_0}}{\nu_0} \sum_{j=1}^{t} W_j^2} \\
\text{CL} &= \lambda_0^{1/\nu_0} - \frac{\nu_0 - 1}{2\nu_0} \\
\text{LCL}_t &= \max\left(0,\ \lambda_0^{1/\nu_0} - \frac{\nu_0 - 1}{2\nu_0} - B_2 \sqrt{\frac{\lambda_0^{1/\nu_0}}{\nu_0} \sum_{j=1}^{t} W_j^2}\right).
\end{aligned} \tag{6.49}$$

The CMP-DGWMA special case where $\nu_0 = 1$ is the PDGWMA chart. Meanwhile, when $q_1 = q_2 = q$ and $\alpha = \beta = 1$, the CMP-DGWMA simplifies to the CMP-DEWMA chart (and, hence, the PDEWMA when $\nu_0 = 1$). More broadly, for $q_1 = q_2 = q$ and $0 < \alpha = \beta < 1$, the CMP-DGWMA chart can be viewed as a scheme where the chart statistic twice applies the CMP-GWMA weighting sequence.

Chen et al. (2008) conducted a series of simulation studies to demonstrate the flexibility and performance of the CMP-GWMA and CMP-DGWMA charts. First, designing Monte Carlo simulations where $q_1 = q_2 = q$ and $\alpha = \beta$ for ease, in-state simulations show that $B_1 > B_2$ when $\nu_0 \leq 1$, given $q$ and $\alpha$, i.e. when the in-control state recognizes the data to be equi- or over-dispersed, the CMP-GWMA bound is consistently greater than the corresponding CMP-DGWMA bound in order to attain the same $\text{ARL}_0$ value. This implies that the CMP-DGWMA chart will more quickly detect significant shifting with respect to $\lambda$ than the CMP-GWMA chart. Further, when the data are equi- or over-dispersed, $B_i$ decreases as $q$ increases, given a fixed value $\alpha$, $i = 1, 2$. When the data are under-dispersed, however, there does not exist a consistent comparison between $B_1$ and $B_2$ except that both bounds increase as $q$ increases for $\alpha \leq 0.2$.

Simulation studies further showed that the CMP-GWMA and CMP-DGWMA charts are more sensitive in their abilities to detect upward shifts than downward shifts in the location parameter $\lambda$, for a fixed level of data equi- or over-dispersion. Under such circumstances, analysts are encouraged to consider a large $q$ and $\alpha$ near 0.5 in order to properly monitor for small location shifts via the CMP-(D)GWMA charts. Meanwhile, the CMP-(D)GWMA charts are more efficient at detecting upward location shifts than downward ones when the in-control data are under-dispersed. Accordingly, analysts are encouraged to use a CMP-DGWMA chart with $q$ and $\alpha$ near 0.9 to detect small upward location shifts, while a large $q$ and $\alpha$ near 0.5 is recommended for small downward shifts. Given a fixed dispersion level and fixed parameters $q$ and $\alpha$, the $\text{ARL}_1$ decreases as the location shift size (whether upward or downward) increases, implying that these charts can more quickly detect location shifts, particularly as the size of the said shift increases. The needed time to detection, however, increases with the size of $\nu_0$; this implies that less underlying dispersion associates with a longer needed time to detect location shifts. Nonetheless, the CMP-DGWMA chart outperforms the CMP-GWMA and CMP-(D)EWMA charts at detecting location shifts of any kind. In studying dispersion parameter shifts given a fixed location $\lambda_0$, simulation studies showed that the CMP-DGWMA chart is unbiased for dispersion parameter shifts in either direction when $q \geq 0.7$ and $\alpha \geq 0.5$ when starting from an over-dispersed in-control state. The CMP-(D)GWMA charts can more efficiently detect downward shifts than upward shifts in $\nu$ for any $\nu_0$. Accordingly, a large $q$ and $\alpha$ around 0.5 is recommended to detect small downward shifts in dispersion via the CMP-DGWMA chart, while a larger $\alpha$ is suggested to detect small upward shifts in $\nu$ via the CMP-GWMA chart when $\nu_0 \leq 1$ (i.e. data are equi- to over-dispersed). For $\nu_0 > 1$ (i.e. under-dispersion exists), the CMP-(D)GWMA charts are meanwhile unbiased when detecting downward shifts in the dispersion parameter, $\nu$. Finally, when studying shifts in both parameters simultaneously, the CMP-(D)GWMA charts are again more efficient than the CMP-(D)EWMA charts with regard to the simultaneous shift detection.

## 6.5 COM–Poisson Chart Via Progressive Mean Statistic

Alevizakos and Koukouvinos (2022) propose a CMP-motivated "memory-type" control chart based on the progressive mean statistic that combines ideas from both the CUSUM and EWMA charts. Letting $X_j$ be CMP($\lambda, \nu$) distributed for all $j$, the CMP progressive mean (CMP-PM) is $M_t^* = \frac{\sum_{j=1}^t X_j}{t}$, and the resulting centerline and control limits associated with the CMP chart via the progressive mean statistic are

$$\begin{aligned} \mathrm{UCL}_t &= \lambda_0 \frac{\partial \ln (Z(\lambda_0, \nu))}{\partial \lambda_0} + \frac{B}{f(t)} \sqrt{\frac{1}{t} \frac{\partial E(X)}{\partial \ln (\lambda_0)}}, \\ \mathrm{CL} &= \lambda_0 \frac{\partial \ln (Z(\lambda_0, \nu))}{\partial \lambda_0}, \\ \mathrm{LCL}_t &= \max \left(0, \ \lambda_0 \frac{\partial \ln (Z(\lambda_0, \nu))}{\partial \lambda_0} - \frac{B}{f(t)} \sqrt{\frac{1}{t} \frac{\partial E(X)}{\partial \ln (\lambda_0)}} \right), \end{aligned} \tag{6.50}$$

where $B$ denotes the control bound width, $f(t)$ is an arbitrary function with respect to $t$ that serves as a penalty term, and $X$ denotes the CMP($\lambda, \nu$) random variable. The process is out of control if $M_t^* > \mathrm{UCL}_t$ or $M_t^* < \mathrm{LCL}_t$; otherwise, the process is in control and thus does not demonstrate any significant shift in the process mean. For $\nu = 1$, these bounds simplify to those of the Poisson progressive mean chart by Abbasi (2017).

Data simulations show that the CMP-PM chart can detect any type of in-control process mean shift and that the chart is unbiased with respect to ARL (i.e. the $\mathrm{ARL}_1$ results are smaller than their $\mathrm{ARL}_0$ counterparts). The run-length distribution is positively skewed where that skewness decreases as the shift with respect to $\lambda$ increases, and its attributes (its average, standard deviation, and percentile points) decrease as the shifts increase. Accordingly, letting $f(t) = t^{0.2}$ is suggested because this optimizes run-length distribution properties (Alevizakos and Koukouvinos, 2022). Simulations further show that the CMP-PM chart can better detect upward (downward) shifts when the data are over-dispersed (under-dispersed).

The CMP-PM chart slows in its ability to detect shifts as $t$ increases, however, if the data have been in control for an extended period, because $M_t^*$ is the average of all $t$ previous observations. Further, given $t$ and prespecified shift in $\lambda$, the percentage of difference between the zero-state and state ARLs decreases as $\nu$ increases.

Monte Carlo-simulated illustrations show that the CMP-PM chart consistently outperforms the CMP Shewhart chart, while the CMP-PM chart often outperforms the CMP-EWMA chart. In particular, when the data are over-dispersed, the CMP-PM chart can better detect small and moderate shifts than the CMP-EWMA, while the CMP-EWMA can better detect large shifts. When the data are equi-dispersed, the CMP-PM chart better detects shifts than the CMP-GWMA chart except when the shift is very large. Finally, when the data are under-dispersed, the CMP-PM chart consistently outperforms the CMP-EWMA chart. Meanwhile, the CMP-PM chart generally outperforms the $\lambda$-CUSUM chart as well. When the data are over-dispersed, the CMP-PM consistently outperforms $\lambda$-CUSUM while, when the data are equi-dispersed, the CMP-PM can better detect shifts except when the shift is very small. When the data are under-dispersed, the CMP-PM chart better detects larger shifts than the $\lambda$-CUSUM chart (Alevizakos and Koukouvinos, 2022).

## 6.6 Summary

The CMP distribution offers added flexibility in the vast array of control charts that exist for discrete attribute data. Whether Shewhart, EWMA, CUSUM charts, their broader representations or even others, each of the aforementioned CMP-inspired charts demonstrates growing efficiency in its ability to detect significant shifts in the CMP parameters (particularly with attention on $\lambda$). Most of the presented works assume that the approximations for the CMP mean and variance hold; thus, analysts should use the presented bounds with caution. If the in-control data are equi- or over-dispersed, then analysts can proceed as presented; however, for under-dispersed data, analysts are instead encouraged to work with the true forms of the CMP expectation and variance as presented in Equations (2.14)–(2.15) in order to gain more precise representations of the resulting control limits. Meanwhile, in all cases, analysts are reminded to Winsorize their lower bound to 0 to address any results that produce a negative value since the bound still reflects the larger context regarding the parameter constraints and attribute representation.

All of the work regarding control charts assume a CMP parametrization of the COM–Poisson distribution. Several of the discussed control charts then rely on the CMP mean and variance approximations for control chart theory development. All of the charts introduced in this chapter, however, can likewise be derived with the other COM–Poisson parametrizations; the various COM–Poisson parametrizations are provided and discussed in Section 2.6. In particular, control chart developments motivated by (say) the MCMP1 distribution should circumvent the issues that currently arise regarding control chart developments that rely on the approximations; the MCMP1 parametrization directly transforms the distribution such that emphasis focuses on the distribution mean. Further, analysts should no longer be constrained as they are with the above approaches that rely on constrained parameter spaces to ensure chart accuracy.

Only the `CMPControl` package appears available for use by analysts; thus, only the CMP-Shewhart chart analysis is readily available. The necessary information is provided, however, for analysts to construct their own `R` codes for the control chart(s) of interest.

# 7

# COM–Poisson Models for Serially Dependent Count Data

In previous chapters, discussion largely assumes a random sample or at least some measures of independence; however, what if dependence exists between variables? This chapter considers various models that focus largely on serially dependent variables and the respective methodologies developed with a COM–Poisson underpinning (in any of its parametrizations). Section 7.1 introduces the reader to the various stochastic processes that have been established, including a homogeneous CMP process, a copula-based CMP Markov model, and a CMP-hidden Markov model (CMP-HMM). Meanwhile, there are two approaches for conducting time series analysis on time-dependent count data. One approach assumes that the time-dependence occurs with respect to the intensity vector; see Section 7.2. Under this framework, the usual time series models that assume a continuous variable can be applied. Alternatively, the time series model can be applied directly to the outcomes themselves. Maintaining the discrete nature of the observations, however, requires a different approach referred to as a thinning-based method. Different thinning-based operators can be considered for such models; see Section 7.3 for details. Section 7.4 broadens the discussion of dependence to consider COM–Poisson-based spatio-temporal models, thus allowing for both serial and spatial dependence among variables. Finally, Section 7.5 concludes the chapter with discussion.

## 7.1 CMP-motivated Stochastic Processes

### *7.1.1 The Homogeneous CMP Process*

Various homogenous discrete-observation processes exist in the literature to model the number of events. Assume that a count outcome of interest $N_t$ over a time period $t$ follows a discrete distribution and has independent increments. The most well-known models for describing such a series of discrete events are the Bernoulli and Poisson processes. A Bernoulli process is a discrete-time discrete-observation process where the success probability $p$ defines the number of successes that occur over $t$ trials with a Binomial$(t, p)$-distributed random variable $N_t = X_1 + \cdots + X_t$, where $X_1, \ldots, X_t$ are independent and identically Bernoulli($p$) distributed random variables. In this context, the independent increments imply that $N_{s+t} - N_s$ also follows a Binomial$(t, p)$ distribution (independent of $s$) for any $s, t \in \mathbb{N}$. The waiting time variable $T_k$ denotes the number of trials it takes to get the $k^{\text{th}}$ success. For a Bernoulli process, the time between successes $T_{k+1} - T_k$ for any $k \in \mathbb{N}$ follows a geometric distribution, i.e. $P(T_{k+1} - T_k = r) = p(1-p)^{r-1}$, $r = 1, 2, \ldots$. Further, for any $k$ and $m$, the interval $T_{k+m} - T_k$ is independent of $k$ (Çinlar, 1975). The homogeneous Poisson process is meanwhile a continuous-time discrete-observation process over a time period $(s, s+t]$, where $N_t$ represents the number of events over a time period $t$ and has a Poisson$(\lambda t)$ distribution such that $N_{s+t} - N_s$ is likewise Poisson$(\lambda t)$, independent of $s$. This definition implies that

$$P(N_{s+t} - N_s = 1) = \lambda t + o(t)$$

$$P(N_{s+t} - N_s \geq 2) = o(t),$$

where a function $f(t)$ of order $o(t)$ implies that $\lim_{t \to 0} \frac{f(t)}{t} = 0$ (Kannan, 1979). The associated waiting time between events $T_k$ for a Poisson process is an exponential$(\lambda)$ distribution with probability density function $f_{T_k}(t) = \lambda e^{-\lambda t}, t \geq 0$.

To generalize these discrete-observation processes, Zhu et al. (2017b) introduced a homogeneous CMP process to describe the number of events with independent increments. Let $N_t$ denote the number of events that occur in a time interval $(s, s+t]$ where this random variable has a sCMP$(\lambda, \nu, t)$ distribution, i.e. a sum-of-COM–Poissons (sCMP) distribution as described in Section 3.2. This CMP process likewise has independent and stationary increments – for ordered time points $t_0, t_1, \ldots, t_n$, the variables $N_{t_1} - N_{t_0}, \ldots, N_{t_n} - N_{t_{n-1}}$ are independent. Special cases of this process include a homogeneous Poisson$(\lambda)$ process (when $\nu = 1$), a Bernoulli process with success probability $\frac{\lambda}{1+\lambda}$ $(\nu \to \infty)$, and what can be termed a "geometric process" with success probability $1 - \lambda$ when $\nu = 0$ and $\lambda < 1$. This process likewise follows a CMP$(\lambda, \nu)$ distribution over a one-unit interval $(t, t+1]$.

The CMP process has a generalized waiting time distribution whose cumulative probability is

$$P(T_t \leq \tau) = 1 - \frac{1}{[Z(\lambda, \nu)]^{\tau}}; \tag{7.1}$$

however, the form is unique because this CMP process can accommodate an underlying discrete or continuous waiting time model. Let $T_t^* = \lceil T_t \rceil$ be a discrete waiting time variable; Equation (7.1) then implies that $T_t^*$ is a geometric$\left(p = 1 - \frac{1}{Z(\lambda,\nu)}\right)$ distribution with probability

$$P(T_t^* = \tau) = \left(\frac{1}{Z(\lambda, \nu)}\right)^{\tau-1}\left(1 - \frac{1}{Z(\lambda, \nu)}\right), \quad \tau = 1, 2, 3, \ldots. \tag{7.2}$$

For a continuous-time process with waiting time $T_t$, Equation (7.1) implies that $T_t$ has the probability density function

$$f(\tau) = [Z(\lambda, \nu)]^{-\tau} \ln(Z(\lambda, \nu)) = \ln(Z(\lambda, \nu))e^{-\tau \ln Z(\lambda,\nu)}, \quad \tau \geq 0, \tag{7.3}$$

namely an exponential($\ln(Z(\lambda, \nu))$) distribution. Zhu et al. (2017b) refer to Equations (7.2) and (7.3) as COM-geometric and COM-exponential distributions, respectively, reflecting the fact that both distributions depend on the CMP($\lambda$, $\nu$) normalizing constant. The relationship between these two distributions is a natural analog to the usual relationship between the geometric and exponential distributions.

*Parameter Estimation*

Three approaches can be considered for conducting parameter estimation and quantifying variability for a CMP process; each is derived under a different scenario, namely (1) for count data regarding a number of events in a single time unit, (2) when information is supplied regarding a single time unit solely based on the wait-time data, and (3) when given a sample of events over an $s$-unit interval. For a CMP process over a single time unit, the number of events is modeled via a CMP($\lambda$, $\nu$) distribution. Thus, for an ordered sequence of count data $x_1, \ldots, x_n$, the method of maximum likelihood can be used to estimate $\lambda$ and $\nu$ as discussed in Section 2.4.2. If instead only waiting time information $t_1, \ldots, t_n$ is supplied, this approach produces the undetermined equation

$$\frac{1}{Z(\lambda, \nu)} = 1 - \frac{1}{\bar{t}}. \tag{7.4}$$

Thus, more information is required in order to obtain unique estimates. Assuming additional knowledge (say) of the observed dispersion index can aid in extrapolating $\hat{\lambda}$ and $\hat{\nu}$. Finally, for a random sample from an sCMP($\lambda, \nu, s$) distribution with $s \geq 1$, the method of maximum likelihood can be achieved as described in Section 3.2 to determine $\hat{\lambda}$ and $\hat{\nu}$ for a given $s$. Under any of these scenarios, the resulting estimates are then used to ascertain the waiting time either as a geometric distribution with success probability $\hat{p} = 1 - \frac{1}{Z(\hat{\lambda},\hat{\nu})}$ or as an exponential $(\ln(Z(\hat{\lambda}, \hat{\nu})))$ distribution. While variation quantification can be determined via the Fisher information matrix, analysts are encouraged to instead utilize nonparametric bootstrapping using the `boot` package (Canty and Ripley, 2020) because the parameters have skewed distributions (Sellers and Shmueli, 2013; Zhu et al., 2017b).

### R *Computing*

The `cmpprocess` (Zhu et al., 2017a) package performs the statistical computations for homogeneous CMP process analysis; Table 7.1 provides the function names that respectively perform parameter estimation based on information provided by the analyst. The `cmpproc` function determines maximum likelihood estimates (MLEs) based on the provided count data (`counts`) over a time interval (`s`, defaulted to equal 1 to assume a one-unit interval) corresponding to the levels associated with the count data vector. This function further contains the `h.out` input that is a logical indicator that determines whether or not R should include the associated Fisher information matrix and thus standard errors among the provided outputs; `h.out` has the default setting equal to `FALSE`. Resulting output from this function reports the respective MLEs $\hat{\lambda}$ (`lambda`) and $\hat{\nu}$ (`nu`), along with their standard errors (`lambda se` and `nu se`) if the analyst updates the input to `h.out=TRUE`. Otherwise, maintaining the default setting `h.out=FALSE` results in the outputs, `lambda se: NA` and `nu se: NA`. Additional outputs that are reported through `cmpproc` include the log-likelihood, Akaike information criterion (`AIC`), and dispersion index values, along with

Table 7.1 R *functions provided in the* `cmpprocess` *package for CMP process analysis. These functions determine (approximate) MLEs based on the information provided by the analyst.*

| Function | Provided information |
|---|---|
| `cmpproc` | Count data |
| `cmpprocwt` | Wait-time data and observed dispersion index value |

an estimated waiting time distributional form assuming discrete time intervals.

The `cmpprocwt` function meanwhile determines approximate MLEs based on the waiting time data (`t`) and an assumed dispersion index measure (`dispersion`). This function likewise reports the estimated parameters for $\lambda$ (`lambda`) and $\nu$ (`nu`), along with convergence information (`convergence`) in order to assess whether the underlying optimization scheme converged. The `cmpprocess` package supplies the real-world data examples discussed in Zhu et al. (2017b), both as count (`fetalcount` and `floodcount`) and waiting time data (`fetalwait` and `floodwait`), to provide analysts with the ability to reproduce the analyses.

The `cmpprocess` package remains accessible through the Comprehensive `R` Archive Network (CRAN) and is searchable online. This package was archived because it relies on `compoisson` that has also since been archived. Nonetheless, the archive supplies the `tar.gz` file associated with the package that can be installed directly into `R`/`RStudio` for use. The `cmpprocess` package likewise requires the `numDeriv` package.

To illustrate the functionality of the `cmpprocess` package, consider a series of count data regarding the number of alpha particles emitted from a disk coated with polonium in successive 7.5-second intervals via the scintillation method (Rutherford et al., 1910); see Table 7.2. They are part of a larger experiment to compare the average number of $\alpha$ particles deduced from an extensive number of scintillations (namely 31.0 per minute) with the average number based on much smaller intervals. The associated discussion regarding this experiment suggests that these data stem from a homogeneous Poisson process (Guttorp, 1995; Rutherford et al., 1910). This exercise serves to analyze the data to see if (in fact) that assumption is reasonable and to see if the hypothesis of 31 alpha particles per minute on average is likewise reasonable.

Table 7.2 *Data (presented in sequential order, left to right) regarding the number of alpha particles emitted in successive 7.5-second intervals from a disk coated with polonium via the scintillation method (Rutherford et al., 1910).*

| | | | | | | | | | |
|---|---|---|---|---|---|---|---|---|---|
| 3 | 7 | 4 | 4 | 2 | 3 | 2 | 0 | 5 | 2 |
| 5 | 4 | 3 | 5 | 4 | 2 | 5 | 4 | 1 | 3 |
| 3 | 1 | 5 | 2 | 8 | 2 | 2 | 2 | 3 | 4 |
| 2 | 6 | 7 | 4 | 2 | 6 | 4 | 5 | 10 | 4 |

The `cmpproc` function calls the object `alphaparticle` that is a single column that lists the respective consecutive recorded number of alphaparticles reported in Table 7.2, and the logical indicator `h.out = TRUE` in order to obtain associated standard errors for $\hat{\lambda}$ and $\hat{\nu}$; see Code 7.1. The resulting MLEs are approximately $\hat{\lambda} = 3.553$ and $\hat{\nu} = 0.963$ with respective approximate standard errors, 1.332 and 0.250. Accordingly, we can infer that $\nu = 1$ is a reasonable assumption, implying that the homogeneous Poisson process is (in fact) represented through these data.

To address the latter question of how the data compare with the hypothesized average number of alpha particles emitted per minute equaling 31, recall that the reported output above is based on an interval equaling 7.5 seconds, where multiple approaches exist for estimating the average number of alpha particles emitted in that time frame. The first approach for estimating the average number of alpha particles per minute is to simply assume that the mean number of particles in a 7.5-second interval is precisely $\hat{\lambda} \approx 3.553$, which thus implies that the average number of particles per minute is approximately 28.4224. Another approach is to take advantage of the mean approximation as discussed in Section 2.3. Since $\hat{\nu} \approx 0.963 < 1$, we can approximate the mean via Equation (2.23), thus obtaining the estimated mean equaling approximately 3.747 alpha particles over 7.5 seconds. Thus, with this estimate, the estimated average number of alpha particles per minute equals approximately 29.9776. Compare these results with the average over the five-minute period (30) and we see that these estimates again seem reasonable.

Additional supplied outputs include the maximized log-likelihood (approximately $-82.6532$) and AIC (169.3065); see Code 7.1. These results are meaningless on their own but can prove valuable if conducting any appropriate model comparisons. Other outputs stemming from

Code 7.1 Code and output analyzing the alpha-particles dataset (Table 7.2) via the `cmpproc` function, contained in the `cmpprocess` package.

```
> alpha.out <- cmpproc(alphaparticle, h.out = TRUE)
log likelihood: -82.65323
lambda: 3.552797
nu: 0.9631337
lambda se: 1.331656
nu se: 0.249702
real dispersion: 1.104274
waiting time distribution is geometric with parameter:
0.9744633
AIC: 169.3065
```

the `cmpproc` function include the observed dispersion level (`real dispersion` $\approx 1.1043$) which is greater than 1 (indicating possible over-dispersion) or which can be viewed as approximately 1 (thus affirming the homogeneous Poisson process assumption). One can use this information to aid in conducting a hypothesis test regarding the dispersion parameter to infer whether any statistically significant data dispersion exists. The last output reports that the `waiting time distribution is geometric with parameter` approximately 0.9745. This makes sense if we view the data as a discrete-time discrete-observation process; the result confirms that the next outcome occurs during the next unit (in this case, the next 7.5-second interval). Viewing this as a homogeneous Poisson process, however, implies that we should estimate the waiting time distribution as an exponential($\ln(Z(\hat{\lambda}, \hat{\nu}))$) $\approx$ exponential(3.553) distribution, if we assume $\nu = 1$.

### *7.1.2 Copula-based CMP Markov Models*

Alqawba and Diawara (2021) developed a Markov model based on a copula-based multivariate ZICMP distribution to model data containing excess zeroes, noting that there is no simple stochastic representation for a discrete $p$-order Markov model of the form $X_t = g(\epsilon_t; X_{t-1}, \ldots, X_{t-p})$ with copula-based transition probabilities, i.e. where $X_t$ depends on its $p$ previous observations, $g(\cdot)$ is an increasing function, and $\epsilon_t$ is an independent and identically distributed (iid) stochastic continuous latent process. Let $\{X_t\}$ be a time series that assumes a ZICMP distribution as defined in Section 5.3 (Equation (5.26)) with a first-order Markov chain such that

$$P(X_1 = x_1, \ldots, X_n = x_n) = P(X_1 = x_1) \prod_{t=2}^{n} P(X_t = x_t \mid X_{t-1} = x_{t-1}) \quad (7.5)$$

$$= P(X_1 = x_1) \prod_{t=2}^{n} \frac{P(X_t = x_t, X_{t-1} = x_{t-1})}{P(X_{t-1} = x_{t-1})}, \quad (7.6)$$

where

$$P(X_t = x_t, X_{t-1} = x_{t-1}) = C(F_t(x_t), F_{t-1}(x_{t-1})) - C(F_t(x_t - 1), F_{t-1}(x_{t-1}))$$
$$- C(F_t(x_t), F_{t-1}(x_{t-1} - 1)) + C(F_t(x_t - 1), F_{t-1}(x_{t-1} - 1)) \quad (7.7)$$

for a copula $C$ as defined in Section 4.4. Analogously, a second-order Markov model has the joint probability

$$P(X_1 = x_1, \ldots, X_n = x_n) = P(X_1 = x_1, X_2 = x_2) \prod_{t=3}^{n} P(X_t = x_t \mid X_{t-1} = x_{t-1}, X_{t-2} = x_{t-2}),$$

where

$$P(X_t = x_t \mid X_{t-1} = x_{t-1}, X_{t-2} = x_{t-2}) = \frac{P(X_t = x_t, X_{t-1} = x_{t-1}, X_{t-2} = x_{t-2})}{P(X_{t-1} = x_{t-1}, X_{t-2} = x_{t-2})} \quad (7.8)$$

such that $P(X_1 = x_1, X_2 = x_2)$ and $P(X_{t-1} = x_{t-1}, X_{t-2} = x_{t-2})$ are determined via Equation (7.7), while

$$P(X_t = x_t, X_{t-1} = x_{t-1}, X_{t-2} = x_{t-2}) = \sum_{j_1=0}^{1} \sum_{j_2=0}^{1} \sum_{j_3=0}^{1} C(F_t(x_t - j_1), F_{t-1}(x_{t-1} - j_2), F_{t-2}(x_{t-2} - j_3)). \quad (7.9)$$

Equations (7.7) and (7.9) incorporate the copula approach for the transition probabilities. Numerous copulas exist for consideration; see Section 4.4 and Table 4.1 for discussion. Alqawba and Diawara (2021) consider bivariate Gaussian, Frank, and Gumbel copulas for the first-order Markov process, while they propose that either a Gaussian or max-id copula be used to determine the necessary joint probabilities in higher order Markov models, because these copulas can fit Markov models whose order is at least 2. Analysts are encouraged to use the max-id copula "when there is stronger dependence for measurements at nearer time points" (Alqawba and Diawara, 2021).

*Statistical Inference*

Parameter estimation can be conducted via the method of maximum likelihood, where the resulting log-likelihood associated with a first-order Markov model has a closed form if the selected copula likewise has one. For the second-order Markov models, the log-likelihood associated with the Gaussian copula produces a form that is not closed; however, the log-likelihood established via a trivariate max-id copula can have a closed form. Per usual, the MLEs are obtained at that multi-dimensional point of estimated parameter values that maximizes the log-likelihood function for a given dataset of observed values. These estimators are typically achieved computationally in R via an optimization procedure, while the corresponding Hessian matrix aids in determining the associated standard errors. Assuming that regularity conditions hold, the asymptotic properties associated with the usual random sample case extend to data assuming a Markov process; hence, the MLEs obtained from the adjusted first-order Markov process log-likelihood

$$\ln L_*((\boldsymbol{\theta}', \boldsymbol{\delta}')';\ \boldsymbol{x}) = \sum_{t=2}^{n} \ln P(X_t = x_t \mid X_{t-1} = x_{t-1};\ (\boldsymbol{\theta}', \boldsymbol{\delta}')') \quad (7.10)$$

are likewise consistent for $(\boldsymbol{\theta}', \boldsymbol{\delta}')'$ and the random vector is asymptotically normally distributed. Further, the associated approximate Fisher information matrix attained here can be used to estimate the covariance matrix associated with the MLEs attained from the random sample case. These ideas further apply for Markov processes of a known higher order (Alqawba and Diawara, 2021; Joe, 1997). Simulation studies validate the theory finding that this method produces sound estimates that appear to converge to the true values as the sample size increases.

### 7.1.3 CMP-Hidden Markov Models

"Hidden Markov models (HMMs) are models in which the distribution that generates an observation depends on the state of an underlying and unobserved Markov process. They provide flexible general-purpose models for univariate and multivariate time series, especially for discrete-valued series, including categorical series and series of counts" (Zucchini et al., 2016, p. 3). This hierarchical structure has the observations $\boldsymbol{X}_t = (X_1, X_2, \ldots, X_t)$ generated from latent variables $\boldsymbol{M}_t = (M_1, M_2, \ldots, M_t)$ defining an irreducible Markov chain where

$$P(M_t \mid \boldsymbol{M}_{t-1}) = P(M_t \mid M_{t-1}), \quad t = 2, 3, 4, \ldots; \text{ and} \quad (7.11)$$

$$P(X_t \mid \boldsymbol{X}_{t-1}, \boldsymbol{M}_t) = P(X_t \mid M_t), \qquad t \in \mathbb{N}; \quad (7.12)$$

for ease, we further assume the Markov chain to be stationary. For example, given an underlying $m$-state Markov chain, $X_t$ may assume a Poisson distribution, i.e. $P(X_t = x \mid M_t = i) = \frac{e^{-\lambda_i} \lambda_i^x}{x!}$, where $i = 1, 2, \ldots, m$ and

$x = 0, 1, 2, \ldots$; this defines a Poisson-hidden Markov model (P-HMM).[1] A general HMM satisfies the equation

$$V(X_t) = \sum_{i=1}^{m} \delta_i \sigma_i^2 + \sum_{i<j} \delta_i \delta_j (\mu_i - \mu_j)^2, \tag{7.13}$$

where $\mu_i$ and $\sigma_i^2$, respectively, denote the conditional mean and variance of $X_t$ given $M_t = i$, and $\boldsymbol{\delta}' = (\delta_1, \ldots, \delta_m)$ is the initial distribution across the $m$ states with elements $\delta_i = P(M_1 = i)$; hence, for a P-HMM, Equation (7.13) becomes

$$V(X_t) = \underbrace{\sum_{i=1}^{m} \delta_i \mu_i}_{E(X_t)} + \sum_{i<j} \delta_i \delta_j (\mu_i - \mu_j)^2.$$

Thus, one can see that the P-HMM allows for equi- or over-dispersion because $\sum_{i<j} \delta_i \delta_j (\mu_i - \mu_j)^2 \geq 0$. This construct, however, is unable to address data under-dispersion. MacDonald and Bhamani (2020) instead develop a stationary HMM where the state-dependent distribution is CMP($\lambda_i, \nu_i$) (i.e. CMP-HMM). Accordingly, the conditional probability is

$$P(X_t = x \mid M_t = i) = \frac{\lambda_i^x}{(\lambda_i)^{\nu_i} Z(\lambda_i, \nu_i)} \tag{7.14}$$

and, given $T$ observations $\boldsymbol{x} = (x_1, \ldots, x_T)$, the likelihood function becomes

$$L(\boldsymbol{\Gamma}, \boldsymbol{\lambda}, \boldsymbol{\nu};\, \boldsymbol{x}) = \boldsymbol{\delta}' \boldsymbol{P}(x_1) \boldsymbol{\Gamma} \boldsymbol{P}(x_2) \boldsymbol{\Gamma} \boldsymbol{P}(x_3) \cdots \boldsymbol{\Gamma} \boldsymbol{P}(x_T) \mathbf{1},$$

where $\boldsymbol{P}(x)$ is the $m \times m$ diagonal matrix of elements $p_i(x) = P(X_t = x \mid M_t = i)$ as defined in Equation (7.14), $\boldsymbol{\Gamma}$ is an $m \times m$ transition matrix with elements $\gamma_{ij} = P(M_t = j \mid M_{t-1} = i)$, and $\mathbf{1}$ is an $m$-length column vector of ones.

Parameter estimation can be performed in R via the method of maximum likelihood to estimate the $m(m+1)$ parameters of interest, namely the $m(m-1)$ elements of $\boldsymbol{\Gamma}$ (because row sums of $\boldsymbol{\Gamma}$ are constrained to equal *1*) along with the $2m$ elements of $\boldsymbol{\lambda} = (\lambda_1, \ldots, \lambda_m)$ and $\boldsymbol{\nu} = (\nu_1, \ldots, \nu_m)$. In order to circumvent the usual computational issues that arise when conducting such statistical computing, analysts are encouraged to take the log-transform of the likelihood function to reduce the computational scale and range and transform any bounded natural parameters to an unconstrained working space. Under this framework, analysts can use the `nlm`

[1] The state-dependent distributions do not all need to be from the same family (MacDonald and Bhamani, 2020).

or optim functions in R to obtain the MLEs, where the likelihood function can be constructed. MacDonald and Bhamani (2020) uses the dcmp function (COMPoissonReg) with its default settings to evaluate the CMP($\lambda_i$, $\nu_i$) probability, including the normalizing constant's infinite sum default approximation that sums the first 101 terms. While the number of terms can be modified by the user, the default setting produces generally sufficient performance; however, complexities can arise when $\lambda$ is large and $\nu$ is small. The nlm and optim functions can further provide the approximate-associated Hessian matrix that is useful in determining parameter estimate standard errors. Analysts are encouraged to consider various starting values for these optimization functions in order to instill confidence regarding their search for the MLEs.

Two-state CMP-HMMs reasonably estimate the sample mean and variance and thus oftentimes adequately model the data; however, the estimated autocorrelation function does not approximate the sample analog. Three or more state CMP-HMMs can likewise be considered and model comparison conducted via the AIC or Bayesian information criterion (BIC). Analysts are cautioned that introducing more states heightens the risks associated with more parameters, e.g. due to multiple local maxima (MacDonald and Bhamani, 2020).

### R *Computing*

While a formal R package to conduct statistical computing for CMP-HMMs does not exist, MacDonald and Bhamani (2020) directly supply the codes and data referenced in their work in order for analysts to reproduce their results.[2] Table 7.3 contains a summary of relevant R codes supplied for statistical computing for CMP-HMMs, including transforming the parameters between their natural and working space, computing the negated log-likelihood associated with parameters under the working or natural space, respectively, and MLE determination for the CMP-HMM. Given the number of states $m$ along with the parameters $\boldsymbol{\lambda}$, $\boldsymbol{\nu}$, and $\boldsymbol{\Gamma}$ (and potentially $\boldsymbol{\delta}$, although this is defaulted as NULL), the CMP.HMM.pn2pw function transforms parameters from their natural to working state. The function performs log transformations on $\boldsymbol{\lambda}$, $\boldsymbol{\nu}$, and the off-diagonal elements of $\boldsymbol{\Gamma}/\text{diag}(\boldsymbol{\Gamma})$ to redefine the parameters from a bounded to an unconstrained space in order to circumvent any potentially associated computational issues that can

[2] R codes and data supplied in the MacDonald and Bhamani (2020) supplementary materials tab online at www.tandfonline.com/doi/suppl/10.1080/00031305.2018.1505656?scroll=top.

Table 7.3 R *functions to conduct statistical computing associated with CMP-hidden Markov modeling. Codes available online as supplementary material associated with MacDonald and Bhamani (2020).*

| Function | Purpose |
|---|---|
| `CMP.HMM.pn2pw` | Transform parameters from natural to working space |
| `CMP.HMM.pw2pn` | Transform parameters from working to natural space |
| `CMP.HMM.mllk` | Computes the negated working parameter log-likelihood |
| `CMP.HMM.mllk_np` | Computes the negated natural parameter log-likelihood |
| `CMP.HMM.mle` | Determines working parameter MLEs |
| `CMP.HMM.mle_np` | Determines natural parameter MLEs |

arise from working within the natural space. The `CMP.HMM.pw2pn` function meanwhile reverses these transformations, returning the parameters from their working, unconstrained space to their natural space and thus their true respective values.

The functions `CMP.HMM.mllk` and `CMP.HMM.mllk_np` compute the negated log-likelihood from the working and natural parameters, respectively. For both functions, the required inputs are the parameter vector of interest on the appropriate scale `parvect`, along with the data `x` and number of states `m`. `CMP.HMM.mle` and `CMP.HMM.mle_np` conduct parameter estimation via maximum likelihood to obtain the MLEs for the CMP-HMM of interest, where the initialized parameters are either working or natural parameters. Given the data `x`, number of states `m`, and starting values `lambda0`, `nu0`, and `gamma0`, these functions use the `nlm` function to optimize the associated likelihood function provided in `CMP.HMM.mllk` or `CMP.HMM.mllk_np`. Along with the MLEs $\hat{\boldsymbol{\lambda}}$, $\hat{\boldsymbol{\nu}}$, $\hat{\boldsymbol{\Gamma}}$, and $\hat{\boldsymbol{\delta}}$, the output further reports the corresponding Hessian matrix and the log-likelihood, AIC and BIC values. The Hessian matrix is used to determine the associated parameter estimate standard errors.

A well-studied dataset originally attributed to Fürth (1918) serves as a nice, illustrative example of this functionality; the data comprise 505 observations that count the number of pedestrians on a city block during five-second increments (Jung and Tremayne, 2006; Mills and Seneta, 1989). The observations range from 0 to 7 with a sample mean and variance equaling 1.59 and 1.51, respectively. This illustration assumes that a two-state CMM-HMM is a reasonable selection to represent the time series. The fitted two-state CMM-HMM has $\hat{\boldsymbol{\lambda}} = (0.886, 9.165)$, $\hat{\boldsymbol{\nu}} = (28.527, 2.400)$, and

$$\hat{\boldsymbol{\Gamma}} = \begin{pmatrix} 0.809 & 0.191 \\ 0.107 & 0.893 \end{pmatrix},$$

thus producing the stationary distribution, $\hat{\boldsymbol{\delta}} = (0.359, 0.641)$. The supplied code verifies that the sample mean and variance are 1.59 and 1.51, while the model mean and variance are approximately 1.5852 and 1.4630, respectively; both result pairs imply that the model addresses data underdispersion. Finally, the model autocorrelation function is $\rho(k) \approx 0.4754 \times 0.7017^k$. As an aside, it is worth noting that the underlying `nlm` function used to determine the MLEs associated with this model reported `code = 1`; thus, the relative gradient associated with the underlying log-likelihood is close to zero, and analysts can feel confident that the resulting output determines the MLEs.

## 7.2 Intensity Parameter Time Series Modeling

To date, two model approaches have been proposed for the flexible modeling of time series data where an underlying COM–Poisson distribution is assumed such that the serial dependence exists and is modeled via the intensity vector. Both approaches respectively operate assuming reparametrized versions of the CMP distribution (see Section 2.6 for distributional discussions) and propose different model structures for analysis.

### *7.2.1 ACMP-INGARCH*

Zhu (2012) uses the ACMP parametrization described in Section 2.6 to develop an integer-valued generalized autoregressive conditional heteroscedastic (INGARCH) model for time series count data. The ACMP-INGARCH($p, q$) model assumes that $X_t$, $t = 1, \ldots, T$, has an ACMP($\mu_{*t}$, $\nu$) distribution with

$$\mu_{*t} = \alpha_0 + \sum_{i=1}^{p} \alpha_i X_{t-i} + \sum_{j=1}^{q} \beta_j \mu_{*,t-j},$$

where $\alpha_0 > 0$, $\alpha_i \geq 0$ for $i = 1, \ldots, p$, and $\beta_j \geq 0$ for $j = 1, \ldots, q$. The model is approximately stationary when $\sum_{i=1}^{p} \alpha_i + \sum_{j=1}^{q} \beta_j < 1$. Special cases of the ACMP-INGARCH($p, q$) model include the ACMP-INARCH($p$) when $q = 0$ and the Poisson INGARCH($p, q$) when $\nu = 1$.

Assuming that the mean and variance of $X_t$ conditioned on previous outcomes can be approximated as Equations (2.49) and (2.50), the unconditional mean and variance for $X_t$ are approximately

$$E(X_t) \approx \frac{\alpha_0 - \left(1 - \sum_{j=1}^{q} \beta_j\right) \frac{\nu-1}{2\nu}}{1 - \sum_{i=1}^{p} \alpha_i - \sum_{j=1}^{q} \beta_j} \tag{7.15}$$

$$V(X_t) \approx \frac{E(\mu_{*t})}{\nu} + V(\mu_{*t}), \tag{7.16}$$

while the auto-covariance function is

$$\gamma_X(h) = \mathrm{Cov}(X_t, X_{t-h}) \approx \sum_{i=1}^{p} \alpha_i \gamma_X(|h-i|)$$
$$+ \sum_{j=1}^{\min(h-1,q)} \beta_j \gamma_X(h-j) + \sum_{j=1}^{q} \beta_j \gamma_{\mu_*}(j-h), \quad h \geq 1,$$

which reduces to $\gamma_X(h) \approx \sum_{i=1}^{p} \alpha_i \gamma_X(|h-i|)$ for the ACMP-INARCH($p$) model. Meanwhile, the special case of the ACMP-INGARCH(1, 1) model has the variance and autocorrelation,

$$V(X_t) \approx \frac{[1 - (\alpha_1 + \beta_1)^2 + \alpha_1^2]\left[E(X_t) + \frac{\nu-1}{2\nu}\right]}{\nu[1 - (\alpha_1 + \beta_1)^2]} \tag{7.17}$$

$$\rho_X(h) \approx \frac{\alpha_1(\alpha_1 + \beta_1)^{h-1}[1 - \beta_1(\alpha_1 + \beta_1)]}{1 - (\alpha_1 + \beta_1)^2 + \alpha_1^2}, \quad h \geq 1. \tag{7.18}$$

Simulation studies illustrate that $\nu > 1$ alone does not guarantee data under-dispersion in ACMP-INGARCH(1, 1) models; larger values of $\nu$ may be required to ensure under-dispersion as $\alpha_0, \alpha_1$, and $\beta_1$ increase. This may, however, demonstrate a repercussion of utilizing the ACMP parametrization. As discussed in Section 2.6, the approximations for this parametrization's mean and variance rely on the data either being over-dispersed (i.e. $\nu < 1$) or $\lambda$ being sufficiently large in relation to $\nu$ such that $\mu_* = \lambda^{1/\nu} \geq 10$. Recall that, while $\mu_*$ denotes a measure of center, it is not necessarily the mean. Accordingly, this approach should be considered with caution, e.g. when the mean and dispersion are both small (Chanialidis et al., 2017).

Parameter estimation can be conducted via the method of maximum likelihood where, given observations $x_1, \ldots, x_T$, the log-likelihood is

$$\ln L(\alpha_0, \alpha_1, \ldots, \alpha_p, \beta_1, \ldots, \beta_q, \nu;\ \boldsymbol{x})$$

$$= \nu \sum_{\max(p,q)}^{T} x_t \ln(\mu_{*t}) - \nu \sum_{\max(p,q)}^{T} \ln(x_t!) - \sum_{\max(p,q)}^{T} Z_1(\mu_{*t}, \nu)$$

and numerical methods are utilized to determine the MLEs $\hat{\alpha}_0$, $\hat{\alpha}_1, \ldots, \hat{\alpha}_p, \hat{\beta}_1, \ldots, \hat{\beta}_q, \hat{\nu}$. Zhu (2012) conducts maximum likelihood estimation in MATLAB via the constrained nonlinear optimization function `fmincon` where the negated log-likelihood function is supplied along with the constraints that $\alpha_0 > 0$, $\nu > 0$, and $\sum_{i=1}^{p} \alpha_i + \sum_{j=1}^{q} \beta_j < 1$. Further, the ACMP normalizing constant $Z_1(\mu_*, \nu)$ is approximated by Winsorizing the summation to the first 101 terms (i.e. from 0 to 100). These ideas, however, are easily transferable to `R`. Performing maximum likelihood estimation via constrained nonlinear optimization can be conducted, say, via the `optim` or `nlminb` functions with the same provided constraints. Meanwhile, the normalizing function can likewise be Winsorized as suggested. The `optim` function with `hessian = TRUE` will include the Hessian matrix among the provided output from which MLE standard errors can be determined. Alternatively (as suggested in Zhu (2012)), analysts can obtain the associated standard errors via the robust sandwich matrix; this too can be achieved in `R`, say, via the `sandwich` package (Zeileis et al., 2021).

### 7.2.2 MCMP1-ARMA

Melo and Alencar (2020) develop an MCMP1($\mu_t, \nu$)-parametrized COM–Poisson autoregressive moving average (MCMP1-ARMA) structure that assumes the mean $\mu_t$ has a temporal relationship involving autoregressive and moving average components. The MCMP1-ARMA($p, q$) model is a special case of the Benjamin et al. (2003) generalized autoregressive moving average (GARMA) construct and has the form

$$g(\mu_t) = \alpha + \boldsymbol{x}_t'\boldsymbol{\beta} + \sum_{i=1}^{p} \phi_i[g(y_{t-i}) - \boldsymbol{x}_{t-i}'\boldsymbol{\beta}] + \sum_{i=1}^{q} \theta_i[g(y_{t-i}) - g(\mu_{t-i})] \tag{7.19}$$

for the observations $y_t$ and the link function $g(\cdot)$, where $\alpha$ and $\boldsymbol{\beta} = (\beta_1, \ldots \beta_r)'$ are coefficients associating $\mu_t$ with the explanatory variables $\boldsymbol{x}_t = (x_{t1}, \ldots, x_{tr})'$, and $\boldsymbol{\phi} = (\phi_1, \ldots, \phi_p)'$ and $\boldsymbol{\theta} = (\theta_1, \ldots, \theta_q)'$ denote the autoregressive and moving average components, respectively. By definition, the MCMP1-ARMA model contains the GARMA models based on the Poisson ($\nu = 1$), geometric ($\nu = 0$), and Bernoulli ($\nu \to \infty$) distributions as special cases.

The serial dependence of the observed outcomes warrants using the method of conditional maximum likelihood to estimate $\alpha, \boldsymbol{\beta}', \boldsymbol{\phi}', \boldsymbol{\theta}'$, and $\nu$. Given $m \geq \max(p, q)$ observations, $\boldsymbol{y} = (y_1, \ldots, y_m)$, the conditional log-likelihood is

$$\begin{aligned} &\ln L(\alpha, \boldsymbol{\beta}', \boldsymbol{\phi}', \boldsymbol{\theta}', \nu) \\ &\quad = \sum_{t=m+1}^{n} y_t \ln(\lambda(\mu_t, \nu)) - \nu \sum_{t=m+1}^{n} \ln(y_t!) - \sum_{t=m+1}^{n} \ln Z(\lambda(\mu_t, \nu), \nu), \end{aligned} \tag{7.20}$$

from which the estimates $\hat{\alpha}, \hat{\boldsymbol{\beta}}', \hat{\boldsymbol{\phi}}', \hat{\boldsymbol{\theta}}'$, and $\hat{\nu}$ are obtained by solving the resulting conditional score equations. When the sample size is large and the appropriate regularity conditions hold, these conditional MLEs are consistent and asymptotically multivariate normal,

$$(\hat{\alpha}, \hat{\boldsymbol{\beta}}', \hat{\boldsymbol{\phi}}', \hat{\boldsymbol{\theta}}', \hat{\nu})' \sim N_{p+q+r+2}((\alpha, \boldsymbol{\beta}', \boldsymbol{\phi}', \boldsymbol{\theta}', \nu)', \boldsymbol{I}^{-1}),$$

where $\boldsymbol{I}$ denotes the Fisher information matrix. This distributional result aids in determining confidence intervals and performing appropriate hypothesis tests associated with the estimators. Simulation studies show that, as the sample size increases, the mean-squared error decreases, demonstrating that the consistency of the conditional MLEs improves with an increased sample size. The autoregressive (moving average) terms are underestimated (overestimated), however; thus, analysts should only include either autoregressive or moving average terms in an initial model (Melo and Alencar, 2020). Simulated and real-data examples further show that, when the data are over-dispersed, the MCMP1-ARMA performs as well as the NB-GARMA model (both showing optimal results), while the MCMP1-ARMA remains an optimal model when data are under-dispersed. Meanwhile, the hypothesis test used to detect statistically significant data dispersion performs effectively, properly rejecting the null hypothesis when significant over- or under-dispersion exists in the data.

## 7.3 Thinning-Based Models

Thinning-based models are an alternative approach to represent serially dependent count data. A benefit to this method is that it directly preserves the discrete nature of the count data; see Weiss (2008) for a comprehensive discussion regarding thinning-based models. This section discusses integer-valued autoregressive (INAR) and moving average (INMA) time series constructions that are derived from the CMP distribution and its related forms, and are based on thinning operations; see Table 7.4 for a section summary. Section 7.3.1 introduces univariate INAR models motivated by CMP and sCMP models, as well as a bivariate CMP autoregression. Section 7.3.2 meanwhile considers univariate INMA models developed assuming a CMP, MCMP2, or sCMP underpinning, along with a bivariate CMP moving average construct. The univariate or bivariate COM–Poisson models (under either CMP or MCMP2 parametrizations) assume aforementioned innovation distributions with binomial thinning operators, while the sCMP-motivated models assume sCMP-distributed innovations with gCMB thinning operations; see Chapter 2 (Sections 2.2 and 2.6 regarding the respective COM–Poisson parametrizations), and Chapter 3 (Sections 3.2 and 3.3, regarding the sCMP and gCMB distributions) for detailed discussion regarding these distributions. These models are special cases of the infinitely divisible convolution-closed class of discrete AR and MA models (Joe, 1996). While there does not currently exist a (univariate or multivariate) CMP- or sCMP-inspired integer-valued autoregressive moving average (INARMA) or integer-valued autoregressive integrated moving average

Table 7.4 *Univariate and multivariate thinning-based time series constructions involving COM–Poisson-motivated distributions. Khan and Jowaheer (2013) and Jowaheer et al. (2018) use a modified CMP notation (namely CMP$(\frac{\mu}{\nu}, \nu)$) that relies on the approximations for the mean $\mu = \lambda^{1/\nu} - \frac{\nu-1}{2\nu}$ and variance $\sigma^2 = \frac{1}{\nu}\lambda^{1/\nu}$. Considered estimation methods are either generalized quasi-likelihood (GQL) or maximum likelihood (ML).*

| Innovations distribution | Thinning operator | Estimation approach | References |
|---|---|---|---|
| CMP | Binomial | GQL | Khan and Jowaheer (2013), Jowaheer et al. (2018), Sunecher et al. (2020) |
| MCMP2 | Binomial | GQL | Sunecher et al. (2020) |
| sCMP | gCMB | ML | Sellers et al. (2020, 2021a) |

(INARIMA) model, the respective components discussed in Sections 7.3.1 and 7.3.2 can aid in developing such forms.

### 7.3.1 Autoregressive Models

INAR models are foundational in the discussion of time series analyses of discrete data and are typically derived via an appropriate thinning operator to maintain discrete observations and outcomes. Al-Osh and Alzaid (1987) introduced a first-order Poisson autoregressive (PAR(1)) model,

$$U_t = \gamma \circ U_{t-1} + \epsilon_t, \tag{7.21}$$

where $\epsilon_t$ is Poisson($\eta$) distributed, $\gamma \in [0,1]$, $U \in \mathbb{N}$ is a random variable, and $\circ$ is a binomial thinning operator such that $\gamma \circ U = \sum_{i=1}^{U} B_i$, where, independent of $U$, $B_i$ is a sequence of iid Bernoulli($\gamma$) random variables. The PAR(1) model has the resulting transition probability

$$P(U_t = u_t \mid U_{t-1} = u_{t-1}) = \sum_{s=0}^{\min(u_t, u_{t-1})} \binom{u_{t-1}}{s} \gamma^s (1-\gamma)^{u_{t-1}-s} \cdot \frac{e^{-\eta}\eta^{u_t - s}}{(u_t - s)!},$$
$$u_t = 0, 1, 2, \ldots \tag{7.22}$$

(Freeland and McCabe, 2004). Letting $\phi_\epsilon(s)$ denote the probability generating function (pgf) of $\epsilon$, $U_t$ has the pgf

$$\phi_{U_t}(s) = \phi_{U_0}(1 - \gamma^t + \gamma^t s) \prod_{k=0}^{t-1} \phi_\epsilon(1 - \gamma^k + \gamma^k s), \quad |s| \leq 1, \tag{7.23}$$

thus producing the measures

$$E(U_t) = \gamma^t E(U_0) + \eta \sum_{j=0}^{t-1} \gamma^j, \tag{7.24}$$

$$V(U_t) = \gamma^{2t} V(U_0) + (1-\gamma) \sum_{j=1}^{t} E(U_{t-j}) + \eta \sum_{j=1}^{t} \gamma^{2(j-1)}, \text{ and} \tag{7.25}$$

$$\text{Cov}(U_t, U_{t-k}) = \gamma^k V(U_t), \tag{7.26}$$

where $E(U_0) = \frac{\eta}{1-\gamma}$ and $V(U_0) = \frac{\eta(1+\gamma)}{1-\gamma^2}$. Jin-Guan and Yuan (1991) generalize the PAR(1) model to a PAR($k$) for general lag $k$.

Various models such as the first-order NB INAR (NB-INAR(1)) (Weiss, 2008) or those derived in Brännäs and Hellström (2001) that relax PAR(1) assumptions give rise to INAR models that can accommodate data over-dispersion; however, they are unable to address discrete time series data that contain data under-dispersion. The first-order generalized Poisson (GPAR(1)) process overcomes this matter, allowing for over- or under-dispersed count data. This model has the form

$$W_t = Q_t(W_{t-1}) + \epsilon_t, \qquad t = 1, 2, \ldots, \tag{7.27}$$

where $\{\epsilon_t\}$ is a series of iid generalized Poisson GP($q_*\lambda_*, \theta_*$) random variables, and $\{Q_t(\cdot) : t = 1, 2, \ldots\}$ is a sequence of independent quasi-binomial QB($p_*, \theta_*/\lambda_*, \cdot$) operators independent of $\{\epsilon_t\}$. This process is an ergodic Markov chain with transition probability

$$\begin{aligned} P(W_t = w_t | W_{t-1} = w_{t-1}) &= \sum_{k=0}^{\min(w_{t-1}, w_t)} \binom{w_{t-1}}{k} \frac{p_* q_* k}{\lambda_* + w_{t-1}\theta_*} \left( \frac{p_*\lambda_* + k\theta_*}{\lambda_* + w_{t-1}\theta_*} \right)^{k-1} \\ &\times \left( \frac{q_*\lambda_* + (w_{t-1} - k)\theta_*}{\lambda_* + w_{t-1}\theta_*} \right)^{w_{t-1}-k-1} \\ &\times \frac{\lambda_* q_* [\lambda_* q_* + \theta_*(w_t - k)]^{w_t - k - 1} e^{-\lambda_* q_* - \lambda_* \theta_* (w_t - k)}}{(w_t - k)!} \end{aligned}$$

and a unique GP($\lambda_*, \theta_*$) stationary distribution with mean and variance $E(W) = \lambda_*/(1 - \theta_*)$ and $V(W) = \lambda_*/(1 - \theta_*)^3$, respectively, and autocorrelation $\rho_W(k) = \text{Corr}(W_t, W_{t-k}) = p_*^{|k|}$, $k = 0, \pm 1, \pm 2, \ldots$ (Alzaid and Al-Osh, 1993). Its pgf is symmetric with the form

$$\phi_{W_{t+1}, W_t}(v, w) = \exp\left[\lambda_* q_* (A_{\theta_*}(v) + A_{\theta_*}(w) - 2) + \lambda_* p_* (A_{\theta_*}(vw) - 1)\right],$$

where $A_{\theta_*}(s)$ is the inverse function of $se^{-\theta_*(s-1)}$, implying that the GPAR(1) process is time reversible. While the GPAR(1) model allows for over- or under-dispersion, some GP distributions can be limited in their ability to properly adhere to probability axioms, e.g. when data are severely under-dispersed (Famoye, 1993).

These issues demonstrate the need for alternative, flexible models to accommodate serially dependent count time series data that express an autoregressive construct. This section introduces the current research for univariate and multivariate CMP-motivated INAR models. While the proposed multivariate INAR model is presented solely in the bivariate form, the approach (in theory) could be generalized for $d \geq 2$ dimensions.

*The CMPAR(1) Model*

Khan and Jowaheer (2013) maintain the CMP parametrization provided in Equation (2.8) yet denote it as $\mathrm{CMP}\left(\frac{\mu}{\nu}, \nu\right)$, assuming that the approximations for the CMP mean and variance (Equations (2.23) and (2.24)) hold and are exact. Given this representation, they maintain the INAR(1) model structure described in Equation (7.21) but assume COM–Poisson innovations represented as described above, thus deriving a first-order CMP-autoregressive (CMPAR(1)) model. Given that $U_t = \gamma \circ U_{t-1} + \epsilon_t$ such that $U_t$ is $\mathrm{CMP}\left(\frac{\mu}{\nu}, \nu\right)$ distributed, the innovations $\epsilon_t$ are $\mathrm{CMP}\left(\frac{\mu_t - \rho\mu_{t-1}}{\nu_*}, \nu_*\right)$ distributed where

$$\nu_* = \frac{(2\mu_t - 2\rho\mu_{t-1} + 1) + \sqrt{(2\mu_t - 2\rho\mu_{t-1} + 1)^2 - 8[(1-\rho^2)\frac{\nu-1}{2\nu^2} + \frac{\mu_t}{\nu} - \frac{\rho^2\mu_{t-1}}{\nu}\rho(1-\rho)\mu_{t-1}]}}{4[(1-\rho^2)\frac{\nu-1}{2\nu^2} + \frac{\mu_t}{\nu} - \frac{\rho^2\mu_{t-1}}{\nu}\rho(1-\rho)\mu_{t-1}]} > 0. \tag{7.28}$$

The nonstationary PAR(1) process is a special case of the CMPAR(1) when $\nu = 1$. The resulting serial correlation between two observations $U_t$ and $U_{t+h}$ is

$$\mathrm{Corr}(U_t, U_{t+h}) = \frac{\rho^h \sqrt{\frac{\mu_t}{\nu} + \frac{\nu-1}{2\nu^2}}}{\sqrt{\frac{\mu_{t+h}}{\nu} + \frac{\nu-1}{2\nu^2}}}. \tag{7.29}$$

Readers should proceed with caution in using this approach for data modeling because the underlying assumptions rely on the CMP approximations for the mean and variance. These approximations generally hold for $\nu \leq 1$ or $\lambda > 10^\nu$; however, the precision of these approximations is debated; see Section 2.8 for details.

*The SCMPAR(1) Model*

Sellers et al. (2020) developed the first-order sCMP-autoregressive (SCMPAR(1)) process as

$$X_t = C_t(X_{t-1}) + \epsilon_t, \qquad t = 1, 2, \ldots, \tag{7.30}$$

where $\epsilon_t$ has an $\mathrm{sCMP}(\lambda, \nu, n_2)$ distribution, and $\{C_t(\bullet): \ t = 1, 2, \ldots\}$ is a sequence of independent generalized COM-Binomial $\mathrm{gCMB}\left(\frac{1}{2}, \nu, \bullet, n_1, n_2\right)$ operators independent of $\{\epsilon_t\}$; see Chapter 3 for a detailed discussion of the sCMP and gCMB distributions. This model satisfies the ergodic Markov property with the transition probability

$$P(X_t = x_t | X_{t-1} = x_{t-1})$$
$$= \sum_{k=0}^{\min(x_t, x_{t-1})} \frac{\binom{x_{t-1}}{k}^{\nu} \left[ \sum_{\substack{a_1,\ldots,a_{n_1}=0 \\ a_1+\cdots+a_{n_1}=k}}^{k} \binom{k}{a_1,\ldots,a_{n_1}}^{\nu} \right] \left[ \sum_{\substack{b_1,\ldots,b_{n_2}=0 \\ b_1+\cdots+b_{n_2}=x_{t-1}-k}}^{x_{t-1}-k} \binom{x_{t-1}-k}{b_1,\ldots,b_{n_2}}^{\nu} \right]}{\sum_{\substack{c_1,\ldots,c_{n_1+n_2}=0 \\ c_1+\cdots+c_{n_1+n_2}=x_{t-1}}}^{x_{t-1}} \binom{x_{t-1}}{c_1 \cdots c_{n_1+n_2}}^{\nu}}$$
$$\times \frac{\lambda^{x_t-k}}{[(x_t-k)!]^{\nu} Z^{n_2}(\lambda, \nu)} \sum_{\substack{d_1,\ldots,d_{n_2}=0 \\ d_1+\cdots+d_{n_2}=x_{t-1}-k}}^{x_t-k} \binom{x_t-k}{d_1,\ldots,d_{n_2}}^{\nu}, \qquad (7.31)$$

implying that the sCMP($\lambda, \nu, n_1 + n_2$) distribution is the unique stationary distribution. Meanwhile, the SCMPAR(1) model is time reversible with the symmetric joint pgf

$$\phi_{X_{t+1},X_t}(u, l) = \frac{(Z(\lambda u, \nu) Z(\lambda l, \nu))^{n_2} Z^{n_1}(\lambda u l, \nu)}{Z^{n_1+2n_2}(\lambda, \nu)} \qquad (7.32)$$

with the general autocorrelation function $\rho_k = \text{Corr}(X_t, X_{t-k}) = \left(\frac{n_1}{n_1+n_2}\right)^k$ for $k = 0, 1, 2, \ldots$. The SCMPAR(1) contains the Al-Osh and Alzaid (1987) PAR(1) ($\nu = 1$), the Al-Osh and Alzaid (1991) Binomial INAR(1) model with a hypergeometric thinning operator ($\nu \to \infty$), and an INAR(1) model with NB marginals and negative hypergeometric thinning operator ($\nu = 0$ and $\lambda < 1$) as special cases.

The conditional maximum likelihood estimation method is used to fit the SCMPAR(1) to integer count data. For proposed discrete values $n_1$ and $n_2$, analysts can numerically maximize $\sum_{i=1}^{N} \ln P(X_i | X_{i-1})$ (where $P(X_t | X_{t-1})$ is given in Equation (7.31)) in `R` (R Core Team, 2014); the `multicool` package (Curran et al., 2015) is used to calculate combinatorial terms, while the `optim` function in the `stats` package determines the MLEs and associated standard errors. Discrete values for $n_1$ and $n_2$ are presumed to be sufficiently effective under this approach; real values for $n_1$ and $n_2$ are not believed to offer significant improvements (Sellers et al., 2020).

### *Bivariate COM–Poisson Autoregressive Model*

Jowaheer et al. (2018) develop a nonstationary bivariate INAR(1) time series model with COM–Poisson-distributed marginal distributions, i.e. a first-order bivariate CMP-autoregressive (BCMPAR(1)) model of the form

$$U_{1t} = \gamma_1 \circ U_{1,t-1} + \epsilon_{1t} \tag{7.33}$$

$$U_{2t} = \gamma_2 \circ U_{2,t-1} + \epsilon_{2t}, \tag{7.34}$$

where $\gamma_i \in (0, 1)$, $i = 1, 2$, and $\boldsymbol{\epsilon}_t = (\epsilon_{1t}, \epsilon_{2t})$ has a bivariate CMP form with $\text{CMP}\left(\frac{\mu_{it} - \gamma_i \mu_{i,t-1}}{\nu_{i*}}, \nu_{i*}\right)$-distributed marginal distributions, and $\text{Corr}(\epsilon_{1t}, \epsilon_{2t'}) = \rho_{12,t}$ for $t = t'$; $0 < \rho_{12,t} < 1$; $\nu_{i*}$ denotes the respective $\nu_*$ calculations (Equation (7.28)) for each dimension $i = 1, 2$. Analogous to the CMPAR(1) model, additional assumptions include independence for $U_{i,t-h}$ and $\epsilon_{it}$ for $h \geq 1$, and

$$\text{Cov}(U_{i,t}, \epsilon_{jt}) = \begin{cases} V(\epsilon_{it}) & i = j \\ \text{Cov}(\epsilon_{it}, \epsilon_{jt}) & i \neq j \end{cases},$$

thus producing $\boldsymbol{U}_t = (U_{1t}, U_{2t})$ with $\text{CMP}\left(\frac{\mu_{it}}{\nu_i}, \nu_i\right)$ marginal distributions for $U_{it}$, $i = 1, 2$, and the lag $h$ correlation for the $i$th series is analogous to that defined in Equation (7.29) for each dimension. The lag $h$ correlation within each of the two respective series $i \in \{1, 2\}$ is

$$\text{Corr}(U_{it}, U_{i,t+h}) = \frac{\gamma_i^h \sqrt{\frac{\mu_{it}}{\nu_i} + \frac{\nu_i - 1}{2\nu_i^2}}}{\sqrt{\frac{\mu_{i,t+h}}{\nu_i} + \frac{\nu_i - 1}{2\nu_i^2}}}, \tag{7.35}$$

while the lag $h$ covariance between observations across the two series is

$$\text{Cov}(U_{it}, U_{j,t+h}) = \gamma_j^h \text{Cov}(U_{1t}, U_{2t}) \quad \text{for } i \neq j \in \{1, 2\}, \tag{7.36}$$

where

$$\text{Cov}(U_{1t}, U_{2t}) = \gamma_1 \gamma_2 \text{Cov}(U_{1,t-1}, U_{2,t-1}) + \rho_{12,t} \sqrt{\frac{\mu_{1t} - \gamma_1 \mu_{1,t-1}}{\nu_{1*}} + \frac{\nu_{1*} - 1}{2\nu_{1*}^2}} \times \sqrt{\frac{\mu_{2t} - \gamma_2 \mu_{2,t-1}}{\nu_{2*}} + \frac{\nu_{2*} - 1}{2\nu_{2*}^2}}.$$

This flexible construct includes the Khan et al. (2016) nonstationary first-order bivariate PAR model as a special case when $\nu_1 = \nu_2 = 1$. Jowaheer et al. (2018) state that they use the bivariate CMP distribution derived via the compounding method (BCMPcomp; see Section 4.2) for their BCMPAR(1) model; however, this conflicts with the desired/claimed marginal CMP representation for $\epsilon_{it}$ and $U_{it}$ noted above. The BCMP-comp distribution does not produce univariate CMP marginal distributions

(Sellers et al., 2016). One can circumvent this matter, for example, however, by instead developing a bivariate CMP distribution via the Sarmanov family or copulas in order to achieve the desired CMP marginal distributions; see Chapter 4 for discussion.

Parameter estimation is conducted via the generalized quasi-likelihood (GQL) approach with the expression $\boldsymbol{D}'\boldsymbol{\Sigma}^{-1}(\boldsymbol{U} - \boldsymbol{\mu}) = \boldsymbol{0}$, where $\boldsymbol{D}$ is a $2T \times 2(p+1)$ block diagonal matrix whose respective blocks contain the partial derivatives of the approximated mean for each of the two series ($i = 1, 2$) at each time $t = 1, \ldots, T$ with respect to each of the $p + 1$ coefficients, $\beta_1, \ldots, \beta_p$ and $\nu_i$; $\boldsymbol{\Sigma}$ is a $2T \times 2T$ matrix comprised of the covariance components (Equation (7.36)), and $\boldsymbol{U} = (\boldsymbol{U}_1, \boldsymbol{U}_2)$ and $\boldsymbol{\mu} = (\boldsymbol{\mu}_1, \boldsymbol{\mu}_2)$ are the vectors of length $2T$ that respectively represent the variable and associated mean of the CMP$\left(\frac{\mu_{it}}{\nu_i}, \nu_i\right)$-distributed random variables, $\boldsymbol{U}_i = (U_{i1}, \ldots, U_{iT})$, $i = 1, 2$. Parameter estimation is thus achieved via an iterative procedure

$$\hat{\boldsymbol{\theta}}_{r+1} = \hat{\boldsymbol{\theta}}_r + [\boldsymbol{D}'\boldsymbol{\Sigma}^{-1}\boldsymbol{D}]_r^{-1}[\boldsymbol{D}'\boldsymbol{\Sigma}^{-1}(\boldsymbol{U} - \boldsymbol{\mu})]_r,$$

where $\hat{\boldsymbol{\theta}}_r$ is the $r$th iteration of $\hat{\boldsymbol{\theta}} = (\hat{\boldsymbol{\beta}}_1, \hat{\boldsymbol{\beta}}_2, \hat{\nu}_1, \hat{\nu}_2)$ that estimates $\boldsymbol{\theta} = (\boldsymbol{\beta}_1, \boldsymbol{\beta}_2, \nu_1, \nu_2)$. More precisely, $\hat{\boldsymbol{\theta}} - \boldsymbol{\theta}$ is asymptotically normally distributed with mean $\boldsymbol{0}$ and variance–covariance matrix

$$[\boldsymbol{D}'\boldsymbol{\Sigma}^{-1}\boldsymbol{D}]^{-1}[\boldsymbol{D}'\boldsymbol{\Sigma}^{-1}(\boldsymbol{Y} - \boldsymbol{\mu})(\boldsymbol{Y} - \boldsymbol{\mu})'\boldsymbol{\Sigma}^{-1}\boldsymbol{D}][\boldsymbol{D}'\boldsymbol{\Sigma}^{-1}\boldsymbol{D}]^{-1}$$

(Jowaheer et al., 2018).

### 7.3.2 Moving Average Models

Integer-valued moving average (INMA) processes have been considered as a discrete analog to the traditional Gaussian moving average (MA) model for continuous data. Similar to the INAR time series, INMA processes rely on thinning operators to maintain discreteness among subsequent observations. For example, Al-Osh and Alzaid (1988) introduced a first-order Poisson moving average (PMA(1)) of the form

$$U_t = \gamma \circ \epsilon_{t-1} + \epsilon_t, \tag{7.37}$$

where $\{\epsilon_t\}$ and $\gamma$ are defined as in Equation (7.21) and $\circ$ remains the binomial thinning operator defined in Section 7.3.1. The pgf of $U_t$ is

$\Phi_{U_t}(u) = e^{-\eta(1+\gamma)(1-u)}$, signifying that its mean and variance are both $(1+\gamma)\eta$. Meanwhile, the correlation is

$$\rho_U(r) = \text{Corr}(U_{t-r}, U_t) = \begin{cases} \frac{\gamma}{1+\gamma} & r = 1 \\ 0 & r > 1 \end{cases}$$

and the joint pgf of $\{U_1, \ldots, U_r\}$ is

$$\Phi_r(u_1, \ldots, u_r) = \exp\left(-\eta[r+\gamma-(1-\gamma)\sum_{i=1}^{r} u_i - \gamma(u_1+u_r) - \gamma\sum_{i=1}^{r-1} u_i u_{i+1}]\right),$$

implying that the PMA process is time reversible. Finally, letting $T_{U,r} = \sum_{i=1}^{r} U_i$ denote the total PMA(1) counts occurring over lag period $r$, its pgf is

$$\Phi_{T_{U,r}}(u) = \exp\left(-\eta[(1-\gamma)r + 2\gamma](1-u) - \eta\gamma(r-1)(1-u^2)\right).$$

This result is significant because, while the MA(1) analog maintains a Gaussian distributional form for $T$, $\Phi_{T_{U,r}}(u)$ does not maintain the pgf structure of a Poisson random variable; hence, the same is not true for the INMA(1) model (Al-Osh and Alzaid, 1988).

The PMA (like the PAR) process assumes an underlying equi-dispersion property; however, real data can contain some form of data dispersion relative to the Poisson model. Thus, it is important to consider a more flexible time series model that can accommodate data dispersion. One such option is the first-order generalized Poisson moving average (GPMA(1)) process whose model form is

$$W_t = Q_t^*(\epsilon_{t-1}^*) + \epsilon_t^*, \quad t = 0, 1, 2, \ldots, \tag{7.38}$$

where $\{\epsilon_t^*\}$ is a sequence of iid GP$(\mu^*, \theta)$ random variables, and $\{Q_t^*(\cdot)\}$ is a sequence of QB$(p^*, \theta/\mu^*, \cdot)$ random operators independent of $\{\epsilon_t^*\}$. The GPMA(1) structure implies that $W_t$ is GP$((1+p^*)\mu^*, \theta)$ distributed with autocorrelation

$$\rho_W(r) = \text{Corr}(W_t, W_{t+r}) = \begin{cases} \frac{p^*}{1+p^*} \in [0, 0.5] & |r| = 1 \\ 0 & |r| > 1. \end{cases}$$

The sequence $(W_t, W_{t-1}, \ldots, W_{t-r+1})$ meanwhile has the joint pgf

$$\Phi(u_1, \ldots, u_r) = \exp\left[\mu q \sum_{i=1}^{r} (A_\theta(u_i) - 1) + \mu p \sum_{i=1}^{r} (A_\theta(u_i u_{i+1}) - 1)\right] \tag{7.39}$$

which implies that the GPMA(1) is time reversible. Finally, the total GPMA(1) counts occurring during time lag $r$ (i.e. $T_{W,r} = \sum_{i=1}^{r} W_{t-r+i}$) has the pgf

$$\Phi_{T_{W,r}}(u) = \exp\left[\mu q r(A_\theta(u) - 1) + \mu p(r-1)(A_\theta(u^2) - 1)\right].$$

While the GPMA allows for over- or under-dispersion, potential scenarios can occur where the underlying GP structure is not viable (Famoye, 1993). This section introduces alternative INMA models motivated by the CMP distribution to address serially dependent count data.

### *INMA(1) Models with COM–Poisson Innovations*

Sunecher et al. (2020) develop an INMA(1) model

$$V_t = \gamma \circ \eta_{t-1} + \eta_t, \quad t = 1, \ldots, T, \tag{7.40}$$

where $\{\eta_t\}$ is MCMP2($\mu_t, \nu = \exp(\phi)$) distributed[3] as defined in Section 2.6 with $\lambda_t = \exp(\boldsymbol{x}_t'\boldsymbol{\beta})$ for covariates $\boldsymbol{x}_t = (x_{t1}, \ldots, x_{tp})'$ and coefficients $\boldsymbol{\beta} = (\beta_1, \ldots, \beta_p)$. The time-dependent covariates induce the nonstationary correlation structure, while $\gamma$ and $\circ$ induce the binomial thinning operation described in Section 7.3.1, such that $\gamma \circ \eta_{t-1} \mid \eta_{t-1}$ has a binomial distribution with $\eta_{t-1}$ trials and success probability $\gamma$. Accordingly, this first-order mean-parametrized (of the second type) CMP moving average (MCMP2MA(1)) model has

$$E(V_t) \approx \gamma \mu_{t-1} + \mu_t \tag{7.41}$$

$$V(V_t) \approx \gamma(1-\gamma)\mu_{t-1} + \frac{\gamma^2\mu_{t-1} + \mu_t}{\nu} + (\gamma^2 + 1)\frac{\nu - 1}{2\nu^2} \tag{7.42}$$

$$\text{Cov}(V_t, V_{t+h}) \approx \begin{cases} \frac{\gamma[2\nu\mu_t + \nu - 1]}{2\nu^2} & h = 1 \\ 0 & h > 1 \end{cases}, \tag{7.43}$$

where, under stationarity, Equations (7.41) and (7.42) simplify to

$$E(V_t) \approx (1+\gamma)\mu_\bullet$$

$$V(V_t) \approx \frac{\mu_\bullet}{\nu}[1 + \gamma^2(1-\nu) + \gamma\nu] + \frac{(1+\gamma^2)(\nu - 1)}{2\nu^2}.$$

[3] This distribution can be viewed as MCMP2($\mu, \phi$), but Sunecher et al. (2020) maintain using the $\nu$ parameter instead of $\phi = \ln(\nu)$; see Section 2.6 for details regarding the MCMP2 distribution.

Mamode Khan et al. (2018) meanwhile develop a longitudinal CMPMA(1) of the form

$$V_{it} = \gamma \circ \eta_{i,t-1} + \eta_{it}, \tag{7.44}$$

for $i = 1, \ldots, I$ and $t = 1, \ldots, T$, where $\{\eta_{it}\}$ is CMP($\lambda_{it}, \nu$) distributed with $\lambda_{it} = \exp(\boldsymbol{X}_t \boldsymbol{\beta})$ for covariates $\boldsymbol{X}_t = (X_{t1}, \ldots, X_{tp})'$ and coefficients $\boldsymbol{\beta} = (\beta_1, \ldots, \beta_p)$; the time-dependent covariates induce the nonstationary correlation structure, while $\gamma$ and $\circ$ remain the binomial thinning operation described in Equation (7.21). In this longitudinal framework, subject observations (i.e. $V_{it}$ and $V_{jt}$ for $i \neq j$) are uncorrelated. Other properties stemming from this moving average include that the innovations are uncorrelated (i.e. $\eta_{it}$ and $\eta_{i,t+h}$ are uncorrelated for $h \neq 0$); $V_{it}$ and $\eta_{i,t+h}$ are independent for $h \neq 0$; and the mean, variance, and covariance results from the binomial thinning operation are, for $i = 1, \ldots, I, t = 1, \ldots, T$,

1. $E(\gamma \circ \eta_{it}) = \gamma E(\eta_{it})$
2. $V(\gamma \circ \eta_{it}) = \gamma(1-\gamma)E(\eta_{it}) + \gamma^2 V(\eta_{it})$
3. $\text{Cov}(\gamma \circ \eta_{it_1}, \eta_{it_2}) = \gamma \text{Cov}(\eta_{it_1}, \eta_{it_2})$, where $t_1, t_2 = 1, \ldots, T$; $t_1 \neq t_2$.

Given that the CMP mean and variance approximations (Equations (2.23) and (2.24)) hold, and $\lambda_{it} = \lambda_{i\bullet}$ (i.e. stationarity holds),

$$E(V_{it}) = (1+\gamma)\left[\lambda_{i\bullet}^{1/\nu} - \left(\frac{\nu-1}{2\nu}\right)\right] \tag{7.45}$$

$$V(V_{it}) = (1+\gamma^2)\left(\frac{\lambda_{i\bullet}^{1/\nu}}{\nu}\right) + \gamma(1-\gamma)\left[\lambda_{i\bullet}^{1/\nu} - \left(\frac{\nu-1}{2\nu}\right)\right] \tag{7.46}$$

$$\text{Cov}(V_{it}, V_{i,t+1}) = \frac{\gamma \lambda_{i\bullet}^{1/\nu}}{\nu}. \tag{7.47}$$

However, Equations (7.45)–(7.47) assume that $\nu \leq 1$ or $\lambda > 10^\nu$ (Minka et al., 2003; Shmueli et al., 2005).

Mamode Khan et al. (2018) conduct parameter estimation via the method of GQL, where the loglinear association $\ln(\lambda_{it}) = \boldsymbol{x}_{it}'\boldsymbol{\beta}$ is assumed for the intensity vector and the dispersion parameter $\nu$ is assumed constant. This approach uses the mean score vector and exact INMA(1) auto-covariance. The GQL estimation requires solving the equation

$$\sum_{i=1}^{I} \boldsymbol{D}_i' \boldsymbol{\Sigma}_i^{-1}(\boldsymbol{y}_i - \boldsymbol{\mu}_i) = \boldsymbol{0}, \tag{7.48}$$

where $\boldsymbol{D}_i'$ is a $(p+2) \times T$ matrix of partial derivatives of the mean response from Subject $i$ with respect to the parameters of interest,

$$\frac{\partial \mu_{it}}{\partial \boldsymbol{\beta}_j} = \frac{\lambda_{it}^{1/\nu} x_{i,t,j} + \gamma \lambda_{i,t-1}^{1/\nu} x_{i,t-1,j}}{\nu} \tag{7.49}$$

$$\frac{\partial \mu_{it}}{\partial \nu} = \lambda_{i,t-1}^{1/\nu} - \frac{\nu - 1}{2\nu} \tag{7.50}$$

$$\frac{\partial \mu_{it}}{\partial \gamma} = \frac{\lambda_{it}^{1/\nu} \ln(\lambda_{it}^{1/\nu})}{\nu} - \gamma \lambda_{i,t-1}^{1/\nu} \ln(\lambda_{i,t-1}^{1/\nu}); \tag{7.51}$$

$\boldsymbol{\Sigma}_i$ is the $T \times T$ symmetric, tridiagonal auto-covariance matrix for subject $i$ with diagonal and off-diagonal terms

$$\sigma_{i,(t,t)} = \gamma(1-\gamma)\left(\lambda_{i,t-1}^{1/\nu} - \frac{\nu - 1}{2\nu}\right) + \frac{\gamma^2 \lambda_{i,t-1}^{1/\nu} + \lambda_{it}^{1/\nu}}{\nu} \tag{7.52}$$

$$\sigma_{i,(t,t+1)} = \sigma_{i,(t+1,t)} = \frac{\gamma \lambda_{it}^{1/\nu}}{\nu} \tag{7.53}$$

$$\sigma_{i,(t,t+h)} = \sigma_{i,(t+h,t)} = 0 \quad \text{for } h > 1. \tag{7.54}$$

Equation (7.48) is solved via the Newton–Raphson method where, for the parameter vector $\boldsymbol{\theta} = (\boldsymbol{\beta}, \nu, \gamma)'$,

$$\hat{\boldsymbol{\theta}}_{r+1} = \hat{\boldsymbol{\theta}}_r + \left(\sum_{i=1}^{I} \boldsymbol{D}_i' \boldsymbol{\Sigma}_i^{-1} \boldsymbol{D}_i\right)_r^{-1} \left(\sum_{i=1}^{I} \boldsymbol{D}_i' \boldsymbol{\Sigma}_i^{-1} (\boldsymbol{y}_i - \boldsymbol{\mu}_i)\right)_r,$$

and the GQL estimator $\tilde{\boldsymbol{\theta}}$ is determined when the difference between consecutive iterations is sufficiently small. This estimator is asymptotically normal with mean $\boldsymbol{\theta}$ and variance–covariance matrix $\sum_{i=1}^{I} \boldsymbol{D}_i' [\boldsymbol{\Sigma}_i(\boldsymbol{\theta})]^{-1} \boldsymbol{D}_i$. Alternatively, parameter estimation can be conducted via the generalized method of moments (GMM) that requires the empirical auto-covariance structure. Estimation is likewise achieved via an analogous Newton-Rhapson iterative scheme, and the resulting estimators are consistent and asymptotically normal; see Mamode Khan et al. (2018) for details. Simulation studies found that, under either protocol, the mean estimates of the model parameters are consistent with their respective true values, and the associated standard errors decrease as the sample size increases. The GQL outperforms the GMM, however, producing better estimates with smaller standard errors.

*The SCMPMA(1) Model*

Sellers et al. (2021a) introduce a first-order sCMP moving average (SCMPMA(1)) process as an alternative construct to model count time series data expressing data dispersion. This process has the form

$$X_t = C_t^*(\epsilon_{t-1}^*) + \epsilon_t^*, \qquad t = 1, 2, \ldots, \tag{7.55}$$

where $\epsilon_t^*$ is a sequence of iid sCMP($\lambda, \nu, m_1 + m_2$) random variables and, independent of $\{\epsilon_t^*\}$, $C_t^*(\,\bullet\,)$ is a sequence of independent gCMB($1/2, \nu, \bullet, m_1, m_2$) operators; see Chapter 3 for details regarding the sCMP and gCMB distributions. Special cases of the SCMPMA(1) include the Al-Osh and Alzaid (1988) PMA(1) ($\nu = 1$), as well as versions of an NB ($\nu = 0, \lambda < 1$) and binomial ($\nu \to \infty$) INMA(1), respectively. The SCMPMA(1) is a stationary but not Markovian process with sCMP($\lambda, \nu, 2m_1 + m_2$) marginals, where $X_{t+r}$ and $X_t$ are independent for $|r| > 1$, and the autocorrelation between consecutive variables $X_t$ and $X_{t+1}$ is

$$0 \leq \rho_1 = \frac{m_1}{2m_1 + m_2} \leq 0.5, \tag{7.56}$$

assuming $m_1, m_2 \in \mathbb{N}$. The SCMPMA(1) joint pgf is

$$\phi_{X_{t+1},X_t}(u, l) = \frac{(Z(\lambda u, \nu)Z(\lambda l, \nu))^{m_1+m_2}(Z(\lambda ul, \nu))^{m_1}}{(Z(\lambda, \nu))^{3m_1+2m_2}}; \tag{7.57}$$

thus, the SCMPMA(1) is likewise time reversible, and the SCMPMA(1) and SCMPAR(1) processes are similar when $m_1 = n_1 = n_2 - m_2$; see Section 7.3.1 for more information regarding SCMPAR(1) processes.

A conditional estimation procedure (e.g. least squares) for any INMA process is difficult "because of the thinning operators, unless randomization is used" (Brännäs and Hall, 2001). Sellers et al. (2021a) propose an ad hoc profile likelihood procedure where they first determine values $m_1, m_2 \in \mathbb{N}$ for which the observed correlation $\rho_1 \approx \frac{m_1}{2m_1+m_2}$ is satisfied and use these values with an assumed sCMP($\lambda, \nu, 2m_1 + m_2$) likelihood function to ascertain $\hat{\lambda}$ and $\hat{\nu}$, while the corresponding standard errors are obtained via the Fisher information matrix or via nonparametric bootstrapping. Given the resulting combinations of $\{m_1, m_2, \hat{\lambda}, \hat{\nu}\}$, the collection that maximizes the likelihood for a given dataset is identified.

*Bivariate MCMP2MA(1) Model*

Sunecher et al. (2020) establish a bivariate analog to the MCMP2MA(1) model described above that has marginal CMP distributions. This first-order bivariate mean-parametrized (of the second type) CMP moving average (BMCMP2MA(1)) model has the forms

$$V_{1t} = \gamma_1 \circ \eta_{1,t-1} + \eta_{1t} \quad \text{and} \tag{7.58}$$

$$V_{2t} = \gamma_2 \circ \eta_{2,t-1} + \eta_{2t}, \tag{7.59}$$

where $\{\eta_{kt}\}$ is CMP$(\mu_{kt}, \nu_k)$ distributed with $\mu_{kt} = \lambda_{kt}^{1/\nu_k} - \frac{\nu_k - 1}{2\nu_k}$, $k = 1, 2$, and $\gamma$ and $\circ$ induce the binomial thinning operation. This construct includes a bivariate PMA(1) as a special case when $\nu_1 = \nu_2 = 1$, and the model form induces the moments

$$E(V_{kt}) \approx \gamma_k \mu_{k,t-1} + \mu_{kt} \tag{7.60}$$

$$V(V_{kt}) \approx \gamma_k(1-\gamma_k)\mu_{k,t-1} + \frac{\gamma^2 \mu_{k,t-1} + \mu_{kt}}{\nu_k} + \frac{(1+\gamma^2)(\nu_k - 1)}{2\nu_k^2} \tag{7.61}$$

$$\text{Cov}(V_{kt}, V_{k,t+h}) \approx \begin{cases} \frac{\gamma_k[2\nu_k\mu_{kt} + \nu_k - 1]}{2\nu_k^2} & h = 1 \\ 0 & h > 1 \end{cases} \tag{7.62}$$

$$\text{Corr}(\eta_{1t}, \eta_{2t'}) = \begin{cases} \rho_{12,t} & t = t' \\ 0 & t \neq t'. \end{cases} \tag{7.63}$$

These results further imply that the cross-series covariances are

$$\text{Cov}(V_{1t}, V_{2t}) = \gamma_1\gamma_2 \text{Cov}(\eta_{1,t-1}\eta_{2,t-1}) + \text{Cov}(\eta_{1t}\eta_{2t}) \tag{7.64}$$

$$\approx \gamma_1\gamma_2\rho_{12,t-1}\sqrt{\frac{2\nu_1\mu_{1,t-1} + \nu_1 - 1}{2\nu_1^2}}\sqrt{\frac{2\nu_2\mu_{2,t-1} + \nu_2 - 1}{2\nu_2^2}} + \rho_{12,t}\sqrt{\frac{2\nu_1\mu_{1t} + \nu_1 - 1}{2\nu_1^2}}\sqrt{\frac{2\nu_2\mu_{2t} + \nu_2 - 1}{2\nu_2^2}} \tag{7.65}$$

$$\text{Cov}(V_{i,t+h}, V_{jt}) \approx \begin{cases} \gamma_i \rho_{12,t}\sqrt{\frac{2\nu_1\mu_{1t} + \nu_1 - 1}{2\nu_1^2}}\sqrt{\frac{2\nu_2\mu_{2t} + \nu_2 - 1}{2\nu_2^2}} & h = 1 \\ 0 & h > 1 \end{cases} \tag{7.66}$$

for $i, j \in \{1, 2\}$, $i \neq j$.

Conducting parameter estimation via a likelihood-based approach is computationally expensive; thus, Sunecher et al. (2020) instead pursue a modified GQL approach solving $\boldsymbol{D}'\boldsymbol{\Sigma}^{-1}(\boldsymbol{V} - \boldsymbol{\mu}) = \boldsymbol{0}$, where $\boldsymbol{V} = (\boldsymbol{V}_1, \boldsymbol{V}_2) = (V_{11}, \ldots, V_{1T}, V_{21}, \ldots, V_{2T})$ and $\boldsymbol{\mu}$ are the vectors of length $2T$

with respective components $V_{kt}$ and $E(V_{kt})$, defined in Equation (7.60); $\boldsymbol{D}$ is a $2T \times 2(p+2)$ block diagonal matrix of components

$$\frac{\partial \boldsymbol{\mu}_k}{\partial \boldsymbol{\beta} \partial \gamma \partial \nu} = \begin{pmatrix} \frac{\partial \mu_{k1}}{\partial \beta_{k1}} & \cdots & \frac{\partial \mu_{k1}}{\partial \beta_{kp}} & \frac{\partial \mu_{k1}}{\partial \gamma_k} & \frac{\partial \mu_{k1}}{\partial \nu_k} \\ \mathrm{d}\,\frac{\partial \mu_{k2}}{\partial \beta_{k1}} & \cdots & \frac{\partial \mu_{k2}}{\partial \beta_{kp}} & \frac{\partial \mu_{k2}}{\partial \gamma_k} & \frac{\partial \mu_{k2}}{\partial \nu_k} \\ \vdots & \ddots & \vdots & \vdots & \vdots \\ \frac{\partial \mu_{kT}}{\partial \beta_{k1}} & \cdots & \frac{\partial \mu_{kT}}{\partial \beta_{kp}} & \frac{\partial \mu_{kT}}{\partial \gamma_k} & \frac{\partial \mu_{kT}}{\partial \nu_k} \end{pmatrix}, \quad k = 1, 2,$$

and

$$\boldsymbol{\Sigma} = \begin{pmatrix} V(\boldsymbol{V}_1) & \mathrm{Cov}(\boldsymbol{V}_1, \boldsymbol{V}_2) \\ \mathrm{Cov}(\boldsymbol{V}_2, \boldsymbol{V}_1) & V(\boldsymbol{V}_2) \end{pmatrix}$$

is the variance–covariance matrix whose components are defined in Equations (7.61)–(7.66). The GQL iteratively updates via the Newton–Raphson method until convergence is reached to determine the estimates for the parameter vector $\boldsymbol{\theta} = (\boldsymbol{\beta}, \gamma_1, \gamma_2, \nu_1, \nu_2)'$. The estimates $\hat{\boldsymbol{\theta}}$ are consistent and asymptotically normal, assuming that regulatory conditions hold.

## 7.4 CMP Spatio-temporal Models

A popular approach to model spatial and spatio-temporal count data structures is to assume an underlying Poisson distribution (e.g. Jackson and Sellers, 2008; Jackson et al., 2010, Waller et al., 1997). The underlying Poisson assumption, however, can be very constricting because of its equi-dispersion constraint where real data rarely demonstrate such a form. Suggested solutions typically address various forms of data over-dispersion (e.g. Arab, 2015; De Oliveira, 2013); yet little attention has been paid on data under-dispersion in a spatial or spatio-temporal setting. For greater flexibility, Wu et al. (2013) develop a hierarchical Bayesian spatio-temporal model based on the CMP parametrization (Equation (2.8)) that allows for dynamic intensity and dispersion.

Let $\boldsymbol{Y}_t = \{Y_t(\boldsymbol{s}_i)\}$ denote a collection of count data across $n$ locations ($\boldsymbol{s}_i$, where $i = 1, \ldots, n$) and $T$ time points, and assume that the data collection across locations $\boldsymbol{Y}_t$ has a CMP($\boldsymbol{K}_t \boldsymbol{\lambda}_t$, $\nu_t$) distribution at time $t$ with intensities $\boldsymbol{\lambda}_t = \{\lambda_t(s_1), \ldots, \lambda_t(s_n)\}$ and dispersion parameter $\nu_t$, respectively, $t = 1, \ldots, T$; $\boldsymbol{K}_t$ is meanwhile an incidence matrix that accounts for any data missingness among the locations over time as each location is not necessarily observed over all time points. Accordingly, $\boldsymbol{Y}_t$ has varying length $m_t \leq n$, where $m_t$ denotes the number of locations observed at time $t$, and $\boldsymbol{K}_t$ is a matrix with dimension $m_t \times n$. The intensity process is represented as

$$\ln(\boldsymbol{\lambda}_t) = \boldsymbol{\mu} + \boldsymbol{\Psi}\boldsymbol{\alpha}_t + \boldsymbol{\epsilon}_t,$$

for $t = 1, \ldots, T$, where $\mu_i = \boldsymbol{X}_i\boldsymbol{\beta} + \eta_i$ is the average intensity at location $i = 1, \ldots, n$, where $\eta_i$ are iid normally distributed random variables with mean 0 and variance $\sigma^2$, $\boldsymbol{\beta}$ assumes a normal prior distribution, and $\sigma^2$ has an inverse Gamma prior. The residuals $\boldsymbol{\epsilon}_t = \{\epsilon_t(s_1), \ldots, \epsilon_t(s_n)\}$ are likewise assumed to be iid normally distributed random variables with mean 0 and variance $\sigma_\epsilon^2$ (also having an inverse Gamma distribution), and $\boldsymbol{\Psi}\boldsymbol{\alpha}_t$ represents the process of transitioning from the physical $n$-dimensional space to a smaller (say) $p$-dimensional space, $p \ll n$ via some matrix $\boldsymbol{\Psi}$ of basis function (e.g. splines and wavelets). Finally, the $p$-dimensional process $\boldsymbol{\alpha}_t$ assumes an autoregressive form $\boldsymbol{\alpha}_t = \boldsymbol{H}\boldsymbol{\alpha}_{t-1} + \boldsymbol{\gamma}_t$, where $\boldsymbol{H}$ is a redistribution matrix[4] in the $p$-dimensional spectral space with a normal prior distribution, and $\boldsymbol{\gamma}_t$ has a multivariate normal distribution with mean $\boldsymbol{0}$ and variance–covariance matrix $\boldsymbol{\Sigma}_\gamma$ (with an assumed Wishart prior distribution) to account for the spatial correlations within the lower-dimensional space; $\boldsymbol{\alpha}_0$ has a normal prior distribution.

While the dispersion can be likewise assumed to be spatial or even spatio-temporal, Wu et al. (2013) assume a time-varying constant dispersion $\nu_t$ to ease computational complexity. Recognizing the nonnegative support space for the dispersion and its dynamic nature, the dispersion is assumed to adhere to the autoregressive model,

$$\ln(\nu_t) = \phi_0 + \phi_1 \ln(\nu_{t-1}) + \xi_t, \quad t = 2, \ldots, T,$$

where $\ln(\nu_1)$ is uniformly distributed over the space between bounds $(\ln(\nu))_*$ and $(\ln(\nu))^*$ to start the autoregressive relationship, and $\phi_0$, $\phi_1$, and $\xi_t$ are themselves random with $\phi_0$ assuming a normal distribution with mean $\mu_{\phi_0}$ and variance $\sigma^2_{\phi_0}$, $\phi_1$ is uniformly distributed in $[-1,1]$, and $\xi_t$ is a random sample of normally distributed random variables with mean 0 and variance $\sigma_\xi^2$.

[4] Wu et al. (2013) actually consider a collection of redistribution matrices $\boldsymbol{H}_i$, $i = 1, 2, 3$ in conjunction with the considered analysis application.

Bayesian estimation is conducted via Markov Chain Monte Carlo (MCMC) using Metropolis–Hastings within Gibbs sampling. Simulation studies showed that respective parameters had posterior distributions whose mean was nearly unbiased, and the respective standard deviations were small such that their 95% credible intervals contained the true parameter in all cases. The estimates improved even further with a considerable increase in the number of timepoints. While the hierarchical structure of this model introduces additional computational complexity, simulation studies never demonstrated any convergence issues (Wu et al., 2013).

## 7.5 Summary

Serially dependent count data require special consideration to ensure proper analysis and modeling that maintains the observations having a discrete form. While stochastic processes, Markov models, time series and spatio-temporal models all exist to model discrete data, historical methods do not necessarily account for inherent data dispersion. This chapter presents those methods motivated by the (univariate or multivariate) COM–Poisson distribution (given various parametrizations) to account for data over- or under-dispersion.

The work surrounding COM-Poisson-motivated stochastic processes considers the matter in vastly diverse ways with broadly varying assumptions and underlying relationships to model the data (e.g. homogeneous or inhomogeneous processes, or (hidden) Markov models). Data analysis assuming a homogeneous CMP process (`cmpprocess`) or an underlying CMP-hidden Markov structure can be conducted in `R` (MacDonald and Bhamani, 2020; Zhu et al., 2017a,b). The other models and processes, however, do not currently offer associated `R` packages for analysts to conduct related statistical computing.

CMP-based time series modeling is considered in two ways: either by applying traditional time series modeling to the intensity parameter $\lambda$ or via integer-based time series modeling where a thinning parameter aids in maintaining discrete outcomes over time. While substantive work has been done regarding the development of COM-Poisson-motivated integer-based time series models, the current research focuses only on (univariate or bivariate COM–Poisson) autoregressive or moving average models. This is presumably true because of the underlying model complexities that exist due to the stricter constraint of maintaining discreteness in the observations over time and modeling the observations directly. Data analysis assuming a time series model on the intensity parameter, however, already allows for more complex structures (e.g. integer-valued generalized autoregressive conditional heteroscedastic (INGARCH) and autoregressive moving average (INARMA)) because the respective underlying model is represented via the intensity parameter that is not restricted itself to be discrete. Considering classical time series models that assume continuous outcomes provides greater insights into COM-Poisson-motivated time series development. Wu et al. (2013) likewise utilized this approach with their spatio-temporal model.

# 8

# COM–Poisson Cure Rate Models

Survival analysis studies the time to event for various subjects and serves as an important tool for study. In the biological sciences and in medicine, for example, interest can focus on patient time to death due to various (competing) causes. Similarly, in engineering reliability, one may study the time-to-component failure due to analogous factors or stimuli. Cure rate (also known as long-term survival) models serve a particular interest because, with advancements in associated disciplines, subjects can be viewed as "cured" meaning that they do not show any recurrence of a disease (in biomedical studies) or subsequent manufacturing error (in engineering) following a treatment.

A mixture cure model (Berkson and Gage, 1952; Boag, 1949) describes the survival of a mixture of cured and susceptible subjects. For a binary variable $Y$, $Y = 0$ denotes those subjects who are "cured"; thus, we assume that these individuals do not observe the event during the observation period. Meanwhile, $Y = 1$ indicates individuals susceptible to any considered causes whose lifetime distributions can be modeled; see Section 8.5 for details regarding lifetime distributions considered in this chapter. The probability that a subject is cured is usually modeled via a logistic regression, i.e. $P(Y = 1) = \frac{\lambda_\circ}{1+\lambda_\circ}$, where $\lambda_\circ = \exp(\boldsymbol{x}'\boldsymbol{\beta})$ denotes the probability of being susceptible to the event(s) (Farewell, 1982). For a mixture cure model, the cumulative distribution function (cdf) and probability of survival at time $y$ for a general population can be represented as

$$F_p(y) = (1 - p_0)F_s(y) \tag{8.1}$$

$$S_p(y) = 1 - (1 - p_0)F_s(y) = p_0 + (1 - p_0)S_s(y), \tag{8.2}$$

where $F_p$ and $S_p$, respectively, denote the cdf and survival function of the overall population, while $F_s$ and $S_s$ denote the cdf and survival function of the susceptible subpopulation; $p_0$ meanwhile denotes the probability of being cured, i.e. the "cure rate." Accordingly, the probability of being alive at time $y$ can be determined based on the probability of being cured or the probability of not being cured and being alive when one is known to be susceptible. Various statistical methods have been proposed assuming right censoring (Berkson and Gage, 1952; Boag, 1949) or interval censoring (Kim and Jhun, 2008).

The promotion time cure model meanwhile presumes $M$ competing causes associated with an event occurrence and assumes that $M$ is Poisson($\lambda_*$) distributed. Let $W_j$ denote the time to event (i.e. the "lifetime") caused by the $j$th competing factor, where $j = 1, \ldots, M$. Given $M = m$, $W_j$ are independent and identically distributed (iid) with cdf and survival function, $F(\cdot)$ and $S(\cdot)$, respectively, such that $F(\cdot) = 1 - S(\cdot)$, and $W_j$ are independent of $M, j = 1, \ldots, m$. Section 8.5 discusses the various distributional models assumed for $W_j$; all of the distributions utilize the associated parameter denotation $\boldsymbol{\gamma} = (\gamma_1, \ldots, \gamma_l)$. By construct, $M$ and $W_j$ are the latent variables while the lifetime of an individual can be observed as $Y = \min(W_0, W_1, W_2, \ldots, W_M)$, where $W_0$ represents the lifetime of one who is immune to the event occurrence (i.e. $P(W_0 = \infty) = 1$). This latter definition implies the existence of a "cure rate," i.e. a proportion of such subjects who are immune denoted $p_0 = P(M = 0) = \exp(-\lambda_*)$. The promotion time cure model long-term survival function, corresponding probability density function (pdf), and hazard function for the population are

$$S_p(y) = \exp(-\lambda_* F(y)) \tag{8.3}$$

$$f_p(y) = \lambda_* f(y) \exp(-\lambda_* F(y)) \tag{8.4}$$

$$h_p(y) = \lambda_* f(y), \tag{8.5}$$

where $f(y) = \frac{d}{dy}F(y)$ is a proper pdf; however, none of the population functions is proper because the population survival function $S_p(y) > 0$. Nonetheless, various properties hold, including that $S_p(y)$ converges to $p_0$ as $y \to \infty$, while $h_p(y)$ converges to 0 as $t \to \infty$ and $\int_0^\infty h_p(y)dy < \infty$ (Chen et al., 1999). The analogous functions for the subpopulation that is susceptible to the event are

$$S_s(y) = \frac{\exp(-\lambda_* F(y)) - \exp(-\lambda_*)}{1 - \exp(-\lambda_*)} \tag{8.6}$$

$$f_s(y) = \left(\frac{\exp(-\lambda_* F(y))}{1 - \exp(\lambda_*)}\right) \lambda_* f(y) \tag{8.7}$$

$$h_s(y) = \left( \frac{\exp(-\lambda_* F(y))}{\exp(-\lambda_* F(y)) - \exp(-\lambda_*)} \right)$$
$$h_p(y) = \left( \frac{\exp(-\lambda_* F(y))}{\exp(-\lambda_* F(y)) - \exp(-\lambda_*)} \right) \lambda_* f(y). \tag{8.8}$$

Analytic methods have been proposed under a right censoring (Chen et al., 2008) or interval censoring (Banerjee and Carlin, 2004) framework, and various lifetime distributions.

These cure rate models are unified via the development of a COM–Poisson cure rate model, where all of the models assume a CMP parametrization as described in Chapter 2. Section 8.1 describes the CMP cure rate model framework and general notation. Sections 8.2 and 8.3 describe the CMP cure rate model framework assuming right and interval censoring, respectively. Section 8.4 describes the broader destructive CMP cure rate model that allows for the number of competing risks to diminish via damage or eradication. Section 8.5 details the various lifetime distributions considered in the literature to date for CMP-based cure rate modeling. Finally, Section 8.6 concludes the chapter with discussion.

## 8.1 Model Background and Notation

The CMP cure rate model serves as a flexible means to describe the number of competing causes contributing to a time to event where the number of competing causes $M$ assumes a CMP($\lambda$, $\nu$) distribution with probability $p_m = P(M = m) = \frac{\lambda^m}{(m!)^\nu}$ for $m = 0, 1, 2, \ldots$ (see Equation (2.8)); the cure rate is

$$p_0 = P(M = 0; \lambda, \nu) = \frac{1}{Z(\lambda, \nu)}. \tag{8.9}$$

By definition, the CMP cure rate model contains the promotion time cure model (Chen et al., 1999; Yakovlev and Tsodikov, 1996) and the mixture cure model (Berkson and Gage, 1952; Boag, 1949) as special cases. The CMP cure rate model reduces to the promotion time cure model when $\nu = 1$, with a cure rate of $p_0 = 1/Z(\lambda, 1) = \exp(-\lambda)$, and represents the mixture cure model special case when $\nu \to \infty$ with a cure rate of $p_0 = \lim_{\nu \to \infty} 1/Z(\lambda, \nu) = 1/(1 + \lambda)$.

Let $W_j$ and $Y$ maintain the same properties as described in the promotion time cure model. Under the CMP cure rate model construct, the long-term survival function for the population is

$$S_p(y) = P(Y \geq y) = \frac{Z(\lambda S(y), \nu)}{Z(\lambda, \nu)}, \tag{8.10}$$

where (by definition) $\lim_{y\to\infty} S_p(y) = p_0$. The corresponding (improper) density function of $Y$ is

$$f_p(y) = \frac{1}{Z(\lambda, \nu)} \frac{f(y)}{S(y)} \sum_{j=1}^{\infty} \frac{j[\lambda S(y)]^j}{(j!)^\nu}, \tag{8.11}$$

where $f(y)$ and $S(y)$, respectively, denote the pdf and survival function of $Y$. This description is a natural extension from the promotion cure time model framework to utilize the flexibility of the CMP distribution, and this approach is noted in several works (e.g. Balakrishnan and Pal, 2015a; Cancho et al., 2012; Rodrigues et al., 2009). Other references (e.g. Balakrishnan and Pal, 2012), however, likewise draw from the mixture cure model as motivation, partitioning the general population into those that are immune versus susceptible to the competing causes as described in Equations (8.1) and (8.2). This decomposition implies that $\lim_{y\to\infty} S_p(y) = p_0$.

Just as the CMP cure rate structure borrows strength from both the promotion time cure model and the mixture cure model, the same is true when considering link functions to associate the covariates. One option (stemming from the promotion time cure model) is to associate $\lambda_i$ with covariates $\boldsymbol{x}_i$ via a loglinear link function, $\ln(\lambda_i) = \boldsymbol{x}_i'\boldsymbol{\beta}$ with coefficients, $\boldsymbol{\beta} = (\beta_0, \beta_1, \ldots, \beta_b)'$ (Balakrishnan and Pal, 2012; Cancho et al., 2012; Rodrigues et al., 2009). While this is a viable approach in general, the loglinear link for $\lambda$ can be problematic in the broader CMP cure rate model because it cannot restrict $\lambda < 1$ for the special case of the geometric cure rate model (i.e. the CMP cure rate where $\nu = 0$). An alternate approach (often utilized in mixture cure models) is to consider a logistic link to associate the cure rate $\boldsymbol{p}_0$ with the covariates $\boldsymbol{x}_i$ (Balakrishnan and Feng, 2018; Balakrishnan and Pal, 2013, 2015a, 2016; Balakrishnan et al., 2017). Setting the logistic model formulation equal to the cure rate definition for the CMP cure rate model (Equation (8.9)) implies

$$p_0 = \frac{1}{1 + \exp(\boldsymbol{x}'\boldsymbol{\beta})} = \frac{1}{Z(\lambda, \nu)} \tag{8.12}$$

such that, for a given $\nu$, $p_0$ only depends on $\lambda$. Accordingly, we have the function $g(\lambda; \nu) = Z(\lambda, \nu) = 1 + \exp(\boldsymbol{x}'\boldsymbol{\beta})$ and $\lambda = g^{-1}(1 + \exp(\boldsymbol{x}'\boldsymbol{\beta}); \nu)$. The logistic link function implies a broader functional relationship between $\lambda$ and the covariates for a given $\nu$, i.e. $g(\boldsymbol{\lambda}; \nu) = 1 + \exp(\boldsymbol{x}'\boldsymbol{\beta})$, where the function $g$ is a monotone increasing function whose inverse is determined numerically. For each of the special CMP cases, however, $g^{-1}(\lambda; \nu)$ has a closed form (Balakrishnan and Pal, 2013); see Table 8.1 for details regarding the Poisson, geometric, and Bernoulli cases.

Table 8.1 *Special-case CMP cure rate models that have closed forms for* $g(\lambda; \nu)$ *and* $\lambda = g^{-1}(1 + \exp(\boldsymbol{x}'\boldsymbol{\beta}); \nu)$.

| Special case | Restriction | $g(\lambda; \nu)$ | $\lambda$ |
|---|---|---|---|
| Poisson | $\nu = 1$ | $g(\lambda; \nu = 1) = \exp(\lambda)$ | $\ln(1 + \exp(\boldsymbol{x}'\boldsymbol{\beta}))$ |
| Geometric | $\nu = 0$ with $\lambda < 1$ | $g(\lambda; \nu = 0) = \frac{1}{1 - \lambda}$ | $\frac{\exp(\boldsymbol{x}'\boldsymbol{\beta})}{1 + \exp(\boldsymbol{x}'\boldsymbol{\beta})}$ |
| Bernoulli | $\nu \to \infty$ | $g(\lambda; \nu \to \infty) = 1 + \lambda$ | $\exp(\boldsymbol{x}'\boldsymbol{\beta})$ |

## 8.2 Right Censoring

Discussion in this section assumes noninformative right censoring where $T_i = \min(Y_i, C_i)$ denotes the observed survival time as $Y_i$ or $C_i$, i.e. either the time to the event or censoring time, respectively. For Subject $i = 1, \ldots, n$, let $\delta_i = I(Y_i \leq C_i)$ denote the indicator variable that equals 1 (0) if the time to event (censoring time) is observed. Given the respective pairs of observed times and associated censor indicators for the $n$ subjects $\{(t_i, \delta_i)\colon i = 1, \ldots, n\}$, the resulting likelihood function is

$$L(\boldsymbol{\theta}; \boldsymbol{t}, \boldsymbol{\delta}) \propto \prod_{i=1}^{n} [f_p(t_i; \boldsymbol{\theta})]^{\delta_i} [S_p(t_i; \boldsymbol{\theta})]^{1-\delta_i} \tag{8.13}$$

for a set of parameters $\boldsymbol{\theta} = (\boldsymbol{\beta}', \nu, \boldsymbol{\gamma})'$.

### 8.2.1 Parameter Estimation Methods

*Maximum Likelihood Estimation*

Rodrigues et al. (2009) utilize the method of maximum likelihood (ML) for parameter estimation, directly optimizing the log-likelihood with respect to the parameters of interest. Substituting the respective definitions for $S_p$ and $f_p$ (Equations (8.10) and (8.11)) into Equation (8.13) produces the likelihood function

$$L(\boldsymbol{\theta}; \boldsymbol{t}, \boldsymbol{\delta}) \propto \prod_{i=1}^{n} \frac{1}{Z(\lambda_i, \nu)} \left\{ \frac{f(t_i; \boldsymbol{\gamma})}{S(t_i; \boldsymbol{\gamma})} \sum_{j=1}^{\infty} \frac{j(\lambda_i S(t_i; \boldsymbol{\gamma}))^j}{(j!)^\nu} \right\}^{\delta_i}$$

$$\times [Z(\lambda_i S(t_i; \boldsymbol{\gamma}), \nu)]^{1-\delta_i}, \tag{8.14}$$

where $\boldsymbol{\theta} = (\boldsymbol{\beta}', \nu, \boldsymbol{\gamma}')$ and $\lambda_i = \exp(\boldsymbol{x}_i'\boldsymbol{\beta})$ (i.e. a loglinear link with $\boldsymbol{\beta}' = (\beta_0, \beta_1, \ldots, \beta_b)$). To obtain the maximum likelihood estimates (MLEs) $\hat{\boldsymbol{\theta}} = (\hat{\boldsymbol{\beta}}', \hat{\nu}, \hat{\boldsymbol{\gamma}}')$, Equation (8.14) is optimized numerically via the Rigby and Stasinopoulos (2005) method contained in the R package gamlss (Stasinopoulos and Rigby, 2007). The normalizing constant $Z(\lambda, \nu) = \sum_{j=0}^{\infty} \frac{\lambda^j}{(j!)^\nu}$ is meanwhile approximated by truncating the infinite sum at some $K > (\lambda/\epsilon)^{1/\nu}$ for small $\epsilon > 0$. Given appropriate regularity conditions, $\hat{\boldsymbol{\theta}}$ has an asymptotic distribution that is multivariate normal with mean $\boldsymbol{\theta}$ and variance–covariance matrix $\boldsymbol{\Sigma}(\hat{\boldsymbol{\theta}})$, estimated as

$$\hat{\boldsymbol{\Sigma}}(\hat{\boldsymbol{\theta}}) = \left[ -\frac{\partial^2 \ln (L(\boldsymbol{\theta}; \boldsymbol{t}, \boldsymbol{\delta}))}{\partial \boldsymbol{\theta} \partial \boldsymbol{\theta}'} \right]^{-1} |_{\boldsymbol{\theta}=\hat{\boldsymbol{\theta}}} . \tag{8.15}$$

A real-data illustration shows that the CMP cure rate model outperforms the promotion time and mixture cure models, achieving the best fit; see Rodrigues et al. (2009) for details.

### *Bayesian Approach*

Cancho et al. (2012) instead take a Bayesian approach toward analyzing the CMP cure rate model whose marginal likelihood function is defined in Equation (8.14). Assume that the parameters $\boldsymbol{\theta} = (\boldsymbol{\beta}', \nu, \boldsymbol{\gamma}')$ are independent with respective prior distributions whose hyperparameters further induce noninformative priors. More precisely, let $\boldsymbol{\beta}$ have a multivariate normal distribution $N_{b+1}(\boldsymbol{0}, \boldsymbol{\Sigma})$ and $\ln(\nu)$ be normally distributed $N(0, \sigma_\nu^2)$, while the selections for $\boldsymbol{\gamma}$ elements are noted in Section 8.5. The resulting posterior distribution $\pi(\boldsymbol{\theta} \mid \boldsymbol{t}, \boldsymbol{\delta}) \propto L(\boldsymbol{\theta}; \boldsymbol{t}, \boldsymbol{\delta})$ is not easily integrable; however, inferences regarding $\boldsymbol{\theta}$ can be attained via Markov Chain Monte Carlo (MCMC) methods.

### *EM Algorithm*

Several works adopt the mixture cure model approach toward defining the likelihood function, partitioning the general population into two groups (i.e. those who are immune versus susceptible) such that Equation (8.13) can be represented as

$$L(\boldsymbol{\theta}; \boldsymbol{t}, \boldsymbol{\delta}) \propto \prod_{\{i:\, \delta_i=1\}} f_p(t_i; \boldsymbol{\theta}) \prod_{\{i:\, \delta_i=0\}} [p_0 + (1 - p_0) S_s(t_i; \boldsymbol{\theta})], \tag{8.16}$$

where $\boldsymbol{\theta} = (\boldsymbol{\beta}', \nu, \boldsymbol{\gamma})'$ and $\boldsymbol{\theta}_* = (\boldsymbol{\beta}', \nu)'$, assuming a particular distribution for the time-to-event variables $W$ (Balakrishnan and Feng, 2018; Balakrishnan and Pal, 2012, 2013, 2015a, 2016; Balakrishnan et al., 2017). They further introduce the binary variable $J$ to denote whether or not a subject experiences the event because of the considered cause(s). By definition, $J = 1$ if the subject experiences a cause(s)-related event (i.e. $J_i = 1$ for

$i \in \{i\colon \delta_i = 1\}$), while $J$ is otherwise unknown (as either 0 or 1) because the subject could either never experience the event because they are immune (thus, $J = 0$) or the subject could still experience the event but it has yet to occur during the span of the experiment or study (i.e $J$ unknown for $i \in \{i\colon \delta_i = 0\}$). Accordingly, this aspect of the work is recognized as a missing data problem, propelling researchers to determine the MLEs $\hat{\boldsymbol{\theta}} = (\hat{\boldsymbol{\beta}}', \hat{\nu}, \hat{\boldsymbol{\gamma}}')$ via the expectation–maximization (EM) algorithm, where the resulting complete data likelihood and corresponding log-likelihood (without the constant) are

$$
\begin{aligned}
L_c(\boldsymbol{\theta};\boldsymbol{t},\boldsymbol{x},\boldsymbol{\delta},\boldsymbol{J}) \propto & \prod_{\{i\colon \delta_i=1\}} [f_p(t_i,\boldsymbol{x}_i;\boldsymbol{\theta})] \\
& \times \prod_{\{i\colon \delta_i=0\}} [p_0(\boldsymbol{\theta}_*,\boldsymbol{x}_i)]^{1-J_i}[(1-p_0(\boldsymbol{\theta}_*,\boldsymbol{x}_i))S_s(t_i,\boldsymbol{x}_i;\boldsymbol{\theta})]^{J_i}
\end{aligned} \tag{8.17}
$$

$$
\begin{aligned}
\ln L_c(\boldsymbol{\theta};\boldsymbol{t},\boldsymbol{x},\boldsymbol{\delta},\boldsymbol{J}) \propto & \sum_{\{i\colon \delta_i=1\}} \ln f_p(t_i,\boldsymbol{x}_i;\boldsymbol{\theta}) + \sum_{\{i\colon \delta_i=0\}} (1-J_i)\ln p_0(\boldsymbol{\theta}_*,\boldsymbol{x}_i) \\
& + \sum_{\{i\colon \delta_i=0\}} J_i \ln[(1-p_0(\boldsymbol{\theta}_*,\boldsymbol{x}_i))S_s(t_i,\boldsymbol{x}_i;\boldsymbol{\theta})].
\end{aligned} \tag{8.18}
$$

For a given $\nu$, let $\boldsymbol{\theta}^\dagger = (\boldsymbol{\beta}', \boldsymbol{\gamma}')'$ and let $\boldsymbol{\theta}^\dagger_{(k)}$ denote its values at the $k$th iteration. Given this framework, the expectation step determines $G(\boldsymbol{\theta}^\dagger, \boldsymbol{\pi}^{(k)}) = E(\ln L_c(\boldsymbol{\theta};\boldsymbol{t},\boldsymbol{x},\boldsymbol{\delta},\boldsymbol{J}) \mid \boldsymbol{\theta}^\dagger_{(k)}, \boldsymbol{O})$ with regard to the distribution of unobserved $J_i$s, given $\boldsymbol{O} = \{\text{observed } J_i\text{s and } (t_i, \delta_i, \boldsymbol{x}_i); i = 1, \ldots, n\}$ for a fixed $\nu$ by replacing $J_i$ with its expected value, namely

$$
\pi_i^{(k)} = \begin{cases} \dfrac{\left[1-p_0\left(\boldsymbol{\theta}^\dagger_{(k)},\boldsymbol{x}_i\right)\right]S_s\left(t_i,\boldsymbol{x}_i;\boldsymbol{\theta}^\dagger_{(k)}\right)}{S_p\left(t_i,\boldsymbol{x}_i;\boldsymbol{\theta}^\dagger_{(k)}\right)} & \text{for } i \in \{i\colon \delta_i = 0\} \\ 1 & \text{otherwise,} \end{cases} \tag{8.19}
$$

where $\boldsymbol{\theta}^\dagger_{(k)} = (\boldsymbol{\beta}', \boldsymbol{\gamma}')'_{(k)}$ denotes values of $\boldsymbol{\theta}^\dagger$ at the $k$th step; Tables 8.2 and 8.3 contain the resulting $G(\boldsymbol{\theta}^\dagger, \boldsymbol{\pi}^{(k)})$ functions for the CMP cure rate model assuming respective lifetime distributions; see Section 8.5 for further discussion regarding the various lifetime distributions featured in this chapter. The respective functions can be refined for the special cases that comprise the CMP cure rate model (i.e. the Bernoulli, Poisson, and geometric cure rate models). Meanwhile, the maximization step identifies $\boldsymbol{\theta}^\dagger_{(k+1)}$ as the value that maximizes $G(\boldsymbol{\theta}^\dagger, \boldsymbol{\pi}^{(k)})$ with respect to the parameter space $\boldsymbol{\Theta}^\dagger$, i.e. $\boldsymbol{\theta}^\dagger_{(k+1)} = \arg\max_{\boldsymbol{\theta}^\dagger \in \boldsymbol{\Theta}^\dagger} G(\boldsymbol{\theta}^\dagger, \boldsymbol{\pi}^{(k)})$. Given the added complexity of some models, the Newton–Raphson or quasi-Newton methods are

Table 8.2 *Expectation step functions for fixed ν from the CMP cure rate model with right censoring and various associated lifetime distributions.*

| Reference | Lifetime distribution | $G(\boldsymbol{\theta}^\dagger, \boldsymbol{\pi}^{(k)}) = \mathbb{E}(\ln L_c(\boldsymbol{\theta}; \boldsymbol{t}, \boldsymbol{x}, \boldsymbol{\delta}, \boldsymbol{J}) \mid \boldsymbol{\theta}^\dagger_{(k)}, \boldsymbol{O})$ |
|---|---|---|
| Balakrishnan and Pal (2012) | Exponential | $n_1 \ln\gamma + \sum_{\{i:\ \delta_i=1\}} \ln z_2(\boldsymbol{\beta}, \gamma, \nu; \boldsymbol{x}_i, t_i) + \sum_{\{i:\ \delta_i=0\}} w_i^{(k)} \ln z_1(\boldsymbol{\beta}, \gamma, \nu; \boldsymbol{x}_i, t_i)$ $- \sum_{I*} \ln z(\boldsymbol{\beta}, \nu; \boldsymbol{x}_i)$, where $z(\boldsymbol{\beta}, \nu; \boldsymbol{x}_i) = \sum_{j=0}^{\infty} \frac{(\exp(\boldsymbol{x}_i'\boldsymbol{\beta}))^j}{(j!)^\nu}$, $z_1(\boldsymbol{\beta}, \gamma, \nu; \boldsymbol{x}_i, t_i) = \sum_{j=1}^{\infty} \frac{(\exp(\boldsymbol{x}_i'\boldsymbol{\beta} - \gamma t_i))^j}{(j!)^\nu}$, $z_2(\boldsymbol{\beta}, \gamma, \nu; \boldsymbol{x}_i, t_i) = \sum_{j=1}^{\infty} \frac{(j \exp(\boldsymbol{x}_i'\boldsymbol{\beta} - \gamma t_i))^j}{(j!)^\nu}$, $w_i^{(k)} = \frac{z_1(\boldsymbol{\beta}^{(k)}, \gamma^{(k)}, \nu; \boldsymbol{x}_i, t_i)}{1 + z_1(\boldsymbol{\beta}^{(k)}, \gamma^{(k)}, \nu; \boldsymbol{x}_i, t_i)}$ |
| Balakrishnan and Pal (2013) | Lognormal | $-n_1 \ln(\sqrt{2\pi}) - n_1 \ln\gamma_1 - \sum_{\{i:\ \delta_i=1\}} \ln t_i - \frac{1}{2} \sum_{\{i:\ \delta_i=1\}} u_i^2 - \sum_{\{i:\ \delta_i=1\}} \ln(1 - \Phi(u_i))$ $+ \sum_{\{i:\ \delta_i=1\}} \ln z_{2i} + \sum_{\{i:\ \delta_i=0\}} \pi_i^{(k)} \ln z_{1i} - \sum_{I*} \ln(1 + \exp(\boldsymbol{x}_i'\boldsymbol{\beta}))$, where $z_1 \doteq z_1(\boldsymbol{\theta}; \boldsymbol{x}_i, t_i) = \sum_{j=1}^{\infty} \frac{[g^{-1}(1 + \exp(\boldsymbol{x}'\boldsymbol{\beta})(1 - \Phi(u_i)); \nu)]^j}{(j!)^\nu}$, $z_2 \doteq z_2(\boldsymbol{\theta}; \boldsymbol{x}_i, t_i) = \sum_{j=1}^{\infty} \frac{j[g^{-1}(1 + \exp(\boldsymbol{x}'\boldsymbol{\beta})(1 - \Phi(u_i)); \nu)]^j}{(j!)^\nu}$, $\pi_i^{(k)} = \frac{z_1(\boldsymbol{\theta}; \boldsymbol{x}_i, t_i)}{1 + z_1(\boldsymbol{\theta}; \boldsymbol{x}_i, t_i)} \Big\vert_{\boldsymbol{\theta}^* = \boldsymbol{\theta}^{*(k)}}$ for censored observation $i$. |
| Balakrishnan and Pal (2015a) | Generalized Gamma | $n_1 \ln(K(\gamma)) + \left(\frac{1}{q\sigma} - 1\right) \sum_{\{i:\ \delta_i=1\}} \ln t_i - \frac{1}{q^2} \sum_{\{i:\ \delta_i=1\}} (\lambda t_i)^{q/\sigma} - \sum_{\{i:\ \delta_i=1\}} \ln(S(t_i; \boldsymbol{\gamma}))$ $+ \sum_{\{i:\ \delta_i=1\}} \ln z_{2i} + \sum_{\{i:\ \delta_i=0\}} \pi_i^{(k)} \ln z_{1i} - \sum_{I*} \ln\left(1 + \exp(\boldsymbol{x}_i'\boldsymbol{\beta})\right)$, where $K(\boldsymbol{\gamma}) = \frac{q}{\sigma} \frac{(\lambda^{q/\sigma} q^2)^{1/q^2}}{\Gamma(1/q^2)}$, $z_1 \doteq z_1(\boldsymbol{\theta}; \boldsymbol{x}, t) = \sum_{j=1}^{\infty} \frac{[g^{-1}(1 + \exp(\boldsymbol{x}'\boldsymbol{\beta}); \nu) \exp(-(\gamma_2 t)^{1/\gamma_1})]^j}{(j!)^\nu}$, $z_2 \doteq z_2(\boldsymbol{\theta}; \boldsymbol{x}, t) = \sum_{j=1}^{\infty} \frac{j[g^{-1}(1 + \exp(\boldsymbol{x}'\boldsymbol{\beta}); \nu) \exp(-(\gamma_2 t)^{1/\gamma_1})]^j}{(j!)^\nu}$, $\pi_i^{(k)} = \frac{z_1(\boldsymbol{\theta}; \boldsymbol{x}_i, t_i)}{1 + z_1(\boldsymbol{\theta}; \boldsymbol{x}_i, t_i)} \Big\vert_{\boldsymbol{\theta}^* = \boldsymbol{\theta}^{(k)}}$ for censored observation $i$. |
| Balakrishnan and Pal (2015b) | Gamma | $n_1 \ln(G(\boldsymbol{\gamma})) - \sum_{\{i:\ \delta_i=1\}} \ln(S(t_i; \boldsymbol{\gamma})) - \left(\frac{\gamma_2}{\gamma_1^2}\right) \sum_{\{i:\ \delta_i=1\}} t_i + \left(\frac{1}{\gamma_1^2} - 1\right) \sum_{\{i:\ \delta_i=1\}} \ln t_i$ $+ \sum_{\{i:\ \delta_i=1\}} \ln z_{2i} + \sum_{\{i:\ \delta_i=0\}} \pi_i^{(k)} \ln z_{1i} - \sum_{I*} \ln(1 + \exp(\boldsymbol{x}_i'\boldsymbol{\beta}))$, where $z_1 \doteq z_1(\boldsymbol{\theta}; \boldsymbol{x}, t) = \sum_{j=1}^{\infty} \frac{[g^{-1}(1 + \exp(\boldsymbol{x}'\boldsymbol{\beta}); \nu) \exp(-(\gamma_2 t)^{1/\gamma_1})]^j}{(j!)^\nu}$, $z_2 \doteq z_2(\boldsymbol{\theta}; \boldsymbol{x}, t) = \sum_{j=1}^{\infty} \frac{j[g^{-1}(1 + \exp(\boldsymbol{x}'\boldsymbol{\beta}); \nu) \exp(-(\gamma_2 t)^{1/\gamma_1})]^j}{(j!)^\nu}$, $\pi_i^{(k)} = \frac{z_1(\boldsymbol{\theta}; \boldsymbol{x}_i, t_i)}{1 + z_1(\boldsymbol{\theta}; \boldsymbol{x}_i, t_i)} \Big\vert_{\boldsymbol{\theta}^* = \boldsymbol{\theta}^{(k)}}$ for censored observation $i$. |

Table 8.3 *Expectation step functions for fixed ν from the CMP cure rate model with right censoring and various associated lifetime distributions (continued).*

| Reference | Lifetime distribution | $G(\boldsymbol{\theta}^\dagger, \boldsymbol{\pi}^{(k)}) = \mathbb{E}(\ln L_c(\boldsymbol{\theta}; \boldsymbol{t}, \boldsymbol{x}, \boldsymbol{\delta}, \boldsymbol{J}) \mid \boldsymbol{\theta}^\dagger_{(k)}, \boldsymbol{O})$ |
|---|---|---|
| Balakrishnan and Pal (2016) | Weibull | $-n \ln \gamma_1 + \frac{n_1}{\gamma_1} \ln \gamma_2 + \frac{1-\gamma_1}{\gamma_1} \sum_{\{i:\ \delta_i=1\}} \ln t_i + \sum_{\{i:\ \delta_i=1\}} \ln z_{2i}$<br>$+ \sum_{\{i:\ \delta_i=0\}} \pi_i^{(k)} \ln z_{1i} - \sum_{\{i:\ \delta_i=0 \cup 1\}} \ln [1 + \exp(\boldsymbol{x}_i' \boldsymbol{\beta})]$, where<br>$z_1 \doteq z_1(\boldsymbol{\theta}; \boldsymbol{x}, t) = \sum_{j=1}^\infty \frac{[g^{-1}(1+\exp(\boldsymbol{x}'\boldsymbol{\beta}); \nu) \exp(-(\gamma_2 t)^{1/\gamma_1})]^j}{(j!)^\nu}$,<br>$z_2 \doteq z_2(\boldsymbol{\theta}; \boldsymbol{x}, t) = \sum_{j=1}^\infty \frac{j[g^{-1}(1+\exp(\boldsymbol{x}'\boldsymbol{\beta}); \nu) \exp(-(\gamma_2 t)^{1/\gamma_1})]^j}{(j!)^\nu}$,<br>$\pi_i^{(k)} = \frac{z_1(\boldsymbol{\theta}; \boldsymbol{x}_i, t_i)}{1+z_1(\boldsymbol{\theta}; \boldsymbol{x}_i, t_i)} \vert_{\boldsymbol{\theta}^* = \boldsymbol{\theta}^{(k)}}$ for censored observation $i$. |
| Balakrishnan et al. (2017) | Proportional hazard (Weibull baseline hazard function) | $n_1 \ln \gamma_0 - n_1 \gamma_0 \ln \gamma_1 + (\gamma_0 - 1) \sum_{\{i:\ \delta_i=1\}} \ln t_i + \sum_{\{i:\ \delta_i=1\}} \boldsymbol{x}_i' \gamma_2$<br>$- \sum_{\{i:\ \delta_i=0 \text{ or } 1\}} \ln(1 + \exp(\boldsymbol{x}_i' \boldsymbol{\beta})) + \sum_{\{i:\ \delta_i=1\}} \ln z_{2i} + \sum_{\{i:\ \delta_i=0\}} \pi_i^{(k)} \ln z_{1i}$, where<br>$z_1 \doteq z_1(\boldsymbol{\theta}; \boldsymbol{x}, t) = \sum_{j=1}^\infty \frac{[g^{-1}(1+\exp(\boldsymbol{x}_i'\boldsymbol{\beta}); \nu) \exp[-(t_i \gamma_1)_0^\gamma \exp(\boldsymbol{x}_{ic}' \gamma_2)]]^j}{(j!)^\nu}$,<br>$z_2 \doteq z_2(\boldsymbol{\theta}; \boldsymbol{x}, t) = \sum_{j=1}^\infty \frac{j[g^{-1}(1+\exp(\boldsymbol{x}_i'\boldsymbol{\beta}); \nu) \exp[-(t_i \gamma_1)_0^\gamma \exp(\boldsymbol{x}_{ic}' \gamma_2)]]^j}{(j!)^\nu}$,<br>$\pi_i^{(k)} = \frac{z_1(\boldsymbol{\theta}; \boldsymbol{x}_i, t_i)}{1+z_1(\boldsymbol{\theta}; \boldsymbol{x}_i, t_i)} \vert_{\boldsymbol{\theta}^* = \boldsymbol{\theta}^{(k)}}$ for censored observation $i$, and $n_1 = \vert\{i\colon \delta_i = 1\}\vert$. |
| Balakrishnan et al. (2017) | Proportional hazard (general form) | $n_1 \ln \gamma_0 - n_1 \gamma_0 \ln \gamma_1 + (\gamma_0 - 1) \sum_{\{i:\ \delta_i=1\}} \ln t_i + \sum_{\{i:\ \delta_i=1\}} \boldsymbol{x}_i' \gamma_2$<br>$- \sum_{\{i:\ \delta_i=0 \text{ or } 1\}} \ln(1 + \exp(\boldsymbol{x}_i' \boldsymbol{\beta})) + \sum_{\{i:\ \delta_i=1\}} \ln z_{2i} + \sum_{\{i:\ \delta_i=0\}} \pi_i^{(k)} \ln z_{1i}$, where<br>$z_{1i} \doteq z_1(\boldsymbol{\theta}; \boldsymbol{x}_i, t_i) = \sum_{j=1}^\infty \frac{[g^{-1}(1+\exp(\boldsymbol{x}_i'\boldsymbol{\beta}); \nu) S(t_i; \boldsymbol{\gamma})]^j}{(j!)^\nu}$,<br>$z_{2i} \doteq z_2(\boldsymbol{\theta}; \boldsymbol{x}_i, t_i) = \sum_{j=1}^\infty \frac{j[g^{-1}(1+\exp(\boldsymbol{x}_i'\boldsymbol{\beta}); \nu) S(t_i; \boldsymbol{\gamma})]^j}{(j!)^\nu}$,<br>$\pi_i^{(k)} = \frac{z_1(\boldsymbol{\theta}; \boldsymbol{x}_i, t_i)}{1+z_1(\boldsymbol{\theta}; \boldsymbol{x}_i, t_i)} \vert_{\boldsymbol{\theta}^* = \boldsymbol{\theta}^{(k)}}$ for censored observation $i$, and $n_1 = \vert\{i\colon \delta_i = 1\}\vert$. |
| Balakrishnan and Feng (2018) | Proportional odds | $- \sum_{\{i:\ \delta_i=0 \text{ or } 1\}} \ln(1 + \exp(\boldsymbol{x}_i' \boldsymbol{\beta})) + \sum_{\{i:\ \delta_i=1\}} \ln f(t_i, \boldsymbol{\gamma}) - \sum_{\{i:\ \delta_i=1\}} \ln S(t_i, \boldsymbol{\gamma})$<br>$+ \sum_{\{i:\ \delta_i=1\}} \ln z_{2i} + \sum_{\{i:\ \delta_i=0\}} \pi_i^{(k)} \ln z_{1i}$, where<br>$z_{1i} \doteq z_1(\boldsymbol{\theta}; \boldsymbol{x}_i, t_i) = \sum_{j=1}^\infty \frac{[g^{-1}(1+\exp(\boldsymbol{x}_i'\boldsymbol{\beta}); \nu) S(t_i; \boldsymbol{\gamma})]^j}{(j!)^\nu}$,<br>$z_{2i} \doteq z_2(\boldsymbol{\theta}; \boldsymbol{x}_i, t_i) = \sum_{j=1}^\infty \frac{j[g^{-1}(1+\exp(\boldsymbol{x}_i'\boldsymbol{\beta}); \nu) S(t_i; \boldsymbol{\gamma})]^j}{(j!)^\nu}$,<br>$\pi_i^{(k)} = \frac{z_1(\boldsymbol{\theta}; \boldsymbol{x}_i, t_i)}{1+z_1(\boldsymbol{\theta}; \boldsymbol{x}_i, t_i)} \vert_{\boldsymbol{\theta}^* = \boldsymbol{\theta}^{(k)}}$ for censored observation $i$. |

used to perform the maximization steps. The expectation and maximization steps are repeated with each iteration until convergence is reached and $\hat{\boldsymbol{\theta}}^{\dagger}$ is determined for a given $\nu$. The optimal $\hat{\nu}$ is then determined via the profile likelihood approach, assessing which log-likelihood is maximum among the collection of results, $(\hat{\boldsymbol{\theta}}^{\dagger}; \nu)$. More broadly, Balakrishnan and Pal (2015a) use the generalized gamma$(g_1, g_2, g_3)$ distribution to describe the time-to-event variable and perform the expectation step for a fixed $\nu$ and $g_2$. As usual, the sequence of expectation and maximization steps repeats until convergence is attained and the conditional MLEs are thus identified for the given $\nu$ and $g_2$. The profile likelihood approach is then applied for $\nu$ and $g_2$ in order to determine their MLEs for which the overall log-likelihood function is maximized. Several works further supply the first- and second-order derivatives of $G(\boldsymbol{\theta}^{\dagger}, \boldsymbol{\pi}^{(k)})$ to aid in constructing the score function and Fisher information matrix associated with the general CMP cure rate model and its special cases; see, for example, Balakrishnan and Pal (2013), Balakrishnan and Pal (2015a), Balakrishnan and Feng (2018). Computational efforts surrounding this method are assisted by truncating any infinite series at some point where its contribution to the summation is sufficiently small (i.e. less than some $\epsilon > 0$; see Section 2.8 for related discussion).

Initial guesses to enact the EM algorithm are determined in various ways. Several works identify initial values for $\boldsymbol{\beta}$ via a $(b+1)$-dimensional grid search where, for each proposed $\boldsymbol{\beta}$, initial values for $\boldsymbol{\gamma}$ are considered by sampling values independently from within a suggested range of respective values. This combination of proposed initial values is finally identified such that the observed log-likelihood is maximum. This approach is recognized as being computationally intensive. Balakrishnan and Pal (2012) determine initial values for $\boldsymbol{\beta}$ based on Kaplan–Meier estimates of the cure rates under each of the considered conditions and then backsolve for $\boldsymbol{\beta}$. These values are then plugged into the observed log-likelihood function in order to produce a function only involving $\gamma$ (recall that this work considers an exponential lifetime distribution, hence $\gamma$ is an unknown constant) that can be maximized. Similarly, Balakrishnan and Feng (2018) determine $\boldsymbol{\beta}$ based on overestimates of the cure rate under each of the considered conditions and then backsolve for $\boldsymbol{\beta}$, while the starting values for $\boldsymbol{\gamma}$ are estimated by considering a linear model between the Nelson–Aalen estimates of the log-odds and $\ln(t)$, given the proportional odds assumption. Balakrishnan et al. (2017) determine initial values for $\boldsymbol{\beta} = (\beta_0, \beta_1)$ assuming that the observed censoring proportion for each group is its cure rate; meanwhile, the $\boldsymbol{\gamma} = (\gamma_0, \gamma_1, \gamma_2)$ initial values are determined from the regression equation

$$\ln\{-\ln[\hat{S}(t;\boldsymbol{\gamma})]\} = \gamma_0 \ln t + \gamma_2 x - \gamma_0 \ln \gamma_1,$$

where $\hat{S}(t;\boldsymbol{\gamma})$ denotes the Kaplan–Meier estimates of $S(t;\boldsymbol{\gamma})$. As a result of the profile likelihood approach, the respective MLEs are determined for each of a range of $\nu$ values (e.g. $\nu \in [0,4]$) taken in increments of 0.1 such that the ultimate determination of the MLEs is based on the collection of MLEs and $\nu$ that produce the ultimate maximum value for the likelihood.

### *8.2.2 Quantifying Variation*

Inverting the corresponding information matrix will provide the asymptotic variances and covariances of $\hat{\boldsymbol{\beta}}$ and $\hat{\boldsymbol{\gamma}}$. This, in turn, determines the standard errors for $(\hat{\boldsymbol{\beta}}, \hat{\boldsymbol{\gamma}})$ which allows analysts to consider asymptotic confidence intervals given the asymptotic normality of the MLEs $(\hat{\boldsymbol{\beta}}, \hat{\boldsymbol{\gamma}})$. The standard error of $\hat{p}_{0i}$ can be determined via the delta method leading to the asymptotic confidence interval for the cure rate $\boldsymbol{p}_0 = \frac{1}{1+\exp(\boldsymbol{x}'\boldsymbol{\beta})}$ given the asymptotic normality of $\hat{\boldsymbol{p}}_0$. More precisely, $\boldsymbol{p}_0$ can be estimated as $\hat{\boldsymbol{p}}_0 = g(\boldsymbol{\beta}) = \frac{1}{1+\exp(\boldsymbol{x}'\hat{\boldsymbol{\beta}})}$, given the MLE $\hat{\boldsymbol{\beta}}$, where $\hat{\boldsymbol{p}}_0$ is asymptotically normally distributed with mean $\boldsymbol{p}_0$ and variance $\boldsymbol{d}'\boldsymbol{\Sigma}\boldsymbol{d}$ where $\boldsymbol{d} = \left(\frac{\partial g(\boldsymbol{\beta})}{\partial \beta_0}, \ldots, \frac{\partial g(\boldsymbol{\beta})}{\partial \beta_b}\right)'$ and $\boldsymbol{\Sigma}$ is the variance–covariance matrix for $\hat{\boldsymbol{\beta}}$. Analysts can establish a confidence interval for $\boldsymbol{p}_0$ based on $\hat{\boldsymbol{p}}_0$ and its asymptotic variance.

### *8.2.3 Simulation Studies*

Simulation studies illustrate the model flexibilities attainable via the CMP cure rate model, assuming any of a number of associated lifetime distributions. Parameter bias, root mean-squared error (RMSE), and confidence interval coverage probabilities are all determined to assess performance. For the special-case cure rate simulation studies, results show that the EM algorithm estimates converge to their true parameters, and the standard error and RMSE, decrease as the sample size increases and the censoring proportion decreases (Balakrishnan and Pal, 2012, 2015b, 2016; Balakrishnan et al., 2017). Analogous results likewise hold for fixed $\nu$ and/or regarding the cure rate proportion – the estimates are close to their true values, and the RMSE is low when the sample size is large and also when the cure rate is small (Balakrishnan and Pal, 2013). The Bernoulli cure rate model consistently produced the best coverage probabilities, and parameter estimates with the smallest bias and RMSE, while the geometric cure rate model showed substantial under-coverage, particularly when partnered with a high censoring proportion and small lifetime parameter values (Balakrishnan et al., 2017). Utilizing EM on a CMP cure rate model with an underlying proportional odds construct meanwhile produces accurate estimates under the general CMP structure as well as its special cases. As the sample size increases, the bias, standard error, and RMSE decrease. These measures likewise decrease when the cure rate is high or when the amount of censoring is small (Balakrishnan and Feng, 2018). Other works however find that, for the general CMP cure rate data simulation, the profile likelihood technique results in estimated values for $\nu$ with large bias and RMSE; this issue, however, diminishes with the censoring proportion and as the sample size increases. Meanwhile, the estimates for $\boldsymbol{\beta}$ and $\boldsymbol{\gamma}$ are close to their true values; yet, their associated standard errors are smaller than expected, resulting in coverage probabilities that are less than their nominal levels. This phenomenon is due to the constrained inaccuracy behind estimating $\nu$ caused by using the profile likelihood approach to estimate $\nu$. The fact that the standard errors wind up estimated to be smaller produces coverage probabilities for the confidence intervals that are less than their respective nominal levels, resulting in model parameter under-coverage (Balakrishnan and Pal, 2015b, 2016).

Methods that a priori consider the loglinear link function $\ln(\lambda) = \boldsymbol{x}'\boldsymbol{\beta}$ can experience convergence issues because, for the geometric cure rate model, the loglinear link does not guarantee that $\lambda < 1$. The EM algorithm can likewise face convergence issues for the general CMP cure rate model, with $\nu$ being close to zero because this scenario closely mimics the geometric cure rate construct. Even for those models that assume $\lambda = g^{-1}(1 + \exp(\boldsymbol{x}'\boldsymbol{\beta}); \nu)$, the EM algorithm may not converge for the general CMP cure rate model; this phenomenon does not occur, however, when estimating parameters under any of the special case (i.e. Poisson, geometric, or Bernoulli) cure rate models (Balakrishnan and Pal, 2013, 2015b, 2016).

### 8.2.4 Hypothesis Testing and Model Discernment

The likelihood ratio test (LRT) is a popular approach for hypothesis testing to suggest one of the parsimonious special-case cure rate models (i.e. $H_0\colon \nu = 1$, $H_0\colon \nu = 0$, or $H_0\colon \nu \to \infty$) versus the general CMP cure rate model ($H_1\colon$ otherwise). Accordingly, the CMP cure rate model serves as a flexible exploration tool for determining if one of the special-case distributions can serve as an appropriate model. For $H_0\colon \nu = 1$, the LRT statistic $-2\ln\Lambda$ has an asymptotic chi-squared distribution with one degree of freedom, where $\Lambda$ equals the ratio of maximized likelihood functions under $H_0$ versus the unrestricted parameter space. Meanwhile, to test $H_0: \nu = 0$ or $H_0: \nu \to \infty$, the corresponding LRT statistic has a null distribution, equaling $0.5 + 0.5\chi_1^2$ because each of these tests considers a boundary value for $\nu$ (Self and Liang, 1987).

The chi-square distribution properly approximates the null distribution of the LRT regarding $H_0: \nu = 1$ versus $H_1: \nu \neq 1$. In fact, whether the underlying proportional odds structure is assumed to be log-logistic or Weibull, the power function increases as the sample size increases, and as the censoring proportion decreases (Balakrishnan and Feng, 2018; Balakrishnan and Pal, 2013, 2016). For small sample sizes, however, analysts are encouraged to consider a parametric bootstrap to perform model discernment. The respective tests about the boundary ($H_0: \nu = 0$, or $H_0: \nu \to \infty$) meanwhile produce disparate results. The resulting mixture distributions involving a point mass and chi-square distribution likewise approximate the null distribution for the LRT associated with $H_0: \nu \to \infty$, i.e. the Bernoulli case. For the geometric case, however, the mixture chi-square distribution does not reasonably approximate the LRT null distribution; the "observed levels are found to be considerably below the nominal level" (Balakrishnan and Pal, 2013, 2016).

The two boundary (i.e. Bernoulli or geometric) cure rate models produce LRTs with high power of rejecting the opposite boundary special case (Balakrishnan and Pal, 2015b, 2016; Balakrishnan et al., 2017). This is expected because these two represent the extreme CMP cases with regard to dispersion. Thus, when the true data stem from one of the boundary special-case cure rate models, the other boundary case exhibits a rejection rate that is considerably higher than other models. The LRT's ability to reject the Poisson model, however, varies. The LRT produces good power for rejecting the Poisson cure rate model when the true model is geometric, but not so much when the true model is Bernoulli; when the true model is Bernoulli, the LRT's statistical power to reject the Poisson distribution is small. When the true cure rate model is Poisson, the LRT has reasonable power to reject the Bernoulli model but less so for the geometric case. The power associated with rejecting the wrong model nonetheless generally increases with the sample size or as the censoring proportion decreases.

Balakrishnan and Pal (2015a) use the LRT to determine whether the CMP cure rate model with the generalized gamma $(g_1, g_2, g_3)$ lifetime can be simplified to consider any of the special-case lifetime distributions (i.e. lognormal, gamma, Weibull, and exponential). For the gamma ($H_0: g_2 = g_3$), Weibull ($H_0: g_2 = 1$), and exponential ($H_0: g_2 = 1, g_3 = 1$)

cases, the respective LRT statistics have a chi-squared distribution whose degrees of freedom equal the difference in the number of parameters between the respective null models and the generalized gamma distribution. The lognormal ($H_0: g_2 = 0$) special case, however, lies on the boundary, and thus the LRT statistic has an asymptotic distribution whose cumulative probability is adjusted as described above. The LRT null distribution closely approximates the respective chi-square distributions for the gamma, Weibull, and exponential lifetime special-case distributions; however, the lognormal case is found to be too conservative. The gamma and Weibull distributions meanwhile do not differ significantly from each other; yet, the LRTs find that the lognormal distribution differs statistically significantly from the Weibull, gamma, and exponential distributions.

The Akaike information criterion (AIC) and the Bayesian information criterion (BIC) are two popular measures used for model comparisons, as noted and discussed in Chapter 1. These information criteria correctly identify the best model with high accuracy when comparing the promotion time cure model (Poisson), mixture cure model (Bernoulli), or the geometric cure rate model. The respective selection rates each increase with the sample size and with a decrease in the censoring proportion (Balakrishnan and Feng, 2018; Balakrishnan and Pal, 2013, 2015b; Balakrishnan et al., 2017). The CMP with $\nu \in [0.5, 2]$ more closely relates to the promotion time cure model rather than the geometric or mixture cure models because these information criteria select the Poisson model at a higher rate than the Bernoulli or geometric distributions (Balakrishnan and Pal, 2016). Using the AIC and/or BIC for model comparison is appealing because these information criteria do not require $\nu$ to be estimated. In contrast, the LRT approach requires estimating $\nu$ which, given the profile likelihood approach, can be a time-consuming process.

Balakrishnan and Feng (2018) meanwhile find that, if the true model assumes a Weibull baseline, then the chances of selecting a cure rate model with the log-logistic baseline are small (and vice versa), and this rate decreases as the sample size increases or censoring decreases. Balakrishnan and Pal (2015a) likewise show that either criterion does a good job of correctly selecting the CMP cure rate model with lognormal or exponential distributions for the lifetime variable when those respectively represent the true form. More broadly, however, the AIC performs better than BIC at properly identifying the correct model. When the true lifetime variable is gamma distributed, the AIC does well given a large sample size or small censoring proportion, while the BIC does a good job only when the sample size is large. Given simulated data assuming a Weibull lifetime variable, the AIC properly selects that model more frequently with an increase in the sample size and a decrease in the proportion of censored observations, while the BIC consistently mis-selected the exponential distribution.

Cancho et al. (2012) analogously propose the deviance information criterion (DIC) as a Bayesian measure for model selection. Let $\boldsymbol{\theta}_1, \boldsymbol{\theta}_2, \ldots, \boldsymbol{\theta}_L$ be an MCMC sample from the posterior distribution $\pi(\boldsymbol{\theta} \mid \mathcal{D})$ for some dataset $\mathcal{D}$, and $D(\boldsymbol{\theta}) = -2\sum_{i=1}^{n} \ln(h(t_i; \boldsymbol{\theta}))$ denote the deviance, where $h(t_i; \boldsymbol{\theta}) = f_p(t_i; \boldsymbol{\theta})$ if the time to event is observed and $h(t_i; \boldsymbol{\theta}) = S_p(t_i; \boldsymbol{\theta})$ if the time to event is censored; $\boldsymbol{\theta} = (\boldsymbol{\beta}', \nu, \boldsymbol{\gamma}')$. The estimated DIC is

$$\widehat{\mathrm{DIC}} = 2\bar{D} - \widehat{D}, \text{ where } \bar{D} = \sum_{\ell=1}^{L} D(\boldsymbol{\theta}_\ell)/L, \text{ and} \tag{8.20}$$

$$\widehat{D} = D\left(\frac{1}{L}\sum_{\ell=1}^{L} \boldsymbol{\theta}_\ell\right) = D\left(\frac{1}{L}\sum_{\ell=1}^{L} \boldsymbol{\beta}'_\ell,\ \frac{1}{L}\sum_{\ell=1}^{L} \nu_\ell,\ \frac{1}{L}\sum_{\ell=1}^{L} \boldsymbol{\gamma}'_\ell\right); \tag{8.21}$$

among a collection of considered models, the selected model is that whose DIC is minimum. The authors also propose conducting model selection via the conditional predictive ordinate (CPO) statistic. For a dataset with the $i$th observation removed (labeled $\mathcal{D}^{(-i)}$) and associated posterior density function $\pi(\boldsymbol{\theta} \mid \mathcal{D}^{(-i)})$, the $i$th CPO is

$$\mathrm{CPO}_i = \int_{\Theta} h(t_i; \boldsymbol{\theta}) \pi(\boldsymbol{\theta} \mid \mathcal{D}^{(-i)}) d\boldsymbol{\theta}$$

with corresponding estimate

$$\widehat{\mathrm{CPO}}_i = \left[\frac{1}{L}\sum_{\ell=1}^{L} \frac{1}{h(t_i; \boldsymbol{\theta}_\ell)}\right]^{-1}$$

and $A^* = \sum_{i=1}^{n} \ln(\widehat{\mathrm{CPO}}_i)/n$. Large values of CPO and $A^*$ imply better model fit. Both measures show that the CMP cure rate model is preferred over the mixture cure and promotion time cure models.

## 8.3 Interval Censoring

Under the interval censoring framework, subjects are only observed during regular intervals; hence, an exact time to event is not known precisely, but rather the event has occurred within a particular interval (i.e. between consecutive interval timepoints). To date, only Pal and Balakrishnan (2017) address interval censoring in a CMP cure rate model; this section summarizes that work with discussion. Maintaining the notation described in Section 8.1, assume noninformative interval censoring where events occur in an interval $[L, R]$, where $L < R$ denote the timepoints before and after an event occurrence. The censoring indicator $\delta = I(R < \infty)$

equals 1 if an event is observed and 0 otherwise; $R = \infty$ represents the case where the event is not observed by the last inspection time. The authors assume a logistic link function for the cure rate $p_0$ (Equation (8.12)) to maintain the satisfied constraints associated with the CMP model for all $\nu$, and let the lifetime random variable $W$ be a Weibull distribution with shape and scale parameters, $1/\gamma_1$ and $1/\gamma_2$, respectively; see Section 8.5 for details regarding possible lifetime distributions, while other links can also be considered for $p_0$. Given the respective data for the $n$ subjects $\{(L_i, R_i, \boldsymbol{x}_i, \delta_i): i = 1, \ldots, n\}$, the resulting likelihood function is

$$L(\boldsymbol{\theta}; \boldsymbol{L}, \boldsymbol{R}, \boldsymbol{x}, \boldsymbol{\delta}) \propto \prod_{i=1}^{n} [S_p(L_i) - S_p(R_i)]^{\delta_i} [S_p(L_i)]^{1-\delta_i} \tag{8.22}$$

for a set of parameters $\boldsymbol{\theta} = (\boldsymbol{\beta}', \nu, \boldsymbol{\gamma}')'$.

### 8.3.1 Parameter Estimation

*EM Algorithm Approach*

The likelihood function can be partitioned based on those subjects whose events are observed versus otherwise, i.e. Equation (8.22) becomes

$$L(\boldsymbol{\theta}; \boldsymbol{L}, \boldsymbol{R}, \boldsymbol{x}, \boldsymbol{\delta}) \propto \prod_{\{i:\, \delta_i=1\}} (1 - p_{0i})[S_s(L_i) - S_s(R_i)] \prod_{\{i:\, \delta_i=0\}} [p_{0i} + (1 - p_{0i}) S_s(L_i)], \tag{8.23}$$

where the added variable $J_i$ is then incorporated as described in Section 8.2.1 to further develop a complete data likelihood function and associated log-likelihood (minus the constant term),

$$\begin{aligned} L_c(\boldsymbol{\theta}; \boldsymbol{L}, \boldsymbol{R}, \boldsymbol{x}, \boldsymbol{\delta}, \boldsymbol{J}) &\propto \prod_{\{i:\, \delta_i=1\}} [(1 - p_{0i})[S_s(L_i) - S_s(R_i)]]^{J_i} \\ &\quad \times \prod_{\{i:\, \delta_i=0\}} p_{0i}^{1-J_i} [(1 - p_{0i}) S_s(L_i)]^{J_i} \\ \ln L_c(\boldsymbol{\theta}; \boldsymbol{L}, \boldsymbol{R}, \boldsymbol{x}, \boldsymbol{\delta}, \boldsymbol{J}) &\propto \sum_{\{i:\, \delta_i=1\}} J_i \ln [(1 - p_{0i})[S_s(L_i) - S_s(R_i)]] \\ &\quad + \sum_{\{i:\, \delta_i=0\}} (1 - J_i) \ln (p_{0i}) \\ &\quad + \sum_{\{i:\, \delta_i=0\}} J_i \ln [(1 - p_{0i}) S_s(L_i)] \end{aligned}$$

on which an EM algorithm is conducted. Analogous to that described in Section 8.2.1, in the expectation step (assuming a fixed $\nu$) with $\boldsymbol{\theta}^{\dagger} = (\boldsymbol{\beta}', \boldsymbol{\gamma}')'$

and $\boldsymbol{\theta}^{\dagger}_{(k)}$ denoting its values at the $k$th iteration, Pal and Balakrishnan (2017) determine $E(\ln L_c(\boldsymbol{\theta}; \boldsymbol{L}, \boldsymbol{R}, \boldsymbol{x}, \boldsymbol{\delta}, \boldsymbol{J} \mid \boldsymbol{\theta}^{\dagger}_{(k)}, \boldsymbol{O})$ with regard to the distribution of unobserved $J_i$s, given $\boldsymbol{O} = \{\text{observed } J_i\text{s and } (L_i, R_i, \boldsymbol{x}_i, \delta_i), i = 1, \ldots, n\}$ by replacing $J_i$ with its expected value, namely

$$\pi_i^{(k)} = \begin{cases} \frac{(1-p_{0i})S_s(L_i)}{S_p(L_i)} \Big|_{\boldsymbol{\theta}_* = \boldsymbol{\theta}_*^{(k)}} & \text{for censored observations} \\ 1 & \text{for uncensored observations} \end{cases}$$

Assuming a Weibull lifetime distribution with shape and scale parameters $1/\gamma_1$ and $1/\gamma_2$, the resulting function

$$\begin{aligned} G(\boldsymbol{\theta}^{\dagger}, \boldsymbol{\pi}^{(k)}) &= E\left(\ln L_c(\boldsymbol{\theta}; \boldsymbol{L}, \boldsymbol{R}, \boldsymbol{x}, \boldsymbol{\delta}, \boldsymbol{J}) \mid \boldsymbol{\theta}^{\dagger}_{(k)}, \boldsymbol{O}\right) \\ &= \sum_{\{i:\, \delta_i = 1\}} \ln(z_{i1L} - z_{i1R}) + \sum_{\{i:\, \delta_i = 0\}} \pi_i^{(k)} \ln(z_{i1L}) \\ &\quad - \sum_{\{i:\, \delta_i = 0 \text{ or } 1\}} \ln(1 + \exp(\boldsymbol{x}_i'\boldsymbol{\beta})), \end{aligned}$$

where

$$z_{i1t} = \sum_{j=1}^{\infty} \frac{[g^{-1}(1 + \exp(\boldsymbol{x}_i'\boldsymbol{\beta}); \nu) S(t_i; \boldsymbol{\gamma})]^j}{(j!)^{\nu}} \quad \text{and}$$

$$\pi_i^{(k)} = \frac{z_{i1L}}{1 + z_{i1L}} \Big|_{\boldsymbol{\theta}^{\dagger} = \boldsymbol{\theta}^{\dagger}_{(k)}} \quad \text{for censored observation } i$$

is then maximized numerically via the EM gradient algorithm (Lange, 1995) in the maximization step in order to determine the next iteration of estimates, i.e.

$$\boldsymbol{\theta}^{\dagger}_{(k+1)} = \arg\max_{\boldsymbol{\theta}^{\dagger} \in \boldsymbol{\Theta}^{\dagger}} G(\boldsymbol{\theta}^{\dagger}, \boldsymbol{\pi}^{(k)}).$$

This procedure continues until convergence is reached such that the MLEs are attained for a fixed $\nu$. The dispersion parameter $\nu$ is then estimated via the profile likelihood method, thus identifying $\hat{\nu}$ that associates with the largest log-likelihood. See Pal and Balakrishnan (2017) for details, including the special case $G(\boldsymbol{\theta}^{\dagger}, \boldsymbol{\pi}^{(k)})$ for the geometric, Bernoulli, and Poisson cure rate models when the data are interval censored.

### 8.3.2 Variation Quantification

The approximate asymptotic variance–covariance matrix associated with the estimates can be obtained by inverting the observed Fisher information matrix for $\boldsymbol{\beta}$ and $\boldsymbol{\gamma}$ (for a fixed $\nu$). Assuming the usual regularity conditions, the MLEs $\hat{\boldsymbol{\beta}}$ and $\hat{\boldsymbol{\gamma}}$ are asymptotically normal; thus, confidence intervals for the parameters $\boldsymbol{\beta}$ and $\boldsymbol{\gamma}$ can be constructed accordingly. Pal and Balakrishnan (2017) supply the component details necessary to determine the information matrix, both for the general CMP cure rate model and its special cases.

### 8.3.3 Simulation Studies

Simulation studies show that this proposed EM algorithm performs well, producing estimates that are close to their true values associated with each of the special-case (i.e. Poisson, geometric, and Bernoulli) cure rate models. The standard errors and RMSE decrease as the sample size increases, and the coverage probabilities closely approximate their respective nominal levels. For the general CMP cure rate model, the profile likelihood approach works well but introduces component-level under-coverage when estimating $\boldsymbol{\gamma}$.

### 8.3.4 Hypothesis Testing and Model Discernment

The LRT as well as the AIC and BIC can be used for model selection and discrimination. As described in Section 8.2.4, an LRT statistic $\Lambda$ and corresponding $p$-value can aid in evaluating each of the respective special-case models (i.e. Poisson with $H_0: \nu = 1$, geometric as $H_0: \nu = 0$, or Bernoulli with $H_0: \nu \to \infty$) in comparison to a general CMP cure rate model where $\nu$ takes a value not considered in the null hypothesis. In all cases, the asymptotic distribution reasonably approximates the LRT null distribution. When the true model is either boundary case (i.e. $H_0: \nu = 0$ or $H_0: \nu \to \infty$), the test has high power to reject the other boundary case while it has less power to reject the Poisson cure rate model. When the true cure rate model is Poisson, the power to reject the other special-case models is relatively low. The AIC and BIC criteria can each tell the difference between the three special-case cure rate models and properly determine which to be the correct model choice. Data simulations further demonstrate the flexibility of the CMP cure rate model. This flexibility serves as a useful tool for model determination in that it helps to avoid potential working model mis-specification which, in turn, can lead to a substantive bias and lack of efficiency.

## 8.4 Destructive CMP Cure Rate Model

A destructive cure rate model offers a more pragmatic representation of the time to event in a competing cause scenario where the number of competing causes can now decrease via a destructive process. Letting $M$ denote the number of competing causes, we now consider the (latent) total damaged variable

$$B = \begin{cases} \sum_{i=1}^{M} X_i & M > 0 \\ 0 & M = 0, \end{cases} \tag{8.24}$$

where $X_i$ are independent and identically Bernoulli($p$)-distributed random variables (independent from $M$) noting the presence (1) or absence (0) of Cause $i = 1, \ldots, M$. By definition, $B$ is the total number of outcomes among $M$ considered competing causes. Equation (8.24) illustrates that the destruction assumes the binomial probability law, i.e. $B \mid M = m_*$ has a Binomial($m_*, p$) distribution. Given $B = b_*$, $W$ represents the (latent) lifetime random variable, $Y = \min(W_0, W_1, \ldots, W_{b_*})$ denotes the lifetime taking into account the $B$ potential combinations of causes that determine the time to event with $Y = \infty$ when $B = 0$, whose probability $p_0$ is the cured fraction. Rodrigues et al. (2011) introduce both a destructive length-biased Poisson and a destructive NB cure rate model as special cases of a destructive weighted Poisson cure rate model. A destructive length-biased Poisson is constructed via $M$ whose associated probability is $P(M = m; \lambda) = e^{-\lambda}\lambda^{m-1}/(m-1)!$ for $m = 1, 2, 3, \ldots$. The resulting destructive length-biased Poisson cure rate population survival and density functions are

$$S_p(y) = [1 - pF(y)] \exp(-\lambda pF(y)) \tag{8.25}$$

$$f_p(y) = p\left(\lambda + \frac{1}{1 - pF(y)}\right)[1 - pF(y)] \exp(-\lambda pF(y))f(y), \tag{8.26}$$

and cure rate, $p_0 = (1-p)\exp(-\lambda p)$. When $p = 1$, $S_p(y)$ is a proper survival function. A destructive NB cure rate model meanwhile is constructed by defining $M$ as having a NB($\eta, \phi$) distribution with probability mass function

$$P(M = m; \eta, \phi) = \frac{\Gamma(m + 1/\phi)}{m!\Gamma(1/\phi)}\left(\frac{\eta\phi}{1 + \eta\phi}\right)^m \left(\frac{1}{1 + \eta\phi}\right)^{1/\phi}, \quad m = 0, 1, 2, \ldots;$$

the geometric distribution with success probability $(1 + \eta\phi)^{-1}$ is a special case where $\phi = 1$. Given this framework, the resulting destructive NB cure rate population survival function and density function are

$$S_p(y) = [1 + \eta\phi pF(y)]^{-1/\phi} \tag{8.27}$$

$$f_p(y) = \frac{\eta p}{1 + \eta\phi pF(y)}[1 + \eta\phi pF(y)]^{-1/\phi} f(y) \tag{8.28}$$

with cure rate, $p_0 = (1 + \eta\phi p)^{-1/\phi}$.

The destructive CMP cure rate model is a more flexible alternative that contains the destructive Poisson and geometric cure rate models as special cases. For the special case where $M$ has a CMP($\lambda$, $\nu$) distribution, the respective survival and density functions, as well as the cure rate for the destructive framework, are

$$S_p(y) = \frac{Z(\lambda[1 - pF(y)], \nu)}{Z(\lambda, \nu)} \tag{8.29}$$

$$f_p(y) = \frac{pf(y)}{Z(\lambda, \nu)[1 - pF(y)]} \sum_{j=1}^{\infty} \frac{j[\lambda(1 - pF(y))]^j}{(j!)^\nu} \tag{8.30}$$

$$p_0 = \frac{Z(\lambda(1 - p), \nu)}{Z(\lambda, \nu)} \tag{8.31}$$

(Pal et al., 2018; Rodrigues et al., 2011, 2012). The cure rate $p_0$ decreases with respect to both $\lambda$ and $p$. For the special case where $\nu \in \mathbb{N}$, the survival function is

$$S_p(y; \lambda, \nu, p) = \exp[-\lambda pF(y)] \frac{{}_1F_\nu(1; 1, \ldots, 1; \lambda[1 - pF(y)])}{{}_1F_\nu(1; 1, \ldots, 1; \lambda)} \tag{8.32}$$

with total damage distribution having the form

$$P(B = b_*; \lambda, \nu, p) = \exp(-\lambda p) \frac{{}_1F_\nu(1; b_* + 1, \ldots, b_* + 1; \lambda(1 - p))}{{}_1F_\nu(1; 1, \ldots, 1; \lambda)},$$
$$b_* = 0, 1, 2, \ldots, \tag{8.33}$$

where

$${}_uF_\nu(\alpha_1, \ldots, \alpha_{u_1}; \beta_1, \ldots, \beta_{u_2}; x) = \sum_{m=0}^{\infty} \frac{\alpha_1^{(m)} \alpha_2^{(m)} \times \cdots \times \alpha_{u_1}^{(m)}}{\beta_1^{(m)} \beta_2^{(m)} \times \cdots \times \beta_{u_2}^{(m)}} \frac{x^m}{m!}$$

is the generalized hypergeometric function with $\alpha^{(m)} = \alpha(\alpha + 1) \times \cdots \times (\alpha + m - 1)$ denoting the rising factorial. The case where $p = 1$ implies that all initial competing causes are possible, i.e. there is no destructive process. The destructive CMP cure rate model is another example of a destructive weighted Poisson cure rate model with weight function $w(m; \nu) = (m!)^{1-\nu}$ (Rodrigues et al., 2011); see Section 1.4 for a general discussion of weighted Poisson distributions (including the weighted Poisson cure rate model) and Section 2.7 for a focused discussion of the COM–Poisson as a particular weighted Poisson distribution.

Research to date only considers the destructive CMP cure rate model under a noninformative construct where $Y$ is subject to right censoring, $C_i$ denotes the censoring time, $T_i = \min(Y_i, C_i)$ denotes the lifetime, and $\delta_i = I(Y_i \leq C_i)$ indicates whether or not the lifetime is observed.

The parameter vectors $\boldsymbol{\lambda}$ and $\boldsymbol{p}$, respectively, have loglinear and logistic link functions $\ln(\lambda_i) = \boldsymbol{x}'_{1i}\boldsymbol{\beta}_1$ and $\ln\left(\frac{p_i}{1-p_i}\right) = \boldsymbol{x}'_{2i}\boldsymbol{\beta}_2$, where $\boldsymbol{x}'_{1i}$ and $\boldsymbol{x}'_{2i}$ have distinct elements and $\boldsymbol{x}'_{1i}$ does not have a column of ones, while $\boldsymbol{\beta}_j, j = 1, 2$ has $b_j$ coefficients (Pal et al., 2018; Rodrigues et al., 2011, 2012). However, the loglinear link for $\lambda$ is concerning because it does not necessarily guarantee $\lambda < 1$ for the geometric (i.e. when $\nu = 0$) competing cause special case. No work considers the alternate parametrization $\lambda_i = g^{-1}(1 + \exp(\boldsymbol{x}'_{1i}\boldsymbol{\beta}_1); \nu)$ described in Section 8.1.

### 8.4.1 Parameter Estimation

Again, three approaches have been considered for parameter estimation: the method of ML, Bayesian estimation, and the EM algorithm. Rodrigues et al. (2011) perform ML estimation to estimate $\boldsymbol{\theta} = (\boldsymbol{\beta}'_1, \boldsymbol{\beta}'_2, \nu, \boldsymbol{\gamma}')'$ by numerically optimizing $\ln(L(\boldsymbol{\theta}; \boldsymbol{t}, \boldsymbol{\delta}))$ via Equation (8.13) in R, where $S_p$ and $f_p$ are defined in Equations (8.29) and (8.30). The resulting MLEs $\hat{\boldsymbol{\theta}} = (\hat{\boldsymbol{\beta}}'_1, \hat{\boldsymbol{\beta}}'_2, \hat{\nu}, \hat{\boldsymbol{\gamma}}')'$ have an asymptotically MVN distribution with mean $\boldsymbol{\theta}$ and approximate variance–covariance matrix $\boldsymbol{\Sigma}(\hat{\boldsymbol{\theta}})$ estimated by Equation (8.15).

Taking a Bayesian perspective, Rodrigues et al. (2012) assume that $\gamma_i$ ($i = 1, \ldots, I$), $\boldsymbol{\beta}_j$ ($j = 1, 2$), and $\nu$ are all independent such that the a priori joint distribution of these parameters equals the product of marginal distributions associated with all of the individual parameters, i.e. for $\boldsymbol{\theta} = (\boldsymbol{\beta}'_1, \boldsymbol{\beta}'_2, \nu, \boldsymbol{\gamma}')$,

$$\pi(\boldsymbol{\theta}) = \pi(\nu) \prod_{i=1}^{I} \pi(\gamma_i) \prod_{j=1}^{2} \prod_{\ell=1}^{b_j} \pi(\beta_{j\ell}). \tag{8.34}$$

The authors let $\beta_{j\ell}$ be $N(0, \sigma^2_{j\ell})$ distributed where $\ell = 1, \ldots, b_j$ for $j = 1, 2$, while $\nu$ assumes a Gamma$(a_1, a_2)$ distribution. Assuming a Weibull$(\gamma_1, \gamma_2)$ distribution as described in Section 8.5 for $Y$ as an illustration (thus $I = 2$), let $\gamma_1$ assume a Gamma$(a_1, a_2)$ distribution and $\gamma_2$ be normally distributed with mean 0 and variance $\sigma^2_{\gamma_2}$. By definition, $\nu$ and $\gamma_1$ each have shape and scale parameters $a_1$ and $a_2$ such that the mean equals $a_1/a_2$. In all cases, the hyperparameters are chosen to have vague prior distributions. The resulting joint posterior distribution's complex structure is difficult to study; thus, a Gibbs sampler can be used to approximate the form of the joint distribution with Metropolis–Hastings steps performed within the Gibbs sampler in order to simulate samples of $\boldsymbol{\theta}$.

Pal et al. (2018) propose an EM algorithm to determine the model parameter MLEs based on noninformative right censored data, motivated by the fact that missingness exists in that censored subjects can either be cured or susceptible without experiencing the event by the study conclusion. Using the complete data log-likelihood (without the constant; see Equation (8.18)), they determine the conditional expectation to be

$$\begin{aligned} G(\boldsymbol{\theta}, \boldsymbol{\pi}^{(k)}) &= \sum_{\{i:\, \delta_i=1\}} \ln f_p(t_i) + \sum_{\{i:\, \delta_i=0\}} (1-\pi_i^{(k)}) \ln p_{0i} \\ &\quad + \sum_{\{i:\, \delta_i=0\}} \pi_i^{(k)} \ln\left[(1-p_{0i})S_s(t_i)\right] \\ &\propto \sum_{\{i:\, \delta_i=1\}} \left[ \ln(p_i f(t_i)) + \ln\left( \sum_{j=1}^{\infty} \frac{j[\eta_i(1-p_iF(t_i))]^j}{(j!)^\nu} \right) - \ln(Z(1-p_iF(t_i),\nu)) \right] \\ &\quad + \sum_{\{i:\, \delta_i=0\}} (1-\pi_i^{(k)}) \left( \ln(Z(\lambda_i(1-p_i),\nu)) - \ln(Z(\lambda_i,\nu)) \right) \\ &\quad + \sum_{\{i:\, \delta_i=0\}} \pi_i^{(k)} \ln\left( \frac{Z(\lambda_i(1-p_iF(t_i)),\nu) - Z(\lambda_i(1-p_i),\nu)}{Z(\lambda_i,\nu)} \right), \end{aligned} \tag{8.35}$$

where $\pi_i^{(k)} = E(J_i \mid \boldsymbol{\theta}_{(k)}^{\dagger}, \boldsymbol{O}) = \frac{(1-p_{0i})S_1(t_i)}{S_p(t_i)} \mid_{\boldsymbol{\theta}=\boldsymbol{\theta}_{(k)}^{\dagger}}$ as discussed in Balakrishnan and Pal (2012), and then maximize it via the Lange (1995) EM gradient algorithm to obtain the next iterated values of the estimated values, $\boldsymbol{\theta}_{(k+1)} = \arg\max_{\boldsymbol{\theta}\in\boldsymbol{\Theta}} G(\boldsymbol{\theta}, \boldsymbol{\pi}^{(k)})$. The EM algorithm is performed until convergence is attained in that the relative difference in successive values is sufficiently small. The corresponding first- and second-order derivatives of $G(\boldsymbol{\theta}, \boldsymbol{\pi}^{(k)})$ are supplied in the appendix of Pal et al. (2018), along with the resulting $G(\boldsymbol{\theta}, \boldsymbol{\pi}^{(k)})$ and its derivatives for the destructive Bernoulli and Poisson cure rate models, respectively. Note that these represent two of the three special-case destructive CMP cure rate models. The destructive geometric cure rate model, however, is concerning with regard to this approach toward parameter estimation because the setup does not guarantee $\lambda < 1$ (which is a required constraint when $\nu = 0$).

Various computational matters are noted under the destructive cure rate framework and addressed in Pal et al. (2018). The likelihood terrain is very flat with respect to $\nu$, so the EM algorithm is conducted for a given $\nu$, and then the profile likelihood approach determines the optimal dispersion parameter. This approach, however, does not account for variation with respect to $\nu$. The normalizing constant $Z(\lambda,\nu) = \sum_{j=0}^{\infty} \frac{\lambda^j}{(j!)^\nu}$ is meanwhile approximated by terminating the summation at the point where the corresponding contribution to the summation is sufficiently small; see Section 2.8 for related discussion. As $\nu$ approaches 0, the required number of terms increases in order to reach convergence. Finally, the associated approximate variance–covariance matrix of the MLEs is determined by inverting its observed Fisher information matrix, where the MLEs and information matrix are used to determine the asymptotic confidence intervals based on the MLEs being asymptotically normal.

Simulation studies showed that the EM algorithm produces estimates that are close to their true parameters when applied to the destructive Bernoulli and Poisson cure rate models, respectively; results showed that the bias, standard error, and RMSE decrease as the sample size increases. The coverage probabilities likewise approached their respective nominal levels as the sample size increased. For the general destructive CMP cure rate model, the profile likelihood approach for $\nu$ likewise worked well with regard to parameter estimation, where the profile likelihood approach considers grid spacings of 0.1 up to a reasonable range for $\nu > 0$. This framework resulted in a small bias, where the bias and RMSE decreased with an increased sample size. The profile likelihood approach, however, resulted in the variation for $\nu$ not being assessed; this produced underestimated coverage probabilities. Alternatively estimating all parameters simultaneously produces larger standard errors (particularly for $\hat{\nu}$) such that the coverage probabilities overshoot their expected nominal levels.

### *8.4.2 Hypothesis Testing and Model Discernment*

The LRT, as well as the AIC and BIC, can be used for hypothesis testing and model discernment, as described in Section 8.2.4. Rodrigues et al. (2012) instead conduct model comparisons via the conditional predictive ordinate (CPO) statistic and the deviance information criterion (DIC) as described in Section 8.2.4, and they further consider the expected AIC ($\widehat{\text{EAIC}} = \bar{D} + 2k$) and expected BIC ($\widehat{\text{EBIC}} = \bar{D} + k\ln(n)$), respectively, where $k$ denotes the number of parameters, and $\bar{D}$ is defined in Equation (8.20). The model with the smallest criterion (whether DIC, EAIC, or EBIC) is the model having the best fit.

## 8.5 Lifetime Distributions

Various lifetime distributions can be assumed for modeling the time-to-event variable $W$. Balakrishnan and Pal (2012) assume that $W$ follows an exponential distribution with pdf and cdf, $f(w;\gamma) = \gamma\exp(-\gamma w)$ and $F(w;\gamma) = 1 - \exp(-\gamma w)$, respectively.

The most popular choice of distribution for $W$ is the Weibull$(s_1, s_2)$ distribution with shape parameter $s_1 > 0$ and scale parameter $s_2 > 0$; its pdf and cdf, respectively, are

$$f(w; s_1, s_2) = \frac{s_1}{s_2}\left(\frac{w}{s_2}\right)^{s_1 - 1} \quad w \geq 0 \tag{8.36}$$

$$F(w; s_1, s_2) = 1 - \exp\left[-\left(\frac{w}{s_2}\right)^{s_1}\right] \tag{8.37}$$

(Jiang and Murthy, 2011). Several works propose a Weibull distribution with shape parameter $s_1 = \gamma_1$ and scale parameter $s_2 = \exp(-\gamma_2/\gamma_1)$ such that $\gamma_1 > 0$, $\gamma_2 \in \mathbb{R}$ (Cancho et al., 2012; Chen et al., 1999; Farewell, 1982; Rodrigues et al., 2009). The Weibull parameters have robust estimates and standard errors associated with the CMP cure rate model (Balakrishnan and Pal, 2016). Cancho et al. (2012) assume a right-censored CMP cure rate model with Weibull lifetimes where, assuming a Bayesian construct, they let $\gamma_1$ be Gamma$(a_1, a_2)$ distributed (i.e. a Gamma distribution with mean $a_1/a_2$) and $\gamma_2$ have a normal distribution with mean 0 and variance $\sigma^2_{\gamma_2}$ define the respective prior distributions, further setting the hyperparameters to generate noninformative priors. See Section 8.2.1 for discussion regarding parameter estimation via the Bayesian paradigm. All of the works regarding the destructive CMP cure rate model likewise assume a Weibull distribution of some form for the lifetime random variable $W$. Rodrigues et al. (2011, 2012) and assume that $W$ has a Weibull distribution with shape parameter $s_1 = \gamma_1$ and scale parameter $s_2 = \exp(-\gamma_2/\gamma_1)$, while Pal et al. (2018) assume a Weibull distribution with shape and scale parameters $s_1 = 1/\gamma_1$ and $s_2 = 1/\gamma_2$, respectively.

One can assess whether a Weibull assumption is reasonable by considering a scatterplot of $\ln[-\ln \hat{S}(t_{(i)})]$ against $\ln(t_{(i)})$ for ordered time to events $t_{(i)}$ and associated Kaplan–Meier survival estimates $\hat{S}(t_{(i)})$, recognizing that the relationship

$$\ln[-\ln S(t; \gamma)] = \frac{1}{\gamma_1}\ln(\gamma_2) + \frac{1}{\gamma_1}\ln(t)$$

holds if the Weibull model is true. Associated tests based on the correlation coefficient or the supremum of the absolute values of the residuals can be conducted to draw conclusions (Balakrishnan and Pal, 2016).

Boag (1949) and Balakrishnan and Pal (2013) instead consider a lognormal distribution to model the lifetime distribution, where the time-to-event data have the pdf and survival function,

$$f(w;\boldsymbol{\gamma}) = \frac{1}{\sqrt{2\pi}\gamma_1 w}\exp\left[-\frac{1}{2}\left(\frac{\ln(\gamma_2 w)}{\gamma_1}\right)^2\right] \tag{8.38}$$

$$S(w;\boldsymbol{\gamma}) = 1 - \Phi\left(\frac{\ln(\gamma_2 w)}{\gamma_1}\right) \tag{8.39}$$

for $w > 0$, $\gamma_1 > 0$, $\gamma_2 > 0$; i.e. $\ln(W)$ is normally distributed with mean $-\ln(\gamma_2)$ and standard deviation $\gamma_1$.

Balakrishnan and Pal (2015b) consider a gamma time-to-event distribution for $W$. For $w > 0$, the density and survival functions are

$$f(w;\boldsymbol{\gamma}) = \frac{\left(\frac{\gamma_2}{\gamma_1^2}^{1/\gamma_1^2}\right)}{\Gamma\left(\frac{1}{\gamma_1^2}\right)} w^{-\left(1-\frac{1}{\gamma_1^2}\right)} \exp\left(-\frac{\gamma_2 w}{\gamma_1^2}\right) \tag{8.40}$$

$$S(w;\boldsymbol{\gamma}) = \frac{\Gamma\left(\frac{1}{\gamma_1^2}, \frac{\gamma_2 w}{\gamma_1^2}\right)}{\Gamma\left(\frac{1}{\gamma_1^2}\right)}, \tag{8.41}$$

where $\gamma_i > 0$ for $i = 1, 2$, and $\Gamma(\alpha,\beta) = \int_\beta^\infty \exp(-x)x^{\alpha-1}\mathrm{d}x$ and $\Gamma(\zeta) = \int_0^\infty \exp(-x)x^{\zeta-1}\mathrm{d}x$ denote the upper incomplete and complete gamma functions, respectively, for some $\alpha$, $\beta$, and $\zeta$. The appropriateness of this time-to-event distribution for the data can be assessed, recognizing that (in truth) $F^{-1}(1-S(t;\boldsymbol{\gamma})) = t$; for the $i$th-ordered observed lifetime and associated Kaplan–Meier estimate $\hat{S}(t_{(i)})$, the scatterplot of $F^{-1}(1-\hat{S}(t_{(i)}))$ versus $t_{(i)}$ values should produce a straight line if the assumed distribution is valid. Accordingly, analysts can consider the resulting correlation between $F^{-1}(1-\hat{S}(t_{(i)}))$ and $t_{(i)}$ or measure $\mathrm{SD} = \frac{1}{n_1}\sum_{i=1}^{n_1}[F^{-1}(1-\hat{S}(t_i)) - t_i]^2$ and consider associated tests for the respective measures.

Another distrbutional option is to assume that the time-to-event random variable $W$ has a generalized gamma$(g_1, g_2, g_3)$ distribution with pdf

$$f(w; g_1, g_2, g_3) = \begin{cases} g_2(g_2^{-2})^{g_2^{-2}}(g_1 w)^{g_2^{-2}(g_2/g_3)}\exp[-g_2^{-2}(g_1 w)^{g_2/g_3}]/[\Gamma(g_2^{-2})g_3 w] & g_2 > 0 \\ (\sqrt{2\pi}g_3 w)^{-1}\exp\{-[\ln(g_1 w)]^2/(2g_3^2)\} & g_2 = 0 \end{cases} \tag{8.42}$$

and a survival function

$$S(w; g_1, g_2, g_3) = \begin{cases} \Gamma(g_2^{-2}, g_2^{-2}(g_1 w)^{g_2/g_3})/\Gamma(g_2^{-2}) & g_2 > 0 \\ 1 - \Phi\left(\frac{\ln(g_1 w)}{g_3}\right) & g_2 = 0 \end{cases} \tag{8.43}$$

with scale parameter $g_1 > 0$ and shape parameters $g_2 \geq 0$ and $g_3 > 0$; $\Gamma(\cdot)$ and $\Gamma(\alpha,\beta) = \int_\beta^\infty e^{-x}x^{\alpha-1}\mathrm{d}x$, respectively, denote the complete and upper incomplete gamma functions, and $\Phi(\cdot)$ denotes the standard normal cdf. The generalized gamma distribution contains several special cases:

- Weibull (when $g_2 = 1$); in particular, $g_1 = \exp(\gamma_2/\gamma_1)$ and $g_3 = 1/\gamma_1$ provide the Rodrigues et al. (2009) and Cancho et al. (2012) parametrization.
- lognormal (when $g_2 = 0$); $g_1 = \gamma_2$ and $g_3 = \gamma_1$ produce the parametrization represented in Equations (8.38) and (8.39).
- Gamma (when $g_2/g_3 = 1$); setting $g_1 = \gamma_2$ and $g_2 = g_3 = \gamma_1$ produces the parametrization used in Balakrishnan and Pal (2015b).

Balakrishnan and Pal (2015a) opt to determine the gamma and incomplete gamma function derivatives numerically, even though explicit expressions exist for each. For the general CMP cure rate model with generalized gamma lifetime distribution, they find that undercoverage exists among the other lifetime parameters because of the two-way profile likelihood approach to estimate $g_2$. For the geometric special case ($\nu = 0$), however, only $g_3$ produces undercoverage.

Balakrishnan and Feng (2018) assume that the time-to-event variables satisfy a proportional odds model, i.e. that $W_j, j = 1, \ldots, M$, have the odds function $O(w; \boldsymbol{x}) = \theta O_B(w)$, where $O_B(w)$ denotes the baseline odds function and $O(w) = S(w)/F(w)$ defines the odds of surviving up to time $w$. The authors further assume that $\theta$ has a loglinear relationship $\ln(\theta) = \boldsymbol{x}'\boldsymbol{\gamma}_*$ with covariates $\boldsymbol{x} = (x_1, \ldots, x_p)'$ and coefficients $\boldsymbol{\gamma}_2 = (\gamma_{21}, \ldots, \gamma_{2p})'$ that further comprise the distributional form of $W$. The proportional odds relationship implies that $W_j$ has the respective survival function and pdf,

$$S(w) = [1 + \exp(-\boldsymbol{x}'\boldsymbol{\gamma}_2)\{S_B^{-1}(w) - 1\}]^{-1}, \tag{8.44}$$

$$f(w) = f_B(w)[\exp(-\boldsymbol{x}'\boldsymbol{\gamma}_2)][\{1 - S_B(w)\}\exp(-\boldsymbol{x}'\boldsymbol{\gamma}_2) + S_B(w)]^{-2}, \tag{8.45}$$

where $S_B(w)$ and $f_B(w)$ denote the baseline survival function and pdf. Assuming a Weibull baseline distribution with shape and scale parameters, $1/\gamma_0$ and $1/\gamma_1$, respectively (Jiang and Murthy, 2011), results in the survival and pdf having the respective forms

$$S(w; \boldsymbol{\gamma}) = [1 + e^{-\boldsymbol{x}'\boldsymbol{\gamma}_2}(e^{(\gamma_1 w)^{1/\gamma_0}} - 1)]^{-1}, \; w > 0, \tag{8.46}$$

$$f(w; \boldsymbol{\gamma}) = \frac{(\gamma_1 w)^{1/\gamma_0}}{\gamma_0 w} e^{\boldsymbol{x}'\boldsymbol{\gamma}_2 - (\gamma_1 w)^{1/\gamma_0}} [e^{-(\gamma_1 w)^{1/\gamma_0}}(e^{\boldsymbol{x}'\boldsymbol{\gamma}_2} - 1) + 1]^{-2}. \tag{8.47}$$

A log-logistic baseline distribution with scale and shape parameters, $\gamma_0 > 0$ and $\gamma_1 > 0$ (Bennett, 1983) meanwhile has the odds function

$$O(w; \boldsymbol{x}_c, \boldsymbol{\gamma}) = \frac{\gamma_0^{\gamma_1} e^{\boldsymbol{x}_c' \boldsymbol{\gamma}_2}}{w^{\gamma_1}}$$

for $W_i$, where $\gamma_0 > 0$ and $\gamma_1 > 0$, such that the resulting survival function has the form

$$S(w; \boldsymbol{\gamma}) = \frac{\gamma_0^{\gamma_1} e^{\boldsymbol{x}_c' \boldsymbol{\gamma}_2}}{\gamma_0^{\gamma_1} e^{\boldsymbol{x}_c' \boldsymbol{\gamma}_2} + w^{\gamma_1}}, \quad w > 0.$$

Assuming a log-logistic baseline odds, the geometric and Bernoulli cure rate models attain equivalent population survival functions (under reparametrization). Balakrishnan and Feng (2018) consider testing the effect of the proportional odds parameter, i.e. $H_0: \gamma_2 = 0$ versus $H_1: \gamma_2 \neq 0$; this assesses whether the constant lifetime model (whether Weibull or log-logistic) provides a sufficient fit or if the proportional odds model is more appropriate.

Balakrishnan et al. (2017) consider the time-to-event variables $W_i$ to have a proportional hazard construct, i.e. a hazard function of the form $h(w; \boldsymbol{x}_c, \boldsymbol{\gamma}) = h_0(w; \gamma_0, \gamma_1) e^{\boldsymbol{x}_c' \boldsymbol{\gamma}_2}$, where $h_0(w; \gamma_0, \gamma_1)$ is the baseline hazard function associated with the Weibull distribution, and $\boldsymbol{x}_c = (x_1, \ldots, x_p)'$ and $\boldsymbol{\gamma}_2 = (\gamma_{21}, \ldots, \gamma_{2p})'$ denote the proportional hazard regression covariates and corresponding coefficients. Assuming a baseline Weibull distribution with shape parameter $\gamma_0$ and scale $\gamma_1$ (Jiang and Murthy, 2011), the baseline hazard function is

$$h_0(w; \gamma_0, \gamma_1) = \frac{\gamma_0}{\gamma_1} \left( \frac{w}{\gamma_1} \right)^{\gamma_0 - 1}$$

with the hazard function

$$h(w; \boldsymbol{x}_c, \boldsymbol{\gamma}) = \frac{\gamma_0}{\gamma_1} \left( \frac{w}{\gamma_1} \right)^{\gamma_0 - 1} e^{\boldsymbol{x}_c' \boldsymbol{\gamma}_2},$$

i.e. the hazard function of a Weibull distribution with shape $\gamma_0$ and scale $\gamma_1 \exp(-\boldsymbol{x}_c' \boldsymbol{\gamma}_2 / \gamma_0)$. By definition, this CMP cure rate model construct reduces to that of Balakrishnan and Pal (2015a) when $\boldsymbol{\gamma}_2 = \boldsymbol{0}$.

## 8.6 Summary

The CMP cure rate model provides a flexible means for describing time-to-event data regarding a population that can be partitioned between those that are susceptible to an event occurrence and those that are cured or immune to any number of competing risks that can influence an event occurrence. This chapter discusses both a simple and destructive CMP cure rate model assuming noninformative right censoring and likewise considers the simple CMP cure rate model under noninformative interval censoring. Given each of the considered frameworks, this chapter discusses parameter estimation via the method of maximum likelihood, Bayesian estimation, and via expectation maximization. Model discernment is then studied through hypothesis testing via the LRT or comparing various appropriate information criteria to assess the appropriateness of a parsimonious special-case cure rate model (e.g. Poisson, Bernoulli, or geometric). Unfortunately, no `R` scripts are readily available in association with any of the aforementioned CMP cure rate models in order to conduct statistical computing.

# References


Abbas, N., Riaz, M., and Does, R. J. M. M. (2013). Mixed exponentially weighted moving average–cumulative sum charts for process monitoring. *Quality and Reliability Engineering International*, 29(3):345–356.

Abbasi, S. A. (2017). Poisson progressive mean control chart. *Quality and Reliability Engineering International*, 33(8):1855–1859.

Adamidis, K. and Loukas, S. (1998). A lifetime distribution with decreasing failure rate. *Statistics & Probability Letters*, 39(1):35–42.

Ahrens, J. H. and Dieter, U. (1982). Computer generation of Poisson deviates from modified normal distributions. *ACM Transactions on Mathematical Software*, 8(2): 163–179.

Airoldi, E. M., Anderson, A. G., Fienberg, S. E., and Skinner, K. K. (2006). Who wrote Ronald Reagan's radio addresses? *Bayesian Analysis*, 1(2):289–319.

Al-Osh, M. A. and Alzaid, A. A. (1987). First-order integer valued autoregressive (INAR(1)) process. *Journal of Time Series Analysis*, 8(3):261–275.

Al-Osh, M. A. and Alzaid, A. A. (1988). Integer-valued moving average (INMA) process. *Statistical Papers*, 29(1):281–300.

Al-Osh, M. A. and Alzaid, A. A. (1991). Binomial autoregressive moving average models. *Communications in Statistics. Stochastic Models*, 7(2):261–282.

Albers, W. (2011). Control charts for health care monitoring under overdispersion. *Metrika*, 74:67–83.

Alevizakos, V. and Koukouvinos, C. (2019). A double exponentially weighted moving average control chart for monitoring COM–Poisson attributes. *Quality and Reliability Engineering International*, 35(7):2130–2151.

Alevizakos, V. and Koukouvinos, C. (2022). A progressive mean control chart for COM–Poisson distribution. *Communications in Statistics – Simulation and Computation*, 51(3):849–867.

Alqawba, M. and Diawara, N. (2021). Copula-based Markov zero-inflated count time series models with application. *Journal of Applied Statistics*, 48(5):786–803.

Alzaid, A. A. and Al-Osh, M. A. (1993). Some autoregressive moving average processes with generalized Poisson marginal distributions. *Annals of the Institute of Statistical Mathematics*, 45(2):223–232.

Anan, O., Böhning, D., and Maruotti, A. (2017). Population size estimation and heterogeneity in capture–recapture data: A linear regression estimator based on the Conway–Maxwell–Poisson distribution. *Statistical Methods & Applications*, 26(1):49–79.

Arab, A. (2015). Spatial and spatio-temporal models for modeling epidemiological data with excess zeros. *International Journal of Environmental Research and Public Health*, 12(9):10536–10548.

Arbous, A. and Kerrick, J. E. (1951). Accident statistics and the concept of accident proneness. *Biometrics*, 7(4):340–432.

Aslam, M., Ahmad, L., Jun, C. H., and Arif, O. H. (2016a). A control chart for COM–Poisson distribution using multiple dependent state sampling. *Quality and Reliability Engineering International*, 32(8):2803–2812.

Aslam, M., Azam, M., and Jun, C.-H. (2016b). A control chart for COM–Poisson distribution using resampling and exponentially weighted moving average. *Quality and Reliability Engineering International*, 32(2):727–735.

Aslam, M., Khan, N., and Jun, C.-H. (2018). A hybrid exponentially weighted moving average chart for COM–Poisson distribution. *Transactions of the Institute of Measurement and Control*, 40(2):456–461.

Aslam, M., Saghir, A., Ahmad, L., Jun, C.-H., and Hussain, J. (2017). A control chart for COM–Poisson distribution using a modified EWMA statistic. *Journal of Statistical Computation and Simulation*, 87(18):3491–3502.

Atkinson, A. C. and Yeh, L. (1982). Inference for Sichel's compound Poisson distribution. *Journal of the American Statistical Association*, 77(377):153–158.

Bailey, B. J. R. (1990). A model for function word counts. *Journal of the Royal Statistical Society. Series C (Applied Statistics)*, 39(1):107–114.

Balakrishnan, N., Barui, S., and Milienos, F. (2017). Proportional hazards under Conway–Maxwell–Poisson cure rate model and associated inference. *Statistical Methods in Medical Research*, 26(5):2055–2077.

Balakrishnan, N. and Feng, T. (2018). Proportional odds under Conway–Maxwell–Poisson cure rate model and associated likelihood inference. *Statistics, Optimization & Information Computing*, 6(3):305–334.

Balakrishnan, N. and Pal, S. (2012). EM algorithm-based likelihood estimation for some cure rate models. *Journal of Statistical Theory and Practice*, 6(4): 698–724.

Balakrishnan, N. and Pal, S. (2013). Lognormal lifetimes and likelihood-based inference for flexible cure rate models based on COM–Poisson family. *Computational Statistics and Data Analysis*, 67:41–67.

Balakrishnan, N. and Pal, S. (2015a). An EM algorithm for the estimation of parameters of a flexible cure rate model with generalized gamma lifetime and model discrimination using likelihood- and information-based methods. *Computational Statistics*, 30(1):151–189.

Balakrishnan, N. and Pal, S. (2015b). Likelihood inference for flexible cure rate models with gamma lifetimes. *Communications in Statistics – Theory and Methods*, 44(19):4007–4048.

Balakrishnan, N. and Pal, S. (2016). Expectation maximization-based likelihood inference for flexible cure rate models with Weibull lifetimes. *Statistical Methods in Medical Research*, 25(4):1535–1563.

Banerjee, S. and Carlin, B. P. (2004). Parametric spatial cure rate models for interval-censored time-to-relapse data. *Biometrics*, 60(1):268–275.

Barlow, R. and Proschan, F. (1965). *Mathematical Theory of Reliability. Classics in Applied Mathematics, 17.* Society for Industrial and Applied Mathematics (SIAM), Philadelphia, PA. Republished in 1996.

Barriga, G. D. and Louzada, F. (2014). The zero-inflated Conway–Maxwell–Poisson distribution: Bayesian inference, regression modeling and influence diagnostic. *Statistical Methodology*, 21:23–34.

Bates, D., Mächler, M., Bolker, B., and Walker, S. (2015). Fitting linear mixed-effects models using lme4. *Journal of Statistical Software*, 67(1):1–48.

Benjamin, M. A., Rigby, R. A., and Stasinopoulos, D. M. (2003). Generalized autoregressive moving average models. *Journal of the American Statistical Association*, 98(461):214–223.

Bennett, S. (1983). Log-logistic regression models for survival data. *Journal of the Royal Statistical Society. Series C (Applied Statistics)*, 32(2):165–171.

Benson, A. and Friel, N. (2017). Bayesian inference, model selection and likelihood estimation using fast rejection sampling: The Conway–Maxwell–Poisson distribution. *Bayesian Analysis*, 16(3):905–931.

Berkson, J. and Gage, R. P. (1952). Survival curve for cancer patients following treatment. *Journal of the American Statistical Association*, 47(259):501–515.

Blocker, A. W. (2018). fastGHQuad: Fast Rcpp implementation of Gauss-Hermite quadrature, version 1.0. https://cran.r-project.org/web/packages/fastGHQuad/index.html.

Boag, J. W. (1949). Maximum likelihood estimates of the proportion of patients cured by cancer therapy. *Journal of the Royal Statistical Society. Series B (Methodological)*, 11(1):15–53.

Boatwright, P., Borle, S., and Kadane, J. B. (2003). A model of the joint distribution of purchase quantity and timing. *Journal of the American Statistical Association*, 98(463):564–572.

Booth, J. G., Casella, G., Friedl, H., and Hobert, J. P. (2003). Negative binomial loglinear mixed models. *Statistical Modelling*, 3(3):179–191.

Borges, P., Rodrigues, J., Balakrishnan, N., and Bazán, J. (2014). A COM–Poisson type generalization of the binomial distribution and its properties and applications. *Statistics and Probability Letters*, 87(1):158–166.

Borle, S., Boatwright, P., and Kadane, J. B. (2006). The timing of bid placement and extent of multiple bidding: An empirical investigation using eBay online auctions. *Statistical Science*, 21(2):194–205.

Borle, S., Boatwright, P., Kadane, J. B., Nunes, J. C., and Shmueli, G. (2005). The effect of product assortment changes on customer retention. *Marketing Science*, 24(4):616–622.

Borle, S., Dholakia, U., Singh, S., and Westbrook, R. (2007). The impact of survey participation on subsequent behavior: An empirical investigation. *Marketing Science*, 26(5):711–726.

Borror, C. M., Champ, C. W., and Rigdon, S. E. (1998). Poisson EWMA control charts. *Journal of Quality Technology*, 30(4):352–361.

Bourke, P. D. (2001). The geometric CUSUM chart with sampling inspection for monitoring fraction defective. *Journal of Applied Statistics*, 28(8):951–972.

Brännäs, K. and Hall, A. (2001). Estimation in integer-valued moving average models. *Applied Stochastic Models in Business and Industry*, 17(3):277–291.

Brännäs, K. and Hellström, J. (2001). Generalized integer-valued autoregression. *Econometric Reviews*, 20(4):425–443.

Breslow, N. E. (1984). Extra-Poisson variation in log-linear models. *Journal of the Royal Statistical Society. Series C (Applied Statistics)*, 33(1):38–44.

Brook, D. and Evans, D. A. (1972). An approach to the probability distribution of CUSUM run length. *Biometrika*, 59(3):539–549.

Brooks, M. E., Kristensen, K., van Benthem, K. J., Magnusson, A., Berg, C. W., Nielsen, A., Skaug, H. J., Mächler, M., and Bolker, B. M. (2017). glmmTMB balances speed and flexibility among packages for zero-inflated generalized linear mixed modeling. *The R Journal*, 9(2):378–400.

Brown, T. C. and Xia, A. (2001). Stein's method and birth-death processes. *Annals of Probability*, 29(3):1373–1403.

Burnham, K. P. and Anderson, D. R. (2002). *Model Selection and Multimodel Inference*. Springer.

Cancho, V., de Castro, M., and Rodrigues, J. (2012). A Bayesian analysis of the Conway–Maxwell–Poisson cure rate model. *Stat Papers*, 53(1): 165–176.

Canty, A. and Ripley, B. D. (2020). *boot: Bootstrap R (S-Plus) Functions*. R package version 1.3-25.

Casella, G. and Berger, R. L. (1990). *Statistical Inference*. Duxbury Press, Belmont, California.

Çinlar, E. (1975). *Introduction to Stochastic Processes*. Prentice-Hall.

Chakraborty, S. and Imoto, T. (2016). Extended Conway–Maxwell–Poisson distribution and its properties and applications. *Journal of Statistical Distributions and Applications*, 3(5):1–19.

Chakraborty, S. and Ong, S. H. (2016). A COM–Poisson-type generalization of the negative binomial distribution. *Communications in Statistics – Theory and Methods*, 45(14):4117–4135.

Chanialidis, C. (2020). combayes: Bayesian regression for COM–Poisson data. version 0.0.1.

Chanialidis, C., Evers, L., Neocleous, T., and Nobile, A. (2014). Retrospective sampling in MCMC with an application to COM–Poisson regression. *Stat*, 3(1): 273–290.

Chanialidis, C., Evers, L., Neocleous, T., and Nobile, A. (2017). Efficient Bayesian inference for COM–Poisson regression models. *Statistics and Computing*, 28(3): 595–608.

Chatla, S. B. and Shmueli, G. (2018). Efficient estimation of COM–Poisson regression and a generalized additive model. *Computational Statistics and Data Analysis*, 121:71–88.

Chen, J.-H. (2020). A double generally weighted moving average chart for monitoring the COM–Poisson processes. *Symmetry*, 12(6):1014.

Chen, M.-H., Ibrahim, J. G., and Sinha, D. (1999). A new Bayesian model for survival data with a surviving fraction. *Journal of the American Statistical Association*, 94(447):909–919.

Chen, N., Zhou, S., Chang, T.-S., and Huang, H. (2008). Attribute control charts using generalized zero-inflated Poisson distribution. *Quality and Reliability Engineering International*, 24(7):793–806.

Chiu, W.-C. and Lu, S.-L. (2015). On the steady-state performance of the Poisson double GWMA control chart. *Quality Technology & Quantitative Management*, 12(2):195–208.

Chiu, W.-C. and Sheu, S.-H. (2008). Fast initial response features for Poisson GWMA control charts. *Communications in Statistics – Simulation and Computation*, 37(7):1422–1439.

Choo-Wosoba, H. and Datta, S. (2018). Analyzing clustered count data with a cluster-specific random effect zero-inflated Conway–Maxwell–Poisson distribution. *Journal of Applied Statistics*, 45(5):799–814.

Choo-Wosoba, H., Gaskins, J., Levy, S., and Datta, S. (2018). A Bayesian approach for analyzing zero-inflated clustered count data with dispersion. *Statistics in Medicine*, 37(5):801–812.

Choo-Wosoba, H., Levy, S. M., and Datta, S. (2016). Marginal regression models for clustered count data based on zero-inflated Conway–Maxwell–Poisson distribution with applications. *Biometrics*, 72(2):606–618.

Clark, S. J. and Perry, J. N. (1989). Estimation of the negative binomial parameter $\kappa$ by maximum quasi-likelihood. *Biometrics*, 45(1):309–316.

Consul, P. and Jain, G. (1973). A generalization of the Poisson distribution. *Technometrics*, 15(4):791–799.

Consul, P. C. (1988). *Generalized Poisson Distributions*. CRC Press.

Conway, R. W. and Maxwell, W. L. (1962). A queuing model with state dependent service rates. *Journal of Industrial Engineering*, 12(2):132–136.

Cordeiro, G. M., Rodrigues, J., and de Castro, M. (2012). The exponential COM–Poisson distribution. *Stat Papers*, 53(3):653–664.

Curran, J., Williams, A., Kelleher, J., and Barber, D. (2015). *The multicool Package: Permutations of Multisets in Cool-Lex Order*, 0.1-6 edition.

Daly, F. and Gaunt, R. (2016). The Conway-Maxwell-Poisson distribution: Distributional theory and approximation. *Alea: Latin American journal of probability and mathematical statistics*, 13(2):635–658.

De Oliveira, V. (2013). Hierarchical Poisson models for spatial count data. *Journal of Multivariate Analysis*, 122:393–408.

del Castillo, J. and Pérez-Casany, M. (2005). Overdispersed and underdispersed Poisson generalizations. *Journal of Statistical Planning and Inference*, 134(2):486–500.

Demirtas, H. (2017). On accurate and precise generation of generalized Poisson variates. *Communications in Statistics – Simulation and Computation*, 46(1):489–499.

Diggle, P. J. and Milne, R. K. (1983). Negative binomial quadrat counts and point processes. *Scandinavian Journal of Statistics*, 10(4):257–267.

Dobson, A. J. and Barnett, A. G. (2018). *An Introduction to Generalized Linear Models*. Chapman & Hall/CRC Press, 4th edition.

Doss, D. C. (1979). Definition and characterization of multivariate negative binomial distribution. *Journal of Multivariate Analysis*, 9(3):460–464.

Dunn, J. (2012). compoisson: Conway–Maxwell–Poisson distribution, version 0.3. https://cran.r-project.org/web/packages/compoisson/index.html.

Efron, B. (1986). Double exponential families and their use in generalized linear regression. *Journal of the American Statistical Association*, 81(395):709–721.

Famoye, F. (1993). Restricted generalized Poisson regression model. *Communications in Statistics – Theory and Methods*, 22(5):1335–1354.

Famoye, F. (2007). Statistical control charts for shifted generalized Poisson distribution. *Statistical Methods and Applications*, 3(3):339–354.

Famoye, F. (2010). On the bivariate negative binomial regression model. *Journal of Applied Statistics*, 37(6):969–981.

Famoye, F. and Consul, P. C. (1995). Bivariate generalized Poisson distribution with some applications. *Metrika*, 42(1):127–138.

Famoye, F. and Singh, K. P. (2006). Zero-inflated generalized Poisson regression model with an application to domestic violence data. *Journal of Data Science*, 4(1): 117–130.

Famoye, F., Wulu Jr., J., and Singh, K. P. (2004). On the generalized Poisson regression model with an application to accident data. *Journal of Data Science*, 2(3): 287–295.

Fang, Y. (2003). C-chart, X-chart, and the Katz family of distributions. *Journal of Quality Technology*, 35(1):104–114.

Farewell, V. T. (1982). The use of mixture models for the analysis of survival data with long-term survivors. *Biometrics*, 38(4):1041–1046.

Francis, R. A., Geedipally, S. R., Guikema, S. D., Dhavala, S. S., Lord, D., and LaRocca, S. (2012). Characterizing the performance of the Conway–Maxwell–Poisson generalized linear model. *Risk Analysis*, 32(1):167–183.

Freeland, R. K. and McCabe, B. (2004). Analysis of low count time series data by Poisson autoregression. *Journal of Time Series Analysis*, 25(5): 701–722.

Fung, T., Alwan, A., Wishart, J., and Huang, A. (2020). mpcmp: Mean-parametrized Conway–Maxwell–Poisson, version 0.3.6. https://cran.r-project.org/web/packages/mpcmp/index.html.

Fürth, R. (1918). Statistik und wahrscheinlichkeitsnachwirkung. *Physikalische Zeitschrift*, 19:421–426.

Gan, F. F. (1990). Monitoring Poisson observations using modified exponentially weighted moving average control charts. *Communications in Statistics – Simulation and Computation*, 19(1):103–124.

Gaunt, R. E., Iyengar, S., Daalhuis, A. B. O., and Simsek, B. (2016). An asymptotic expansion for the normalizing constant of the Conway–Maxwell–Poisson distribution. arXiv:1612.06618.

Genest, C. and Nešlehová, J. (2007). A primer on copulas for count data. *ASTIN Bulletin*, 37(2):475–515.

Gilbert, P. and Varadhan, R. (2016). *numDeriv: Accurate Numerical Derivatives*. R package version 2016.8-1.

Gillispie, S. B. and Green, C. G. (2015). Approximating the Conway–Maxwell–Poisson distribution normalization constant. *Statistics*, 49(5):1062–1073.

Greene, W. H. (1994). Accounting for Excess Zeros and Sample Selection in Poisson and Negative Binomial Regression Models. *SSRN eLibrary*.

Guikema, S. D. and Coffelt, J. P. (2008). A flexible count data regression model for risk analysis. *Risk Analysis*, 28(1):213–223.

Gupta, R. C. (1974). Modified power series distributions and some of its applications. *Sankhya Series B*, 36:288–298.

Gupta, R. C. (1975). Maximum likelihood estimation of a modified power series distribution and some of its applications. *Communications in Statistics – Theory and Methods*, 4(7):689–697.

Gupta, R. C. and Huang, J. (2017). The Weibull–Conway–Maxwell–Poisson distribution to analyze survival data. *Journal of Computational and Applied Mathematics*, 311:171–182.

Gupta, R. C. and Ong, S. (2004). A new generalization of the negative binomial distribution. *Computational Statistics & Data Analysis*, 45(2):287–300.

Gupta, R. C. and Ong, S. (2005). Analysis of long-tailed count data by Poisson mixtures. *Communications in Statistics – Theory and Methods*, 34(3):557–573.

Gupta, R. C., Sim, S. Z., and Ong, S. H. (2014). Analysis of discrete data by Conway–Maxwell–Poisson distribution. *AStA Advances in Statistical Analysis*, 98(4): 327–343.

Guttorp, P. (1995). *Stochastic Modeling of Scientific Data*. Chapman & Hall/CRC Press.

Hilbe, J. M. (2007). *Negative Binomial Regression*. Cambridge University Press, 5th edition.

Hilbe, J. M. (2014). *Modeling Count Data*. Cambridge University Press.

Hinde, J. (1982). Compound Poisson regression models. In Gilchrist, R., editor, *GLIM 82: Proc. Internat. Conf. Generalized Linear Models*, pages 109–121, Berlin. Springer.

Holgate, P. (1964). Estimation for the bivariate Poisson distribution. *Biometrika*, 51 (1–2):241–245.

Huang, A. (2017). Mean-parametrized Conway–Maxwell–Poisson regression models for dispersed counts. *Statistical Modelling*, 17(6):359–380.

Huang, A. and Kim, A. S. I. (2019). Bayesian Conway–Maxwell–Poisson regression models for overdispersed and underdispersed counts. *Communications in Statistics – Theory and Methods*, 50(13):1–12.

Hwang, Y., Joo, H., Kim, J.-S., and Kweon, I.-S. (2007). Statistical background subtraction based on the exact per-pixel distributions. In *MVA2007 IAPR Conference on Machine Vision Applications, May 16–18, 2007*, pages 540–543, Tokyo, Japan.

Imoto, T. (2014). A generalized Conway-Maxwell-Poisson distribution which includes the negative binomial distribution. *Applied Mathematics and Computation*, 247(C):824–834.

Jackman, S., Tahk, A., Zeileis, A., Maimone, C., Fearon, J., and Meers, Z. (2017). pscl: Political Science Computational Laboratory, version 1.5.2. https://cran.r-project.org/web/packages/pscl/index.html.

Jackson, M. C., Johansen, L., Furlong, C., Colson, A., and Sellers, K. F. (2010). Modelling the effect of climate change on prevalence of malaria in western Africa. *Statistica Neerlandica*, 64(4):388–400.

Jackson, M. C. and Sellers, K. F. (2008). Simulating discrete spatially correlated Poisson data on a lattice. *International Journal of Pure and Applied Mathematics*, 46(1): 137–154.

Jiang, R. and Murthy, D. (2011). A study of Weibull shape parameter: Properties and significance. *Reliability Engineering & System Safety*, 96(12):1619–1626.

Jin-Guan, D. and Yuan, L. (1991). The integer-valued autoregressive (INAR($p$)) model. *Journal of Time Series Analysis*, 12(2):129–142.

Joe, H. (1996). Time series models with univariate margins in the convolution-closed infinitely divisible class. *Journal of Applied Probability*, 33(3):664–677.

Joe, H. (1997). *Multivariate Models and Multivariate Dependence Concepts*. Chapman & Hall/CRC Press, 1st edition.

Joe, H. (2014). *Dependence Modeling with Copulas*. Chapman & Hall/CRC Monographs on Statistics & Applied Probability. CRC Press.

Johnson, N., Kotz, S., and Balakrishnan, N. (1997). *Discrete Multivariate Distributions*. John Wiley & Sons.

Jowaheer, V., Khan, N. M., and Sunecher, Y. (2018). A BINAR(1) time-series model with cross-correlated COM—Poisson innovations. *Communications in Statistics – Theory and Methods*, 47(5):1133–1154.

Jowaheer, V. and Mamode Khan, N. A. (2009). Estimating regression effects in COM–Poisson generalized linear model. *International Journal of Computational and Mathematical Sciences*, 3(4):168–173.

Jung, R. C. and Tremayne, A. R. (2006). Binomial thinning models for integer time series. *Statistical Modelling*, 6(2):81–96.

Kadane, J. B. (2016). Sums of possibly associated Bernoulli variables: The Conway–Maxwell-Binomial distribution. *Bayesian Analysis*, 11(2):403–420.

Kadane, J. B., Krishnan, R., and Shmueli, G. (2006a). A data disclosure policy for count data based on the COM–Poisson distribution. *Management Science*, 52(10): 1610–1617.

Kadane, J. B., Shmueli, G., Minka, T. P., Borle, S., and Boatwright, P. (2006b). Conjugate analysis of the Conway–Maxwell–Poisson distribution. *Bayesian Analysis*, 1(2):363–374.

Kadane, J. B. and Wang, Z. (2018). Sums of possibly associated multivariate indicator functions: The Conway–Maxwell-multinomial distribution. *Brazilian Journal of Probability and Statistics*, 32(3):583–596.

Kaminsky, F. C., Benneyan, J. C., Davis, R. D., and Burke, R. J. (1992). Statistical control charts based on a geometric distribution. *Journal of Quality Technology*, 24(2):63–69.

Kannan, D. (1979). *An Introduction to Stochastic Processes*. Elsevier North Holland.

Karlis, D. and Ntzoufras, I. (2005). *The bivpois package: Bivariate Poisson models using the EM algorithm*, 0.50-2 edition.

Karlis, D. and Ntzoufras, I. (2008). Bayesian modelling of football outcomes: Using the Skellam's distribution for the goal difference. *IMA Journal of Management Mathematics*, 20(2):133–145.

Kemp, C. D. and Kemp, A. W. (1956). Generalized hypergeometric distributions. *Journal of the Royal Statistical Society Series B*, 18(2):202–211.

Khan, N. M. and Jowaheer, V. (2013). Comparing joint GQL estimation and GMM adaptive estimation in COM–Poisson longitudinal regression model. *Communications in Statistics – Simulation and Computation*, 42(4):755–770.

Khan, N. M., Sunecher, Y., and Jowaheer, V. (2016). Modelling a non-stationary BINAR(1) Poisson process. *Journal of Statistical Computation and Simulation*, 86(15):3106–3126.

Kim, Y. J. and Jhun, M. (2008). Cure rate model with interval censored data. *Statistics in Medicine*, 27(1):3–14.

Kocherlakota, S. and Kocherlakota, K. (1992). *Bivariate Discrete Distributions*. Marcel Dekker.

Kokonendji, C. C. (2014). Over- and underdispersion models. In Balakrishnan, N., editor, *Methods and Applications of Statistics in Clinical Trials: Planning, Analysis, and Inferential Methods*, pages 506–526. John Wiley & Sons.

Kokonendji, C. C., Mizere, D., and Balakrishnan, N. (2008). Connections of the Poisson weight function to overdispersion and underdispersion. *Journal of Statistical Planning and Inference*, 138(5):1287–1296.

Kokonendji, C. C. and Puig, P. (2018). Fisher dispersion index for multivariate count distributions: A review and a new proposal. *Journal of Multivariate Analysis*, 165(C):180–193.

Krishnamoorthy, A. S. (1951). Multivariate binomial and Poisson distributions. *Sankhyā: The Indian Journal of Statistics (1933–1960)*, 11(2):117–124.

Kuş, C. (2007). A new lifetime distribution. *Computational Statistics and Data Analysis*, 51(9):4497–4509.

Kutner, M. H., Nachtsheim, C. J., and Neter, J. (2003). *Applied Linear Regression Models, 4th edition*. McGraw-Hill.

Lambert, D. (1992). Zero-inflated Poisson regression, with an application to defects in manufacturing. *Technometrics*, 34(1):1–14.

Lange, K. (1995). A gradient algorithm locally equivalent to the EM algorithm. *Journal of the Royal Statistical Society: Series B (Methodological)*, 57(2):425–437.

Lee, J., Wang, N., Xu, L., Schuh, A., and Woodall, H. W. (2013). The effect of parameter estimation on upper-sided Bernoulli cumulative sum charts. *Quality and Reliability Engineering International*, 29(5):639–651.

Lee, M.-L. T. (1996). Properties and applications of the Sarmanov family of bivariate distributions. *Communications in Statistics – Theory and Methods*, 25(6):1207–1222.

Li, H., Chen, R., Nguyen, H., Chung, Y.-C., Gao, R., and Demirtas, H. (2020). *RNGforGPD: Random Number Generation for Generalized Poisson Distribution*. R package version 1.1.0.

Lindsey, J. K. and Mersch, G. (1992). Fitting and comparing probability distributions with log linear models. *Computational Statistics and Data Analysis*, 13(4):373--384.

Loeys, T., Moerkerke, B., De Smet, O., and Buysse, A. (2012). The analysis of zero-inflated count data: Beyond zero-inflated Poisson regression. *British Journal of Mathematical and Statistical Psychology*, 65(1):163–180.

Lord, D. (2006). Modeling motor vehicle crashes using Poisson-gamma models: Examining the effects of low sample mean values and small sample size on the estimation of the fixed dispersion parameter. *Accident Analysis & Prevention*, 38(4):751–766.

Lord, D., Geedipally, S. R., and Guikema, S. D. (2010). Extension of the application of Conway–Maxwell–Poisson models: Analyzing traffic crash data exhibiting underdispersion. *Risk Analysis*, 30(8):1268–1276.

Lord, D. and Guikema, S. D. (2012). The Conway–Maxwell–Poisson model for analyzing crash data. *Applied Stochastic Models in Business and Industry*, 28(2):122–127.

Lord, D., Guikema, S. D., and Geedipally, S. R. (2008). Application of the Conway–Maxwell–Poisson generalized linear model for analyzing motor vehicle crashes. *Accident Analysis & Prevention*, 40(3):1123–1134.

Lord, D. and Miranda-Moreno, L. F. (2008). Effects of low sample mean values and small sample size on the estimation of the fixed dispersion parameter of Poisson-gamma models for modeling motor vehicle crashes: A Bayesian perspective. *Safety Science*, 46(5):751–770.

Lucas, J. M. (1985). Counted data CUSUM's. *Technometrics*, 27(2):199–244.

Lucas, J. M. and Crosier, R. B. (1982). Fast initial response for CUSUM quality control schemes: Give your CUSUM a head start. *Technometrics*, 24(3):199–205.

Macaulay, T. B. (1923). *Literary Essays Contributed to the Edinburgh Review*. Oxford University Press.

MacDonald, I. L. and Bhamani, F. (2020). A time-series model for underdispersed or overdispersed counts. *The American Statistician*, 74(4):317–328.

Magnusson, A., Skaug, H., Nielsen, A., Berg, C., Kristensen, K., Maechler, M., van Bentham, K., Bolker, B., Sadat, N., Lüdecke, D., Lenth, R., O'Brien, J., and Brooks, M. (2020). glmmTMB: Generalized Linear Mixed Models using Template Model Builder, version 1.0.2.1. https://cran.r-project.org/web/packages/glmmTMB/index.html.

Mamode Khan, N. and Jowaheer, V. (2010). A comparison of marginal and joint generalized quasi-likelihood estimating equations based on the COM–Poisson GLM: Application to car breakdowns data. *International Journal of Mathematical and Computational Sciences*, 4(1):23–26.

Mamode Khan, N., Jowaheer, V., Sunecher, Y., and Bourguignon, M. (2018). Modeling longitudinal INMA(1) with COM–Poisson innovation under non-stationarity: Application to medical data. *Computational and Applied Mathematics*, 37(12):5217–5238.

Marschner, I. C. (2011). glm2: Fitting generalized linear models with convergence problems. *R Journal*, 3(2):12–15.

Marshall, A. W. and Olkin, I. (1985). A family of bivariate distributions generated by the bivariate Bernoulli distribution. *Journal of the American Statistical Association*, 80(390):332–338.

McCullagh, P. and Nelder, J. A. (1997). *Generalized Linear Models*, 2nd edition. Chapman & Hall/CRC Press.

Melo, M. and Alencar, A. (2020). Conway–Maxwell–Poisson autoregressive moving average model for equidispersed, underdispersed, and overdispersed count data. *Journal of Time Series Analysis*, 41(6):830–857.

Miaou, S.-P. and Lord, D. (2003). Modeling traffic crash-flow relationships for intersections: Dispersion parameter, functional form, and Bayes versus empirical Bayes. *Transportation Research Record*, 1840(1):31–40.

Mills, T. and Seneta, E. (1989). Goodness-of-fit for a branching process with immigration using sample partial autocorrelations. *Stochastic Processes and their Applications*, 33(1):151–161.

Minka, T. P., Shmueli, G., Kadane, J. B., Borle, S., and Boatwright, P. (2003). Computing with the COM–Poisson distribution. Technical Report 776, Dept. of Statistics, Carnegie Mellon University.

Mohammed, M. and Laney, D. (2006). Overdispersion in health care performance data: Laney's approach. *Quality and Safety in Health Care*, 15(5):383–384.

Molenberghs, G., Verbeke, G., and Demétrio, C. G. (2007). An extended random-effects approach to modeling repeated, overdispersed count data. *Lifetime Data Analysis*, 13(4):513–531.

Montgomery, D. C. (2001). *Design and Analysis of Experiments*. John Wiley & Sons, 5th edition.

Morris, D. S., Raim, A. M., and Sellers, K. F. (2020). A Conway–Maxwell-multinomial distribution for flexible modeling of clustered categorical data. *Journal of Multivariate Analysis*, 179:104651.

Morris, D. S. and Sellers, K. F. (2022). A flexible mixed model for clustered count data. *Stats*, 5(1):52–69.

Morris, D. S., Sellers, K. F., and Menger, A. (2017). Fitting a flexible model for longitudinal count data using the NLMIXED procedure. In *SAS Global Forum Proceedings Paper 202-2017*, pages 1–6, Cary, NC. SAS Institute. http://support.sas.com/resources/papers/proceedings17/0202-2017.pdf.

Nadarajah, S. (2009). Useful moment and CDF formulations for the COM–Poisson distribution. *Statistics Papers*, 50(3):617–622.

Nelsen, R. B. (2010). *An Introduction to Copulas*. Springer Publishing Company.

Ntzoufras, I., Katsis, A., and Karlis, D. (2005). Bayesian assessment of the distribution of insurance claim counts using reversible jump MCMC. *North American Actuarial Journal*, 9(3):90–108.

Oh, J., Washington, S., and Nam, D. (2006). Accident prediction model for railway-highway interfaces. *Accident Analysis & Prevention*, 38(2):346–356.

Olver, F. W. J. (1974). *Asymptotics and Special Functions*. Academic Press.

Ong, S. H., Gupta, R. C., Ma, T., and Sim, S. Z. (2021). Bivariate Conway–Maxwell–Poisson distributions with given marginals and correlation. *Journal of Statistical Theory and Practice*, 15(10):1–19.

Ord, J. K. (1967). Graphical methods for a class of discrete distributions. *Journal of the Royal Statistical Society, Series A*, 130(2):232–238.

Ord, J. K. and Whitmore, G. A. (1986). The Poisson-inverse Gaussian distribution as a model for species abundance. *Communications in Statistics – Theory and Methods*, 15(3):853–871.

Ötting, M., Langrock, R., and Maruotti, A. (2021). A copula-based multivariate hidden Markov model for modelling momentum in football. *AStA Advances in Statistical Analysis*.

Pal, S. and Balakrishnan, N. (2017). Likelihood inference for COM–Poisson cure rate model with interval-censored data and Weibull lifetimes. *Statistical Methods in Medical Research*, 26(5):2093–2113.

Pal, S., Majakwara, J., and Balakrishnan, N. (2018). An EM algorithm for the destructive COM–Poisson regression cure rate model. *Metrika*, 81(2):143–171.

Piegorsch, W. W. (1990). Maximum likelihood estimation for the negative binomial dispersion parameter. *Biometrics*, 46(3):863–867.

Pollock, J. (2014a). CompGLM: Conway–Maxwell–Poisson GLM and distribution functions, version 1.0. https://cran.r-project.org/web/packages/CompGLM/index.html.

Pollock, J. (2014b). *The CompGLM Package: Conway–Maxwell–Poisson GLM and distribution functions*, 1.0 edition.

Puig, P., Valero, J., and Fernández-Fontelo, A. (2016). Some mechanisms leading to underdispersion.

R Core Team (2014). *R: A Language and Environment for Statistical Computing*. R Foundation for Statistical Computing, Vienna, Austria.

Raim, A. M. and Morris, D. S. (2020). COMMultReg: Conway–Maxwell multinomial regression, version 0.1.0. https://cran.r-project.org/web/packages/commultreg/index.html.

Rao, G. S., Aslam, M., Rasheed, U., and Jun, C.-H. (2020). Mixed EWMA–CUSUM chart for COM–Poisson distribution. *Journal of Statistics and Management Systems*, 23(3):511–527.

Reynolds, M. R. J. R. (2013). The Bernoulli CUSUM chart for detecting decreases in a proportion. *Quality and Reliability Engineering International*, 29(4):529–534.

Ribeiro Jr., E. E., Zeviani, W. M., Bonat, W. H., Demetrio, C. G., and Hinde, J. (2019). Reparametrization of COM–Poisson regression models with applications in the analysis of experimental data. *Statistical Modelling*. 20. 10.1177/1471082X19838651.

Ridout, M. S. and Besbeas, P. (2004). An empirical model for underdispersed count data. *Statistical Modelling*, 4(1):77–89.

Rigby, R. A. and Stasinopoulos, D. M. (2005). Generalized additive models for location, scale and shape (with discussion). *Applied Statistics*, 54:507–554.

Rodrigues, J., Cancho, V. G., de Castro, M., and Balakrishnan, N. (2012). A Bayesian destructive weighted Poisson cure rate model and an application to a cutaneous melanoma data. *Statistical methods in medical research*, 21(6):585–597.

Rodrigues, J., de Castro, M., Balakrishnan, N., and Cancho, V. G. (2011). Destructive weighted Poisson cure rate models. *Lifetime Data Analysis*, 17:333–346.

Rodrigues, J., de Castro, M., Cancho, V. G., and Balakrishnan, N. (2009). COM–Poisson cure rate survival models and an application to a cutaneous melanoma data. *Journal of Statistical Planning and Inference*, 139(10):3605 – 3611.

Rosen, O., Jiang, W., and Tanner, M. A. (2000). Mixtures of marginal models. *Biometrika*, 87(2):391–404.

Roy, S., Tripathi, R. C., and Balakrishnan, N. (2020). A Conway–Maxwell–Poisson type generalization of the negative hypergeometric distribution. *Communications in Statistics – Theory and Methods*, 49(10):2410–2428.

Rutherford, E., Geiger, H., and Bateman, H. (1910). The probability variations in the distribution of $\alpha$ particles. *The London, Edinburgh, and Dublin Philosophical Magazine and Journal of Science*, 20(118):698–707.

Saez-Castillo, A. J., Conde-Sanchez, A., and Martinez, F. (2020). DGLMExtPois: Double Generalized Linear Models Extending Poisson Regression, version 0.1.3.

Saghir, A. and Lin, Z. (2014a). Control chart for monitoring multivariate COM–Poisson attributes. *Journal of Applied Statistics*, 41(1):200–214.

Saghir, A. and Lin, Z. (2014b). Cumulative sum charts for monitoring the COM–Poisson processes. *Computers and Industrial Engineering*, 68(1):65–77.

Saghir, A. and Lin, Z. (2014c). A flexible and generalized exponentially weighted moving average control chart for count data. *Quality and Reliability Engineering International*, 30(8):1427–1443.

Saghir, A., Lin, Z., Abbasi, S., and Ahmad, S. (2013). The use of probability limits of COM–Poisson charts and their applications. *Quality and Reliability Engineering International*, 29(5):759–770.

Santana, R. A., Conceição, K. S., Diniz, C. A. R., and Andrade, M. G. (2021). Type I multivariate zero-inflated COM–Poisson regression model. *Biometrical Journal*.

Sarmanov, O. V. (1966). Generalized normal correlation and two-dimensional Fréchet classes. *Dokl. Akad. Nauk SSSR*, 168(1):32–35.

SAS Institute. (2014). SAS/ETS(R) 13.1 User's Guide. http://www.sas.com/.

Self, S. G. and Liang, K.-Y. (1987). Asymptotic properties of maximum likelihood estimators and likelihood ratio tests under nonstandard conditions. *Journal of the American Statistical Association*, 82(398):605–610.

Sellers, K. (2012a). A distribution describing differences in count data containing common dispersion levels. *Advances and Applications in Statistical Sciences*, 7(3):35–46.

Sellers, K. and Costa, L. (2014). CMPControl: Control Charts for Conway–Maxwell–Poisson Distribution, version 1.0. https://cran.r-project.org/web/packages/CMPControl/index.html.

Sellers, K., Lotze, T., and Raim, A. (2019). COMPoissonReg: Conway–Maxwell–Poisson (COM–Poisson) Regression, version 0.7.0. https://cran.r-project.org/web/packages/COMPoissonReg/index.html.

Sellers, K., Morris, D. S., Balakrishnan, N., and Davenport, D. (2017a). multicmp: Flexible Modeling of Multivariate Count Data via the Multivariate Conway–Maxwell–Poisson Distribution, version 1.0. https://cran.r-project.org/web/packages/multicmp/index.html.

Sellers, K. and Shmueli, G. (2009). A regression model for count data with observation-level dispersion. In Booth, J. G., editor, *Proceedings of the 24th International Workshop on Statistical Modelling*, pages 337–344, Ithaca, NY.

Sellers, K. and Shmueli, G. (2013). Data dispersion: Now you see it... now you don't. *Communications in Statistics – Theory and Methods*, 42(17):1–14.

Sellers, K. F. (2012b). A generalized statistical control chart for over- or under-dispersed data. *Quality and Reliability Engineering International*, 28(1):59–65.

Sellers, K. F., Arab, A., Melville, S., and Cui, F. (2021a). A flexible univariate moving average time-series model for dispersed count data. *Journal of Statistical Distributions and Applications*, 8(1):1–12.

Sellers, K. F., Li, T., Wu, Y., and Balakrishnan, N. (2021b). A flexible multivariate distribution for correlated count data. *Stats*, 4(2):308–326.

Sellers, K. F. and Morris, D. S. (2017). Underdispersion models: Models that are "under the radar". *Communications in Statistics – Theory and Methods*, 46(24):12075–12086.

Sellers, K. F., Morris, D. S., and Balakrishnan, N. (2016). Bivariate Conway–Maxwell–Poisson distribution: Formulation, properties, and inference. *Journal of Multivariate Analysis*, 150:152–168.

Sellers, K. F., Peng, S. J., and Arab, A. (2020). A flexible univariate autoregressive time-series model for dispersed count data. *Journal of Time Series Analysis*, 41(3): 436–453.

Sellers, K. F. and Raim, A. (2016). A flexible zero-inflated model to address data dispersion. *Computational Statistics and Data Analysis*, 99:68–80.

Sellers, K. F. and Shmueli, G. (2010). A flexible regression model for count data. *Annals of Applied Statistics*, 4(2):943–961.

Sellers, K. F., Shmueli, G., and Borle, S. (2011). The COM–Poisson model for count data: A survey of methods and applications. *Applied Stochastic Models in Business and Industry*, 28(2):104–116.

Sellers, K. F., Swift, A. W., and Weems, K. S. (2017b). A flexible distribution class for count data. *Journal of Statistical Distributions and Applications*, 4(22):1–21.

Sellers, K. F. and Young, D. S. (2019). Zero-inflated sum of Conway–Maxwell–Poissons (ZISCMP) regression. *Journal of Statistical Computation and Simulation*, 89(9):1649–1673.

Sheu, S.-H. and Chiu, W.-C. (2007). Poisson GWMA control chart. *Communications in Statistics – Simulation and Computation*, 36(5):1099–1114.

Shmueli, G., Minka, T. P., Kadane, J. B., Borle, S., and Boatwright, P. (2005). A useful distribution for fitting discrete data: Revival of the Conway–Maxwell–Poisson distribution. *Applied Statistics*, 54:127–142.

Şimşek, B. and Iyengar, S. (2016). Approximating the Conway–Maxwell–Poisson normalizing constant. *Filomat*, 30(4):953–960.

Skellam, J. G. (1946). The frequency distribution of the difference between two Poisson variates belonging to different populations. *Journal of the Royal Statistical Society Series A*, 109(3):296.

Spiegelhalter, D. (2005). Handling over-dispersion of performance indicators. *Quality and Safety in Health Care*, 14(5):347–351.

Stasinopoulos, D. M. and Rigby, R. A. (2007). Generalized additive models for location scale and shape (gamlss) in R. *Journal of Statistical Software*, 23(7):1–46.

Statisticat and LLC. (2021). *LaplacesDemon: Complete Environment for Bayesian Inference*. R package version 16.1.6.

Sunecher, Y., Khan, N. M., and Jowaheer, V. (2020). BINMA(1) model with COM–Poisson innovations: Estimation and application. *Communications in Statistics – Simulation and Computation*, 49(6):1631–1652.

Thall, P. F. and Vail, S. C. (1990). Some covariance models for longitudinal count data with overdispersion. *Biometrics*, 46(3):657–671.

Trivedi, P. and Zimmer, D. (2017). A note on identification of bivariate copulas for discrete count data. *Econometrics*, 5(1):1–11.

Waller, L. A., Carlin, B. P., Xia, H., and Gelfand, A. E. (1997). Hierarchical spatio-temporal mapping of disease rates. *Journal of the American Statistical Association*, 92(438):607–617.

Washburn, A. (1996). Katz distributions, with applications to minefield clearance. Technical Report ADA307317, Naval Postgraduate School Moterey CA Department of Operations Research.

Weems, K. S., Sellers, K. F., and Li, T. (2021). A flexible bivariate distribution for count data expressing data dispersion. *Communications in Statistics – Theory and Methods*.

Weiss, C. H. (2008). Thinning operations for modeling time series of counts—a survey. *Advances in Statistical Analysis*, 92(3):319–341.

Winkelmann, R. (1995). Duration dependence and dispersion in count-data models. *Journal of Business & Economic Statistics*, 13(4):467–474.

Winkelmann, R. and Zimmermann, K. F. (1995). Recent developments in count data modelling: Theory and application. *Journal of Economic Surveys*, 9(1):1–24.

Witowski, V. and Foraita, R. (2018). *HMMpa: Analysing Accelerometer Data Using Hidden Markov Models*. R package version 1.0.1.

Wood, S. N. (2012). On p-values for smooth components of an extended generalized additive model. *Biometrika*, 100(1):221–228.

Wu, G., Holan, S. H., Nilon, C. H., and Wikle, C. K. (2015). Bayesian binomial mixture models for estimating abundance in ecological monitoring studies. *The Annals of Applied Statistics*, 9(1):1–26.

Wu, G., Holan, S. H., and Wikle, C. K. (2013). Hierarchical Bayesian spatio-temporal Conway–Maxwell–Poisson models with dynamic dispersion. *Journal of Agricultural, Biological, and Environmental Statistics*, 18(3):335–356.

Yakovlev, A. Y. and Tsodikov, A. D. (1996). *Stochastic Models of Tumor Latency and Their Biostatistical Applications*. World Scientific.

Yee, T. W. (2008). The VGAM package. *R News*, 8(2):28–39.

Yee, T. W. (2010). The VGAM package for categorical data analysis. *Journal of Statistical Software*, 32(10):1–34.

Zeileis, A., Lumley, T., Graham, N., and Koell, S. (2021). sandwich: Robust Covariance Matrix Estimators, version 3.0-1. https://cran.r-project.org/web/packages/sandwich/sandwich.pdf.

Zhang, H. (2015). Characterizations and infinite divisibility of extended COM–Poisson distribution. *International Journal of Statistical Distributions and Applications*, 1(1):1–4.

Zhang, H., Tan, K., and Li, B. (2018). The COM-negative binomial distribution: modeling overdispersion and ultrahigh zero-inflated count data. *Frontiers of Mathematics in China*, 13:967–998.

Zhang, L., Govindaraju, K., Lai, C. D., and Bebbington, M. S. (2003). Poisson DEWMA control chart. *Communications in Statistics – Simulation and Computation*, 32(4):1265–1283.

Zhu, F. (2012). Modeling time series of counts with COM–Poisson INGARCH models. *Mathematical and Computer Modelling*, 56(9):191–203.

Zhu, L., Sellers, K., Morris, D., Shmueli, G., and Davenport, D. (2017a). cmpprocess: Flexible modeling of count processes, version 1.0. https://cran.r-project.org/web/packages/cmpprocess/index.html.

Zhu, L., Sellers, K. F., Morris, D. S., and Shmuéli, G. (2017b). Bridging the gap: A generalized stochastic process for count data. *The American Statistician*, 71(1): 71–80.

Zucchini, W., MacDonald, I. L., and Langrock, R. (2016). *Hidden Markov Models for Time Series: An Introduction Using R*. Chapman & Hall/CRC Press, 2nd edition.

# Index

Printed in the United States
by Baker & Taylor Publisher Services